CAMBRIDGE TRACTS IN MATHEMATICS

General Editors

B. BOLLOBÁS, H. HALBERSTAM, C.T.C. WALL

89: *Irregularities of distribution*

JÓZSEF BECK

Associate Professor of Mathematics, Eötvös Loránd University, Budapest and Research Fellow Mathematical Institute of the Hungarian Academy of Sciences, Budapest

WILLIAM W.L.CHEN

Lecturer in Pure Mathematics
Imperial College of Science and Technology, London

Irregularities of distribution

The right of the University of Cambridge to print and sell all manner of books was granted by Henry VIII in 1534. The University has printed and published continuously since 1584.

CAMBRIDGE UNIVERSITY PRESS

Cambridge

New York New Rochelle

Melbourne Sydney

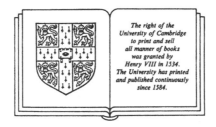

CAMBRIDGE UNIVERSITY PRESS
Cambridge, New York, Melbourne, Madrid, Cape Town, Singapore, São Paulo, Delhi

Cambridge University Press
The Edinburgh Building, Cambridge CB2 8RU, UK

Published in the United States of America by Cambridge University Press, New York

www.cambridge.org
Information on this title: www.cambridge.org/9780521307925

First published 1987
This digitally printed version 2008

A catalogue record for this publication is available from the British Library

Library of Congress Cataloguing in Publication data

Beck, József.
Irregularities of distribution.

Bibliography: p.
Includes indexes
1. Distribution (Probability theory) I. Chen,
William W. L. II. Title
QA273.6.B45 1987 519.2 86-24419

ISBN 978-0-521-30792-5 hardback
ISBN 978-0-521-09300-2 paperback

To
KLAUS ROTH
and
WOLFGANG SCHMIDT

Contents

Preface

Although parts of it border on combinatorics and probability theory, irregularities of distribution began as a branch of the theory of uniform distribution, and may sometimes be described as a quantitative form of the theory. It originated from a conjecture of van der Corput and the work of van Aardenne-Ehrenfest, and owes its current prominence to the contribution of Roth and Schmidt.

Students and researchers in this area have benefited greatly from the excellent monograph of Schmidt published by the Tata Institute of Fundamental Research in 1977. However, Professor Schmidt pointed out to the second author in 1981 that his monograph was out of date and indicated that there was a need for one that would include the more recent work of Halász and Roth. When Professor Schmidt repeated this point in 1983, it was time to do something about the situation. With generous financial support from the Science and Engineering Research Council in Great Britain, the first author was able to spend the academic year 1984–85 at Imperial College London, and we made it our primary task to put Professor Schmidt's suggestion into reality.

In the last few years, much progress has been made in irregularities of distribution. From its early days where Roth's orthogonal function method played a crucial role, the subject has developed through an ever greater dependence on tools in harmonic analysis, first introduced by Halász in his use of Riesz products. In particular, the most recent work depends heavily on summability kernels (de la Vallée-Poussin kernels, Gauss kernels, Fejér kernels) and other standard tools of Fourier analysis. And one of the aims of this monograph is to demonstrate the common role played by all these different approaches – to 'blow up the trivial error'.

In Part A, we study the classical problem of irregularities of point distribution in the unit cube with respect to rectangular boxes with sides parallel to the coordinate axes. This is essentially a reformulation of the problem of van der Corput as studied by van Aardenne-Ehrenfest in 1949. We also study the problem in van Aardenne-Ehrenfest's formulation. In Chapter 1, we list most of the key results; these are either directly concerned with the problem, or applications and variations of the methods involved.

In Chapter 2, we prove lower bound results using Roth's orthogonal function method and Halász's variation of the method. We complement these results with upper bound theorems in Chapter 3, where we attempt to explain the step-by-step development of the probabilistic ideas involved in these rather complicated proofs. In Chapter 4, we study the problem in van Aardenne-Ehrenfest's formulation. We end the chapter and this part by studying Wagner's solution to a problem of Erdös on diophantine approximation.

In Part B, we study generalizations of the classical problem pioneered by Schmidt. We no longer restrict ourselves to the case of rectangular boxes with sides parallel to the coordinate axes. We allow rotation, and we also study more general shapes. Chapter 5 is basically a description of Schmidt's integral equation method. We investigate firstly the simple case of rotated boxes, as the method is rather complicated and mysterious at first sight. We then study the case of balls and prove an essentially sharp lower bound theorem. Chapters 6 and 7 are concerned with a recent Fourier transform approach to the subject. In Chapter 6, we study the effects of rotation and surface area on the discrepancy. In Chapter 7, we study the 2-dimensional case more closely and introduce the idea of an approximability number which is a measure of how well a convex region can be approximated by an inscribed polygon of few sides. We also discuss Roth's problem on disc-segments and an application of the method to a problem in discrete geometry concerning the sums of distances between points on the surface of a sphere. In Chapter 8, we complement the lower bounds results in this part by proving some upper bound theorems with tools in combinatorics and probability theory.

In the final chapter, we prove Roth's famous '$\frac{1}{4}$-theorem' on irregularities of integer sequences relative to arithmetic progressions and show that this is essentially best possible. We conclude the chapter and the book with a most important section, where we mention some well-known open problems and also discuss without proof the recent advances.

Many of the proofs, especially in Part B, are not necessarily the simplest, as a considerable amount of material represents very recent work. Our object here is to present the ideas behind these results as we know them at this point in time.

Our interest in the theory of irregularities of distribution is due almost entirely to the fundamental work of Klaus Roth and Wolfgang Schmidt, their help, encouragement, guidance and friendship in the last few years. We are greatly indebted to them both. We also owe a word of thanks to many of our colleagues and friends – Paul Erdös, Gábor Halász, Hugh Montgomery, András Sárközy, Vera Sós, Robert Tijdeman and Bob

Vaughan, to name but a few – for their encouragement and support, and particularly to Heini Halberstam for recommending our project to the Cambridge University Press. The first author would especially like to thank the members of the Department of Mathematics at Imperial College for their hospitality during his stay in London.

Lastly, but definitely not least, we would like to thank our wives Mária and Lily for their unfailing support.

J.B.
W.W.L.C.
1985

List of notation

(i) For sets, we adopt the following notation:

$\mathbb{R}$	set of all real numbers
$\mathbb{R}^+$	set of all positive real numbers
$\mathbb{Z}$	set of all integers
$\mathbb{N}$	set of all natural numbers
$\mathbb{N}^*$	set of all non-negative integers
U_0	$\{x\in\mathbb{R}: 0\leqslant x<1\}$
U_1	$\{x\in\mathbb{R}: 0<x\leqslant 1\}$
U	$\{x\in\mathbb{R}: 0\leqslant x\leqslant 1\}$
U_*	$\{x\in\mathbb{R}: 0<x<1\}$

U is also used to denote the unit torus.

(ii) The numbers i and e have their usual meanings, and $e(x)=\mathrm{e}^{2\pi i x}$

(iii) We adopt the following standard notation:

[ ]	integral part
{ }	fractional part
$\|\ \|$	distance to the nearest integer
$\varnothing$	empty set
$A\times B$	cartesian product of A and B
$A\backslash B$	difference of A and B
$A\Delta B$	symmetric difference of A and B
$x+S$	$\{x+v: v\in S\}$
λS	$\{\lambda v: v\in S\}$

(iv) For functions f and non-negative functions g, we adopt the following notation:

$f=\mathrm{O}(g)$	$\lvert f\rvert\leqslant cg$ for some positive constant c
$f=\mathrm{o}(g)$	$f/g\to 0$ in the limit
$f\ll g$	$\lvert f\rvert\leqslant cg$ for some positive constant c
$f\gg g$	$f\geqslant cg$ for some positive constant c
$f\gg\ll g$	$f\ll g$ and $g\ll f$

Furthermore, if the notation is followed by a parameter in subscript position, then the constant c may depend on the value of the parameter.

(v) K is reserved to indicate dimension. For a set A in $\mathbb{R}^K$, we adopt the following notation:

$\mu(A)$	measure of A
$\# A$	cardinality of A
$d(A)$	diameter of A
∂A	boundary surface of A
$\sigma(\partial A)$	(K-1)-dimensional volume of ∂A
$r(A)$	radius of the largest inscribed ball in A

(vi) We adopt the following notation only in the parts indicated:

A	area (only in § 3.1)
V	volume (only in § 3.1)
l	length (only in Chapter 4)
$\lvert\mathscr{S}\rvert$	measure of $\mathscr{S}$ (only in Part A)
$\lvert\mathbf{x}\rvert$	euclidean distance (except in Chapter 2)

(vii) The constants $c_1, c_2, c_3, \ldots$ are positive constants (unless otherwise stated) whose values depend at most on the values of the parameters in the attached brackets. If $\mathbf{x} = (x_1, \ldots, x_K)$, we use $\mathrm{d}\mathbf{x}$ to denote $\mathrm{d}x_1 \cdots \mathrm{d}x_K$. Finally, we also use the convention

$$\int_B \int_A f(x, y)\,\mathrm{d}x\,\mathrm{d}y = \int_B \left(\int_A f(x, y)\,\mathrm{d}x \right) \mathrm{d}y.$$

Part A: The classical problem

1

Van der Corput's conjecture

The study of irregularities of distribution of sequences began with the work of van der Corput (1935*a*, *b*) on distribution functions. In 1935, he made the following conjecture.

Conjecture. *If* $s_1, s_2, s_3, \ldots$ *is an infinite sequence of real numbers lying between* 0 *and* 1, *then corresponding to any arbitrarily large* κ, *there exist a positive integer* n *and two subintervals, of equal length, of* (0, 1), *such that the number of* s_ν ($\nu = 1, \ldots, n$) *that lie in one of the subintervals differs from the number of such* s_ν *that lie in the other subinterval by more than* κ.

In short, the conjecture expresses the fact that no sequence can, in a certain sense, be too evenly distributed.

The conjecture was proved in 1945 by van Aardenne-Ehrenfest (1945). In 1949, van Aardenne-Ehrenfest (1949) proved the following refinement.

Theorem. *Let* N *be a large integer, and let* $s_1, \ldots, s_N$ *be a sequence of* N *real numbers satisfying* $0 \leqslant s_\nu < 1$ *for* $\nu = 1, \ldots, N$. *For any integer* n *satisfying* $1 \leqslant n \leqslant N$ *and for any real number* α *satisfying* $0 < \alpha \leqslant 1$, *we denote by* $Z(n, \alpha)$ *the number of* s_ν ($\nu = 1, \ldots, n$) *that satisfy* $0 \leqslant s_\nu < \alpha$. *Then there exist* n *and* α *such that*

$$|Z(n, \alpha) - n\alpha| > c_1 \frac{\log \log N}{\log \log \log N}. \tag{1}$$

In 1954, Roth (1954) considered the problem in a more symmetrical form. He showed that the above theorem is equivalent to the following.

Theorem. *Let* N *be a large integer, and let* $\mathscr{P}$ *be a distribution of* N *points, not necessarily distinct, in the square* $0 \leqslant y_i < 1$ ($i = 1, 2$). *For any point* $\mathbf{x} = (x_1, x_2)$ *satisfying* $0 < x_i \leqslant 1$ ($i = 1, 2$), *let* $B(\mathbf{x})$ *denote the box consisting of all* $\mathbf{y} = (y_1, y_2)$ *such that* $0 \leqslant y_i < x_i$ ($i = 1, 2$). *Denote by* $Z[\mathscr{P}; B(\mathbf{x})]$ *the number of points of* $\mathscr{P}$ *which fall into* $B(\mathbf{x})$, *and let*

$$D[\mathscr{P}; B(\mathbf{x})] = Z[\mathscr{P}; B(\mathbf{x})] - N x_1 x_2.$$

Then

$$\sup_{x_1,x_2\in(0,1]} |D[\mathscr{P}; B(\mathbf{x})]| > c_2 \frac{\log\log N}{\log\log\log N}. \tag{2}$$

Roth was also able to replace the factor $(\log\log N)/(\log\log\log N)$ in (1) and (2) by the factor $(\log N)^{\frac{1}{2}}$. In fact, he proved a stronger theorem, a special case of which is the following.

Theorem. *Let $N > 1$ be an integer, and let $\mathscr{P}$ be a distribution of N points, not necessarily distinct, in the square $[0, 1)^2$. Then*

$$\int_0^1 \int_0^1 |D[\mathscr{P}; B(\mathbf{x})]|^2 \,\mathrm{d}x_1 \,\mathrm{d}x_2 > c_3 \log N,$$

where $D[\mathscr{P}; B(\mathbf{x})]$ is defined as in the previous theorem.

To state Roth's theorem in its generality, and Schmidt's generalization of it, we need to define a multidimensional discrepancy function $D[\mathscr{P}; B(\mathbf{x})]$.

1.1 Roth's formulation of the problem

Let $U_0 = [0, 1)$, $U_1 = (0, 1]$ and $U = [0, 1]$. Suppose that we have a distribution $\mathscr{P} = \mathscr{P}(K, N)$ of N points, not necessarily distinct, in U_0^K, where, for every natural number $K \geqslant 2$, U_0^K is the unit cube consisting of the points $\mathbf{y} = (y_1, \ldots, y_K)$ with $0 \leqslant y_i < 1$ $(i = 1, \ldots, K)$. For $\mathbf{x} = (x_1, \ldots, x_K)$ in U_1^K, let $B(\mathbf{x})$ denote the box consisting of all $\mathbf{y}$ such that $0 \leqslant y_i < x_i$ $(i = 1, \ldots, K)$, and let $Z[\mathscr{P}; B(\mathbf{x})]$ denote the number of points of $\mathscr{P}$ which lie in $B(\mathbf{x})$. Write

$$D[\mathscr{P}; B(\mathbf{x})] = Z[\mathscr{P}; B(\mathbf{x})] - N x_1 \cdots x_K.$$

The irregularity of the distribution $\mathscr{P}$ can be measured in a number of ways by the behaviour of the function $D[\mathscr{P}; B(\mathbf{x})]$. One may consider, for $0 < W < \infty$, the L^W-norm

$$\|D(\mathscr{P})\|_W = \left(\int_{U^K} |D[\mathscr{P}; B(\mathbf{x})]|^W \,\mathrm{d}\mathbf{x} \right)^{1/W}.$$

One may also consider the L^∞-norm

$$\|D(\mathscr{P})\|_\infty = \sup_{\mathbf{x}\in U_1^K} |D[\mathscr{P}; B(\mathbf{x})]|.$$

We write

$$D(K, W, N) = \inf_{\mathscr{P}(K,N)} \|D(\mathscr{P})\|_W$$

and

$$D(K, \infty, N) = \inf_{\mathscr{P}(K,N)} \|D(\mathscr{P})\|_\infty,$$

where, as indicated, the infimum is taken over all distributions $\mathscr{P}$ of N points in U_0^K.

For the lower bound of $D(K, W, N)$, the following are known.

Theorem 1A (Roth (1954)). *We have*

$$D(K, 2, N) > c_4(K)(\log N)^{\frac{1}{2}(K-1)}.$$

Theorem 1B (Schmidt (1977*a*)). *Let* $W > 1$. *Then*

$$D(K, W, N) > c_5(K, W)(\log N)^{\frac{1}{2}(K-1)}.$$

Theorem 1C (Halász (1981)). *We have*

$$D(K, 1, N) > c_6(K)(\log N)^{\frac{1}{2}}.$$

These are complemented by the following upper bound results.

Theorem 2A (Davenport (1956)). *We have, for* $N \geqslant 2$,

$$D(2, 2, N) < c_7(\log N)^{\frac{1}{2}}.$$

Different proofs of Theorem 2A were later given by Vilenkin (1967), Halton and Zaremba (1969) and Roth (1976*b*).

Theorem 2B (Roth (1979)). *We have, for* $N \geqslant 2$,

$$D(3, 2, N) < c_8 \log N.$$

Theorem 2C (Roth (1980)). *We have, for* $N \geqslant 2$,

$$D(K, 2, N) < c_9(K)(\log N)^{\frac{1}{2}(K-1)}.$$

Theorem 2D (Chen (1980)). *Let* $W > 0$. *Then for* $N \geqslant 2$,

$$D(K, W, N) < c_{10}(K, W)(\log N)^{\frac{1}{2}(K-1)}.$$

By Theorem 1B and Theorem 2D, the estimates for $D(K, W, N)$ for $W > 1$ are essentially sharp. On the other hand, it is unlikely that Theorem 1C is sharp for $K \geqslant 3$. It is easily deduced from Theorem 1A that

Theorem 3A (Roth (1954)). *We have*

$$D(K, \infty, N) > c_{11}(K)(\log N)^{\frac{1}{2}(K-1)}.$$

In 1972, Schmidt, using a completely different method, proved

Theorem 3B (Schmidt (1972*b*)). *We have*

$$D(2, \infty, N) > c_{12} \log N.$$

An alternative proof of Theorem 3B was given by Halász (1981). On the other hand, Schmidt's result is sharp, for

Theorem 4 (Halton (1960)). *We have, for* $N \geqslant 2$,

$$D(K, \infty, N) < c_{13}(K)(\log N)^{K-1}.$$

An alternative proof of Theorem 4 was given by Faure (1982). This led also to an alternative proof of Theorem 2D by Chen (1983).

Note that Theorem 3B shows that for any distribution of N points in the unit square, there is an aligned rectangle with discrepancy of order $(\log N)$. By a variation of his own ideas in Halász (1981), Halász was able to show that there is an aligned square with discrepancy of order $(\log N)$. More precisely, we have

Theorem 3C (Halász (1985)). *Let* $\mathscr{P}$ *be a distribution of* N *points in* U_0^2. *For any aligned square* Q *in* U_1^2, *let* $Z[\mathscr{P}; Q]$ *denote the number of points of* $\mathscr{P}$ *which lie in* Q, *and write* $D[\mathscr{P}; Q] = Z[\mathscr{P}; Q] - N\mu(Q)$, *where* $\mu(Q)$ *denotes the area of* Q. *Then*

$$\sup_Q |D[\mathscr{P}; Q]| > c_{14} \log N,$$

where the supremum is taken over all aligned squares in U_1^2.

The following is an excruciatingly difficult problem.

Conjecture (Great open problem). *For* $K > 2$, *the exponent* $\frac{1}{2}(K-1)$ *in Theorem* 3A *can be replaced by the exponent* $(K-1)$.

For $K = 2$, this is, of course, true, in view of Theorem 3B. However, for $K > 2$, Theorem 3A has so far remained the best result known.

We shall discuss the lower bound results in Chapter 2 and the upper bound results in Chapter 3. Also, in Chapter 3, it is convenient to write $K = k + 1$, where $k \geqslant 1$.

We conclude this section by proving the equivalence of the two formulations of the original problem by van Aardenne-Ehrenfest and by Roth.

It is very simple to show that (2) implies (1). We simply take $\mathscr{P}$ to consist of the points $(s_\nu, (\nu - 1)/N)$ for $\nu = 1, \dots N$. Then there exists $\mathbf{x} = (x_1, x_2)$ such that

$$|D[\mathscr{P}; B(\mathbf{x})]| > c_2 \frac{\log \log N}{\log \log \log N}.$$

We obtain (1) with $\alpha = x_1$ and $n = -[-Nx_2]$ (upper integral part).

To show that (1) implies (2), let $(y_1(\nu), y_2(\nu))$ $(\nu = 1, \ldots, N)$ be the points of $\mathscr{P}$ with

$$y_2(1) \leqslant y_2(2) \leqslant \cdots \leqslant y_2(N). \tag{3}$$

Let $\mathscr{P}^*$ consist of the points $(y_1(\nu), (\nu - 1)/N)$ $(\nu = 1, \ldots, N)$. Write

$$M = \sup_{x_1, x_2 \in (0,1]} |D[\mathscr{P}; B(\mathbf{x})]| \quad \text{and} \quad M^* = \sup_{x_1, x_2 \in (0,1]} |D[\mathscr{P}^*; B(\mathbf{x})]|. \tag{4}$$

Roth showed that $M^* \leqslant 7M$ as follows: By (4),

$$|Z[\mathscr{P}; B((1, y_2(\nu)))] - Ny_2(\nu)| \leqslant M;$$

also (3) can have at most $(2M - 1)$ consecutive equalities, so that by (3),

$$|Z[\mathscr{P}; B((1, y_2(\nu)))] - \nu| \leqslant 2M - 1.$$

Hence

$$|(\nu - 1) - Ny_2(\nu)| \leqslant 3M. \tag{5}$$

On the other hand, $|Z[\mathscr{P}; B(\mathbf{x})] - Z[\mathscr{P}^*; B(\mathbf{x})]|$ does not exceed the number of ν for which exactly one of the two numbers $y_2(\nu)$ and $(\nu - 1)/N$ is strictly less than x_2. By (5), there are at most $6M$ such numbers ν. Hence

$$|Z[\mathscr{P}^*; B(\mathbf{x})] - Nx_1x_2| \leqslant |Z[\mathscr{P}; B(\mathbf{x})] - Z[\mathscr{P}^*; B(\mathbf{x})]| + |Z[\mathscr{P}; B(\mathbf{x})] - Nx_1x_2|,$$

and the result follows.

1.2 Application to approximate evaluation of certain functions

Roth's formulation of the original problem enables one to investigate the following question.

Suppose that g is a Lebesgue-integrable function, not necessarily bounded, in U^K, where $U = [0, 1]$. Let μ denote the Lebesgue measure in U^K. We are interested in functions of the type

$$C + \int_{B(\mathbf{x})} g(\mathbf{y})\, \mathrm{d}\mu. \tag{6}$$

Clearly, if g vanishes almost everywhere in U^K, then the integral in (6) vanishes for all $\mathbf{x} \in U^K$. We therefore consider only the following.

Definition. We denote by $\mathscr{F}(K)$ the class of all functions of the type (6) in U^K, where C is a real constant, and where $g(\mathbf{y}) \neq 0$ in a subset $S \subset U^K$ where $\mu(S) > 0$.

We shall show that functions in $\mathscr{F}(K)$ cannot be approximated very well by certain 'simple' functions.

Definition. By an M-simple function in U^K, we mean a function ψ defined by

$$\psi(\mathbf{x}) = \sum_{j=1}^{M} m_j \chi_{B_j}(\mathbf{x}) \tag{7}$$

for all $\mathbf{x} \in U^K$, where for each $j = 1, \dots, M$, B_j denotes a box of the type $(u_1^{(j)}, w_1^{(j)}] \times \cdots \times (u_K^{(j)}, w_K^{(j)}]$ in U^K, χ_{B_j} denotes the characteristic function of the box B_j, and the coefficients m_j are real.

Theorem 5A (Chen (1985)). *Suppose that $f \in \mathscr{F}(K)$. Then for every real number $W > 1$, there exists a positive constant $c_{15}(f, W)$ such that for every M-simple function ψ in U^K,*

$$\int_{U^K} |\psi(\mathbf{x}) - f(\mathbf{x})|^W \, d\mu > c_{15}(f, W) M^{-W} (\log M)^{\frac{1}{2}(K-1)W}.$$

As a simple corollary, we have

Theorem 5B (Chen (1985)). *Suppose that $f \in \mathscr{F}(K)$. Then there exists a positive constant $c_{16}(f)$ such that for every M-simple function ψ in U^K,*

$$\sup_{\mathbf{x} \in U^K} |\psi(\mathbf{x}) - f(\mathbf{x})| > c_{16}(f) M^{-1} (\log M)^{\frac{1}{2}(K-1)}.$$

Theorem 5B can be improved when $K = 2$.

Theorem 5C (Chen (TA)). *Suppose that $f \in \mathscr{F}(2)$. Then there exists a positive constant $c_{17}(f)$ such that for every M-simple function ψ in U^2,*

$$\sup_{\mathbf{x} \in U^2} |\psi(\mathbf{x}) - f(\mathbf{x})| > c_{17}(f) M^{-1} \log M.$$

As before, let $\mathscr{P} = \mathscr{P}(K, N)$ be a distribution of N points in U^K. Let h be any function bounded in U^K. For any $\mathbf{x} \in U^K$, let

$$Z[\mathscr{P}; h; B(\mathbf{x})] = \sum_{\mathbf{y} \in \mathscr{P} \cap B(\mathbf{x})} h(\mathbf{y}),$$

and write

$$D[\mathscr{P}; h; g; B(\mathbf{x})] = Z[\mathscr{P}; h; B(\mathbf{x})] - N \int_{B(\mathbf{x})} g(\mathbf{y}) \, d\mu.$$

Note that the term

$$N \int_{B(\mathbf{x})} g(\mathbf{y}) \, d\mu$$

can be interpreted as the product of the 'expected' number of points of $\mathscr{P}$ in $B(\mathbf{x})$ and the 'average' value of g in $B(\mathbf{x})$. It follows that the term $D[\mathscr{P}; g; g; B(\mathbf{x})]$ measures the 'discrepancy' between a sum of g over points of $\mathscr{P}$ in $B(\mathbf{x})$ and its expected value. In the case $h = g = 1$, the problem reduces to one purely concerned with the irregularities of distribution of the set $\mathscr{P}$ as introduced in §1.1.

We have the following generalizations of Theorems 1B, 3A and 3B.

Theorem 6A (Chen (1985)). *Let $W > 1$. There exists a positive constant $c_{18}(g, W)$ such that for every distribution $\mathscr{P}$ of N points in U^K and for every function h bounded in U^K,*

$$\int_{U^K} |D[\mathscr{P}; h; g; B(\mathbf{x})]|^W \,\mathrm{d}\mu > c_{18}(g, W)(\log N)^{\frac{1}{2}(K-1)W}.$$

Theorem 6B (Chen (1985)). *There exists a positive constant $c_{19}(g)$ such that for every distribution $\mathscr{P}$ of N points in U^K and for every function h bounded in U^K,*

$$\sup_{\mathbf{x} \in U^K} |D[\mathscr{P}; h; g; B(\mathbf{x})]| > c_{19}(g)(\log N)^{\frac{1}{2}(K-1)}.$$

Theorem 6C (Chen (TA)). *There exists a positive constant $c_{20}(g)$ such that for every distribution $\mathscr{P}$ of N points in U^2 and for every function h bounded in U^2,*

$$\sup_{\mathbf{x} \in U^2} |D[\mathscr{P}; h; g; B(\mathbf{x})]| > c_{20}(g) \log N.$$

We conclude this section by deducing Theorem 5C from Theorem 6C. The argument works similarly for deducing Theorems 5A and 5B from Theorems 6A and 6B respectively.

Let $f \in \mathscr{F}(2)$. Then f is of the form (6). Clearly, we may assume, without loss of generality, that $C = 0$. Let ψ be of the form (7), where, for each $j = 1, \ldots, M$,

$$B_j = (u_1^{(j)}, u_1^{(j)} + v_1^{(j)}] \times (u_2^{(j)}, u_2^{(j)} + v_2^{(j)}] \subset U^2.$$

For each $j = 1, \ldots, M$, let

$$\mathscr{P}_j = \{(u_1^{(j)} + \alpha_1 v_1^{(j)}, u_2^{(j)} + \alpha_2 v_2^{(j)}) : \alpha_1, \alpha_2 \in \{0, 1\}\};$$

in other words, $\mathscr{P}_j$ is the set of vertices of the rectangle B_j. For $\mathbf{y} \in U^2$ and $j = 1, \ldots, M$, let

$$h_j(\mathbf{y}) = \begin{cases} 1 & (\mathbf{y} \in \mathscr{P}_j \text{ and } \alpha_1 + \alpha_2 \text{ is even}); \\ -1 & (\mathbf{y} \in \mathscr{P}_j \text{ and } \alpha_1 + \alpha_2 \text{ is odd}); \\ 0 & (\mathbf{y} \notin \mathscr{P}_j). \end{cases}$$

It is quite easy to see that

$$\chi_{B_j}(\mathbf{x}) = \sum_{\mathbf{y} \in \mathscr{P}_j \cap B(\mathbf{x})} h_j(\mathbf{y})$$

in U^2. Let

$$\mathscr{P} = \bigcup_{j=1}^{M} \mathscr{P}_j,$$

and let H be defined for all $\mathbf{y} \in U^2$ by

$$H(\mathbf{y}) = \sum_{j=1}^{M} m_j h_j(\mathbf{y}).$$

Then clearly $\mathscr{P}$ is a distribution of $N \leqslant 4M$ points in U^2, and

$$\psi(\mathbf{x}) = \sum_{\mathbf{y} \in \mathscr{P} \cap B(\mathbf{x})} H(\mathbf{y}) = Z[\mathscr{P}; H; B(\mathbf{x})]$$

in U^2. Applying Theorem 6C with $h = MH$ and $N \leqslant 4M$, we get Theorem 5C.

1.3 A question of Erdös

We return to a formulation of the problem similar to that given by van Aardenne-Ehrenfest. Let $s_1, s_2, s_3, \ldots$ be an infinite sequence of real numbers satisfying $0 \leqslant s_\nu < 1$ for all $\nu \in \mathbb{N}$. For any $n \in \mathbb{N}$ and for any real number α satisfying $0 < \alpha \leqslant 1$, let $Z(n, \alpha)$ denote the number of s_ν $(\nu = 1, \ldots, n)$ that satisfy $0 \leqslant s_\nu < \alpha$, and write

$$D(n, \alpha) = Z(n, \alpha) - n\alpha.$$

Erdös (1964) asked the following question: Does there exist α satisfying $0 < \alpha \leqslant 1$ such that $D(n, \alpha)$ is unbounded?

The problem was solved by Schmidt (1968), who showed that such values of α exist. We shall prove the stronger

Theorem 7A (Schmidt (1977*b*)). *There exists α such that $0 < \alpha \leqslant 1$ and*

$$\limsup_{n \to \infty} \frac{|D(n, \alpha)|}{\log n} > c_{21}.$$

Theorem 7B (Schmidt (1977*b*)). *For almost all (in the sense of Lebesgue) α satisfying $0 < \alpha \leqslant 1$,*

$$\limsup_{n \to \infty} \frac{|D(n, \alpha)|}{\log \log n} > c_{22}.$$

Schmidt asked whether the factor ($\log\log n$) in Theorem 7B could be improved. This was answered in the affirmative by Tijdeman and Wagner, who proved the following improvement of Theorems 7A and 7B.

Theorem 7C (Tijdeman and Wagner (1980)). *For almost all (in the sense of Lebesgue) α satisfying* $0 < \alpha \leqslant 1$,

$$\limsup_{n\to\infty} \frac{|D(n,\alpha)|}{\log n} > c_{23}.$$

The method of proof of Theorem 7C is a variation of the method used by Schmidt (1977*b*) to prove Theorems 7A and 7B. To allow comparison, we prove Theorems 7A and 7B in §4.2 and Theorem 7C in §4.4.

Here, we should mention that using a variation of Roth's method, Halász (1981) showed that the set of values $\alpha \in U_1$ satisfying

$$D(n,\alpha) = \mathrm{o}(\log n)$$

has Hausdorff dimension 0.

A special case of Theorem 7C was studied earlier by Sós (1983) who showed that for the special sequence $s_\nu = \{\nu\theta\}$, where θ is some fixed number, we have that

$$\limsup_{n\to\infty} \frac{|D(n,\alpha)|}{\log n} > 0$$

for almost all α. Sós (1983) also showed that for this special sequence, we have that

$$\sup_{\alpha \in U_1} |D(n,\alpha)| > c_{24} \log n$$

for almost all n.

Schmidt also considered the function

$$E(\alpha) = \sup_{n\in\mathbb{N}} |D(n,\alpha)|.$$

For any $\kappa \in \mathbb{N}^*$, let

$$S(\kappa) = \{0 < \alpha \leqslant 1 : E(\alpha) \leqslant \kappa\},$$

and let

$$S(\infty) = \{0 < \alpha \leqslant 1 : E(\alpha) < \infty\}.$$

Theorem 8A (Schmidt (1972*a*)). *For any $\kappa \in \mathbb{N}^*$, $S(\kappa)$ is countable (i.e. finite or countably infinite) and nowhere dense. The set $S(\infty)$ is countable.*

Suppose now, instead of considering intervals of the form $[0,\alpha)$, we consider arbitrary intervals $\mathscr{I} \subset [0,1)$. For a given sequence $s_1, s_2, s_3, \ldots$ of real numbers in $[0,1)$, let $Z(n, \mathscr{I})$ denote the number of s_ν $(\nu = 1, \ldots, n)$ satisfying $s_\nu \in \mathscr{I}$, and write

$$D(n, \mathscr{I}) = Z(n, \mathscr{I}) - n|\mathscr{I}|,$$

where $|\mathscr{I}|$ denotes the length of $\mathscr{I}$. Let

$$E(\mathscr{I}) = \sup_{n \in \mathbb{N}} |D(n, \mathscr{I})|.$$

Theorem 8B (Schmidt (1974)). *The set* $\{|\mathscr{I}| : E(\mathscr{I}) < \infty\}$ *is countable.*

This generalization of Theorem 8A to arbitrary intervals causes considerable extra difficulties in the proof, which is too long to be included here. The interested reader is referred to Schmidt (1974), where he not only proved Theorem 8B, but also proved a generalization to rectangular boxes $\mathscr{B}$ with sides parallel to the coordinate axes in higher dimensions. Let $\mathbf{s}_1, \mathbf{s}_2, \mathbf{s}_3, \ldots$ be a sequence in U_0^k. For any $n \in \mathbb{N}$ and any set $\mathscr{S}$ in U_0^k, let $Z(n, \mathscr{S})$ denote the number of $\mathbf{s}_\nu$ $(\nu = 1, \ldots, n)$ satisfying $\mathbf{s}_\nu \in \mathscr{S}$, and let

$$D(n, \mathscr{S}) = Z(n, \mathscr{S}) - n|\mathscr{S}|,$$

where $|\mathscr{S}|$ denotes the measure of $\mathscr{S}$. Let

$$E(\mathscr{S}) = \sup_{n \in \mathbb{N}} |D(n, \mathscr{S})|.$$

Schmidt proved

Theorem 8C (Schmidt (1974)). *The set* $\{|\mathscr{B}| : E(\mathscr{B}) < \infty\}$ *is countable.*

In the opposite direction, we state but do not prove

Theorem 8D (Schmidt (1974)). *Suppose that* $k > 1$. *There exists a sequence* $\mathbf{s}_1, \mathbf{s}_2, \mathbf{s}_3, \ldots$ *in* U_0^k *such that for every* μ *satisfying* $0 \leqslant \mu \leqslant 1$, *there is a convex set* $\mathscr{S}$ *in* U_0^k *with* $|\mathscr{S}| = \mu$ *such that* $E(\mathscr{S}) \leqslant \frac{1}{2}$.

We conclude this chapter by mentioning a related result. Suppose that f is a bounded measurable function on the real line, having period 1. Let $s_1, s_2, s_3, \ldots$ be a sequence of real numbers. Let

$$D(n, f) = \sum_{i=1}^{n} f(s_i) - n \int_0^1 f(x)\,\mathrm{d}x.$$

Let f_t denote the translated function $f_t(x) = f(x - t)$, and write

$$\Delta(n, f) = \sup_t |D(n, f_t)|.$$

Schmidt (1977*b*) asked whether there exist functions f such that

$$\Delta(n, f) \to \infty \quad \text{as} \quad n \to \infty \tag{8}$$

no matter what the given sequence $s_1, s_2, s_3, \ldots$. This was answered in the affirmative.

Theorem 9 (Baker (1978)). *Suppose that f has Fourier series*

$$f(x) \sim \sum_{m=-\infty}^{\infty} a_m e(mx) \quad (a_{-m} = \overline{a_m}),$$

where $e(x) = e^{2\pi i x}$, and suppose that

$$\lim_{m \to \infty} m|a_m| = \infty.$$

Then (8) *holds no matter what the given sequence $s_1, s_2, s_3, \ldots$.*

We shall not prove Theorem 9. In fact, the method of proof is closer to material covered in Part B.

2
Lower bounds – Roth's method

2.1 Roth's orthogonal function method

The method of Roth (1954) to prove Theorem 1A is dependent on Schwarz's inequality. Corresponding to every distribution $\mathscr{P}$ of N points in U_0^K, Roth constructed an auxiliary function $F[\mathscr{P};\mathbf{x}]$ such that, writing $D(\mathbf{x})$ and $F(\mathbf{x})$ in place of $D[\mathscr{P};B(\mathbf{x})]$ and $F[\mathscr{P};\mathbf{x}]$ respectively,

$$\int_{U^K} F(\mathbf{x})D(\mathbf{x})\,d\mathbf{x} > c_1(K)(\log N)^{K-1} \tag{1}$$

and

$$\int_{U^K} F^2(\mathbf{x})\,d\mathbf{x} < c_2(K)(\log N)^{K-1}. \tag{2}$$

These, together with Schwarz's inequality, give

$$\int_{U^K} |D(\mathbf{x})|^2\,d\mathbf{x} > c_3(K)(\log N)^{K-1},$$

so that Theorem 1A follows easily.

The idea of Schmidt (1977*a*) is to use Hölder's inequality instead of Schwarz's inequality. He showed that Roth's auxiliary function $F(\mathbf{x})$ also satisfies

$$\int_{U^K} F^{2m}(\mathbf{x})\,d\mathbf{x} < c_4(K,m)(\log N)^{m(K-1)} \quad (m = 1, 2, \ldots). \tag{3}$$

Since $(\int_{U^K} |F(\mathbf{x})|^r\,d\mathbf{x})^{1/r}$ is an increasing function of $r > 0$ for every fixed function F, we have, for every $r > 0$, that

$$\int_{U^K} |F(\mathbf{x})|^r\,d\mathbf{x} < c_5(K,r)(\log N)^{\frac{1}{2}(K-1)r}. \tag{4}$$

(1) and (4), together with Hölder's inequality, give Theorem 1B.

In fact, the auxiliary function that Schmidt constructed is slightly different from that constructed by Roth, although Roth's auxiliary function also satisfies (3). We shall follow closely Schmidt's argument and indicate Roth's ideas.

Any $x \in U_0$ can be written in the form

$$x = \sum_{j=0}^{\infty} \beta_j(x) 2^{-j-1},$$

where $\beta_j(x) = 0$ or 1 and such that the sequence $\beta_j(x)$ does not end with $1, 1, \ldots$. For $r = 0, 1, 2, \ldots$, let

$$R_r(x) = (-1)^{\beta_r(x)}$$

(these are called the Rademacher functions).

Definition. By an r-interval, we mean an interval of the form $[m2^{-r}, (m+1)2^{-r})$, where the integer m satisfies $0 \leqslant m < 2^r$.

Definition. By an r-function, we mean a function $f(x)$ defined in U_0 such that in every r-interval, $f(x) = R_r(x)$ or $f(x) = -R_r(x)$.

Clearly, if $f(x)$ is an r-function, then

$$\int_U f(x)\,\mathrm{d}x = 0.$$

Suppose that $\mathbf{r} = (r_1, \ldots, r_K)$ is a K-tuple of non-negative integers. Let

$$|\mathbf{r}| = r_1 + \cdots + r_K; \tag{5}$$

and for any $\mathbf{x} \in U_0^K$, let

$$R_{\mathbf{r}}(\mathbf{x}) = R_{r_1}(x_1) \cdots R_{r_K}(x_K).$$

Definition. By an $\mathbf{r}$-box, we mean a set of the form $I_1 \times \cdots \times I_K$, where, for $j = 1, \ldots, K, I_j$ is an r_j-interval.

Definition. By an $\mathbf{r}$-function, we mean a function $f(\mathbf{x})$ defined in U_0^K such that in every $\mathbf{r}$-box, $f(\mathbf{x}) = R_{\mathbf{r}}(\mathbf{x})$ or $f(\mathbf{x}) = -R_{\mathbf{r}}(\mathbf{x})$.

The following three lemmas are useful.

Lemma 2.1. *Suppose that $f_1, \ldots, f_t$ are r_1-, $\ldots, r_t$-functions respectively. If an odd number among $r_1, \ldots, r_t$ are equal to $r = \max\{r_1, \ldots, r_t\}$, then the function $f_1(x) \cdots f_t(x)$ is an r-function, and so*

$$\int_U f_1(x) \cdots f_t(x)\,\mathrm{d}x = 0.$$

Proof. Note simply that the product of an odd number of r-functions is an r-function, and the product of an r-function and an s-function is an r-function if $s < r$. ■

Given $x_1,\dots,x_{j-1},x_{j+1},\dots,x_K$, an $\mathbf{r}$-function is an r_j-function in the variable x_j. It follows that if we single out the last coordinate, then we have

Lemma 2.2. *Suppose that $f_1,\dots,f_t$ are $\mathbf{r}_1$-$,\dots,\mathbf{r}_t$-functions respectively. For $i=1,\dots t$, write $\mathbf{r}_i=(r_{i1},\dots,r_{iK})$. Suppose further that an odd number among $r_{1K},\dots,r_{tK}$ are equal to $r=\max\{r_{1K},\dots,r_{tK}\}$. Then*

$$\int_{U^K} f_1(\mathbf{x})\cdots f_t(\mathbf{x})\,d\mathbf{x}=0.$$

For the case $K=2$, we have the following

Lemma 2.3. *Suppose that $K=2$ and $n\geqslant 0$. For each $\mathbf{r}$ with $|\mathbf{r}|=n$, let $f_{\mathbf{r}}$ be an $\mathbf{r}$-function. For $\mathbf{r}_1,\dots,\mathbf{r}_{2m}$ with $|\mathbf{r}_1|=\cdots=|\mathbf{r}_{2m}|=n$, write*

$$\mathscr{I}(\mathbf{r}_1,\dots,\mathbf{r}_{2m})=\int_{U^K} f_{\mathbf{r}_1}(\mathbf{x})\cdots f_{\mathbf{r}_{2m}}(\mathbf{x})\,d\mathbf{x}. \tag{6}$$

Then $\mathscr{I}(\mathbf{r}_1,\dots,\mathbf{r}_{2m})=0$ unless $\mathbf{r}_1,\dots,\mathbf{r}_{2m}$ form m pairs of equal K-tuples.

Proof. For $m=1$, it follows from Lemma 2.2 that $\mathscr{I}(\mathbf{r}_1,\mathbf{r}_2)=0$ unless $r_{12}=r_{22}$, i.e. unless $\mathbf{r}_1=\mathbf{r}_2$ since $|\mathbf{r}_1|=|\mathbf{r}_2|$. For $m>1$, the integral is again 0 unless there exist t and T satisfying $1\leqslant t<T\leqslant 2m$ such that $\mathbf{r}_t=\mathbf{r}_T$. Since the square of an $\mathbf{r}$-function is 1, it follows that

$$\mathscr{I}(\mathbf{r}_1,\dots,\mathbf{r}_{2m})=\mathscr{I}(\mathbf{r}_1,\dots,\mathbf{r}_{t-1},\mathbf{r}_{t+1},\dots,\mathbf{r}_{T-1},\mathbf{r}_{T+1},\dots,\mathbf{r}_{2m}),$$

and the result follows by induction on m. ■

It is because of the case $m=1$ of Lemma 2.3 that we sometimes call the functions $f_{\mathbf{r}}(\mathbf{x})$ orthogonal functions.

In Roth (1954), Roth considered the following: For any $\mathbf{r}$-box B, let

$$g_{\mathbf{r}}(\mathbf{x})=\begin{cases}(-1)^{K+1}R_{\mathbf{r}}(\mathbf{x}) & (B \text{ contains no points of } \mathscr{P}),\\ 0 & (B \text{ contains a point of } \mathscr{P});\end{cases}$$

and write

$$F(\mathbf{x})=\sum_{|\mathbf{r}|=n} g_{\mathbf{r}}(\mathbf{x}), \tag{7}$$

where $n \gg \ll \log N$ is a suitably chosen integer. Schmidt (1977*a*) considered instead

$$F(\mathbf{x})=\sum_{|\mathbf{r}|=n} f_{\mathbf{r}}(\mathbf{x}),$$

where, for each $\mathbf{r}$, $f_{\mathbf{r}}(\mathbf{x})$ is a suitably chosen $\mathbf{r}$-function.

Schmidt proved the following form of (3).

Lemma 2.4. *Suppose that $K\geqslant 2$, $m\geqslant 1$ and $n\geqslant 0$. Suppose that for every $\mathbf{r}$*

with $|\mathbf{r}| = n$, $f_{\mathbf{r}}$ *is an* **r**-*function. If*

$$F(\mathbf{x}) = \sum_{|\mathbf{r}|=n} f_{\mathbf{r}}(\mathbf{x}), \tag{8}$$

then

$$\int_{U^K} F^{2m}(\mathbf{x})\,d\mathbf{x} < (2m)^{m(2K-3)}(n+1)^{m(K-1)}. \tag{9}$$

Remark. It can be shown that Roth's function (7) satisfies

$$\int_{U^K} F^{2m}(\mathbf{x})\,d\mathbf{x} < (2m)^{m(2K-2)}(n+1)^{m(K-1)}.$$

Note also that in Lemma 2.4, we have not chosen our auxiliary function yet, as (9) holds for every function of the form (8).

Let n be chosen to satisfy

$$2N \leqslant 2^n < 4N. \tag{10}$$

Then Theorem 1B follows from Lemma 2.4 and

Lemma 2.5. *Suppose that* $2^n \geqslant 2N$. *Then for every* **r** *satisfying* $|\mathbf{r}| = n$, *there is an* **r**-*function* $f_{\mathbf{r}}$ *satisfying*

$$\int_{U^K} f_{\mathbf{r}}(\mathbf{x})D(\mathbf{x})\,d\mathbf{x} \geqslant 2^{-n-2K-1}N. \tag{11}$$

For given n satisfying (10), we can construct $F(\mathbf{x})$ by (8), where for every **r** satisfying $|\mathbf{r}| = n$, $f_{\mathbf{r}}$ is chosen to satisfy (11). Now the number of K-tuples **r** satisfying $|\mathbf{r}| = n$ is greater than $c_6(K)(n+1)^{K-1}$, so that

$$\int_{U^K} F(\mathbf{x})D(\mathbf{x})\,d\mathbf{x} > c_7(K)(n+1)^{K-1}2^{-n}N,$$

a suitable form of (1). It remains to prove Lemmas 2.4 and 2.5.

Proof of Lemma 2.4. Let $M = M(K, m, n)$ be the maximum of $\int_{U^K} F^{2m}(\mathbf{x})\,d\mathbf{x}$, where the maximum is taken over all functions $F(\mathbf{x})$ of the type (8), where each $f_{\mathbf{r}}(\mathbf{x})$ is either an **r**-function or is identically zero. To prove Lemma 2.4, it suffices to prove that

$$M < (2m)^{m(2K-3)}(n+1)^{m(K-1)}. \tag{12}$$

Let $f_{\mathbf{r}}$ be chosen so that

$$\int_{U^K} F^{2m}(\mathbf{x})\,d\mathbf{x} = M.$$

Then

$$M = \sum_{|\mathbf{r}_1|=n} \cdots \sum_{|\mathbf{r}_{2m}|=n} \mathscr{I}(\mathbf{r}_1,\ldots,\mathbf{r}_{2m}). \tag{13}$$

We shall prove (12) by induction on K. Suppose that $K=2$. Then by Lemma 2.3, we can restrict the sum (13) to a sum over those $\mathbf{r}_1,\ldots,\mathbf{r}_{2m}$ which form m pairs of equal K-tuples. Then the summand is 0 if any of the functions $f_{\mathbf{r}_1},\ldots,f_{\mathbf{r}_{2m}}$ is identically zero, and 1 otherwise. The number of divisions of $\{1,\ldots,2m\}$ into pairs is $(2m-1)(2m-3)\cdots 3\cdot 1<(2m)^m$, and the number of possibilities for $\mathbf{r}_{i_1},\ldots,\mathbf{r}_{i_m}$ is $(n+1)^m$. Hence for $K=2$,

$$M = \int_{U^2} F^{2m}(\mathbf{x})\,d\mathbf{x} < (2m)^m(n+1)^m,$$

which is (12) with $K=2$. Suppose now that $K>2$. We make use of Lemma 2.2 with $t=2m$. For each $\mathscr{I}(\mathbf{r}_1,\ldots,\mathbf{r}_{2m})$ on the right-hand side of (13), let $r=\max\{r_{1K},\ldots,r_{(2m)K}\}$ and let h be the number of r_{iK} $(i=1,\ldots,2m)$ equal to r. By Lemma 2.2, $\mathscr{I}(\mathbf{r}_1,\ldots,\mathbf{r}_{2m})=0$ unless h is even, i.e. unless $h=2u$ where $1\leqslant u\leqslant m$. Note that there are $\binom{2m}{2u}$ ways of choosing $2u$ elements out of $\{1,\ldots,2m\}$. Note also that $\mathscr{I}(\mathbf{r}_1,\ldots,\mathbf{r}_{2m})$ is invariant under permutations of $\mathbf{r}_1,\ldots,\mathbf{r}_{2m}$. It follows that

$$M = \sum_{u=1}^{m}\binom{2m}{2u}\sum_{r=0}^{n}\sum_{u+v=m}\sum\nolimits_{2u}'\sum\nolimits_{2v}''\mathscr{I}(\mathbf{r}_1,\ldots,\mathbf{r}_{2u},\mathbf{s}_1,\ldots,\mathbf{s}_{2v}),$$

where

$$\sum\nolimits_{2u}' = \sum_{\substack{\mathbf{r}_1,\ldots,\mathbf{r}_{2u}\\ |\mathbf{r}_1|=\cdots=|\mathbf{r}_{2u}|=n\\ r_{1K}=\cdots=r_{(2u)K}=r}} \quad\text{and}\quad \sum\nolimits_{2v}'' = \sum_{\substack{\mathbf{s}_1,\ldots,\mathbf{s}_{2v}\\ |\mathbf{s}_1|=\cdots=|\mathbf{s}_{2v}|=n\\ s_{1K},\ldots,s_{(2v)K}<r}}.$$

Writing

$$A_r(\mathbf{x}) = \sum_{\substack{|\mathbf{r}|=n\\ r_K=r}} f_{\mathbf{r}}(\mathbf{x}) \quad\text{and}\quad B_r(\mathbf{x}) = \sum_{\substack{|\mathbf{r}|=n\\ r_K<r}} f_{\mathbf{r}}(\mathbf{x}),$$

we have, with the convention $u+v=m$, that

$$\begin{aligned} M &= \sum_{u=1}^{m}\binom{2m}{2u}\sum_{r=0}^{n}\int_{U^K} A_r^{2u}(\mathbf{x})B_r^{2v}(\mathbf{x})\,d\mathbf{x}\\ &\leqslant \sum_{u=1}^{m}\binom{2m}{2u}\sum_{r=0}^{n}\left(\int_{U^K} A_r^{2m}(\mathbf{x})\,d\mathbf{x}\right)^{u/m}\left(\int_{U^K} B_r^{2m}(\mathbf{x})\,d\mathbf{x}\right)^{v/m}. \end{aligned} \tag{14}$$

If we write $\bar{\mathbf{r}}=(r_1,\ldots,r_{K-1})$, then

$$A_r(\mathbf{x}) = \sum_{|\bar{\mathbf{r}}|=n-r} f_{(\bar{\mathbf{r}},r)}(\mathbf{x}).$$

Hence for given x_K, $A_r(\mathbf{x})$ is of the form $F(\mathbf{x})$ with $(K-1)$ and $(n-r)$ in place of K and n respectively. By the induction hypothesis,

$$\int_{U^K} A_{\mathbf{r}}^{2m}(\mathbf{x})\,d\mathbf{x} < (2m)^{m(2K-5)}(n+1)^{m(K-2)}. \tag{15}$$

The function $B_{\mathbf{r}}(\mathbf{x})$ is of the form $F(\mathbf{x})$, except that certain $f_{\mathbf{r}}(\mathbf{x})$ are replaced by 0. Hence

$$\int_{U^K} B_{\mathbf{r}}^{2m}(\mathbf{x})\,d\mathbf{x} \leqslant M. \tag{16}$$

Combining (14)–(16), we have

$$M < (n+1)\sum_{u=1}^{m}\binom{2m}{2u}((2m)^{2K-5}(n+1)^{K-2})^u M^{v/m},$$

so that

$$(n+1)^{-1} < \sum_{u=1}^{m}\binom{2m}{2u}((2m)^{2K-5}(n+1)^{K-2}M^{-1/m})^u$$

$$\leqslant \sum_{u=1}^{m}\frac{1}{(2u)!}((2m)^{2K-3}(n+1)^{K-2}M^{-1/m})^u.$$

Since

$$(n+1)^{-1} > \sum_{u=1}^{m}\frac{1}{(2u)!}(n+1)^{-u},$$

we have

$$(n+1)^{-1} < (2m)^{2K-3}(n+1)^{K-2}M^{-1/m},$$

which gives (12). ■

Proof of Lemma 2.5. We decompose the integral (11) into integrals over **r**-boxes. We choose $f_{\mathbf{r}}(\mathbf{x})$ such that the integral $\int f_{\mathbf{r}}(\mathbf{x})D(\mathbf{x})\,d\mathbf{x}$ over every **r**-box is non-negative. Let B be an **r**-box given by

$$B = [m_1 2^{-r_1}, (m_1+1)2^{-r_1}) \times \cdots \times [m_K 2^{-r_K}, (m_K+1)2^{-r_K}),$$

and let B' be the box

$$B' = [m_1 2^{-r_1}, (m_1+\tfrac{1}{2})2^{-r_1}) \times \cdots \times [m_K 2^{-r_K}, (m_K+\tfrac{1}{2})2^{-r_K}).$$

Then it is not difficult to see that $\int_B R_{\mathbf{r}}(\mathbf{x})D(\mathbf{x})\,d\mathbf{x}$ is equal to

$$\int_{B'} \sum_{\alpha_1=0}^{1}\cdots\sum_{\alpha_K=0}^{1}(-1)^{\alpha_1+\cdots+\alpha_K}D((y_1+\alpha_1 2^{-r_1-1},\ldots,y_K+\alpha_K 2^{-r_K-1}))\,d\mathbf{y}.$$

Note that the sum

$$\left|\sum_{\alpha_1=0}^{1}\cdots\sum_{\alpha_K=0}^{1}(-1)^{\alpha_1+\cdots+\alpha_K}Z[\mathscr{P};B((y_1+\alpha_1 2^{-r_1-1},\ldots,y_K+\alpha_K 2^{-r_K-1}))]\right|$$

is the number of points of $\mathscr{P}$ in $[y_1, y_1 + 2^{-r_1-1}) \times \cdots \times [y_K, y_K + 2^{-r_K-1})$. This box is contained in B. Hence if B contains no points of $\mathscr{P}$, the sum is 0. Note also that

$$\sum_{\alpha_1=0}^{1} \cdots \sum_{\alpha_K=0}^{1} (-1)^{\alpha_1+\cdots+\alpha_K}(y_1 + \alpha_1 2^{-r_1-1}) \cdots (y_K + \alpha_K 2^{-r_K-1})$$
$$= (-1)^K 2^{-|\mathbf{r}|-K}.$$

It follows from the definition of $D(\mathbf{x})$ that if B contains no points of $\mathscr{P}$, then since $|\mathbf{r}| = n$, we have

$$\int_B R_{\mathbf{r}}(\mathbf{x}) D(\mathbf{x})\, d\mathbf{x} = (-1)^{K+1} 2^{-2n-2K} N.$$

There are 2^n **r**-boxes B with $|\mathbf{r}| = n$, but only $N \leqslant 2^{n-1}$ points. It follows that at least half of the **r**-boxes contain no points of $\mathscr{P}$. Since $f_{\mathbf{r}}(\mathbf{x})$ is chosen to make the integral $\int f_{\mathbf{r}}(\mathbf{x}) D(\mathbf{x})\, d\mathbf{x}$ over any **r**-box non-negative, it follows that

$$\int_{U^K} f_{\mathbf{r}}(\mathbf{x}) D(\mathbf{x})\, d\mathbf{x} \geqslant (2^n - N) 2^{-2n-2K} N \geqslant 2^{-n-2K-1} N. \qquad \blacksquare$$

Remark. In Roth (1954), the function $R_{\mathbf{r}}(\mathbf{x})$ offers no choice for the sign of $\int R_{\mathbf{r}}(\mathbf{x}) D(\mathbf{x})\, d\mathbf{x}$ over an **r**-box, but gives the same sign for $\int R_{\mathbf{r}}(\mathbf{x}) D(\mathbf{x})\, d\mathbf{x}$ over any **r**-box not containing any points of $\mathscr{P}$. Roth therefore replaced $R_{\mathbf{r}}(\mathbf{x})$ by 0 in those **r**-boxes containing a point of $\mathscr{P}$ so that there is no contribution from these boxes where the contribution to the integral might otherwise be of the 'wrong' sign.

2.2 Halász's variation of Roth's method

Motivated by the fact that Roth's analytic method in Roth (1954) works efficiently in higher dimensions, Halász (1981) devised an ingenious variation of the method, whereby he replaced Roth's auxiliary function $F(\mathbf{x})$ by a quite different one, albeit only in the case $K = 2$. By this approach, Halász was able to give an alternative proof of Theorem 3B. Unfortunately, his argument does not seem to generalize to higher dimensions.

Let α satisfy $0 < \alpha < \frac{1}{2}$, and let $K = 2$. Instead of defining the auxiliary function $F(\mathbf{x})$ by (7), Halász considered, with suitably chosen $n \gg \ll \log N$,

$$F(\mathbf{x}) = \prod_{|\mathbf{r}|=n} (1 + \alpha g_{\mathbf{r}}(\mathbf{x})) - 1.$$

Then, in view of a variation of Lemma 2.2,

$$\int_{U^2} |F(\mathbf{x})|\, d\mathbf{x} \leqslant \int_{U^2} \prod_{|\mathbf{r}|=n} (1 + \alpha g_{\mathbf{r}}(\mathbf{x}))\, d\mathbf{x} + 1 = 2,$$

so that

$$\left|\int_{U^2} F(\mathbf{x})D(\mathbf{x})\,d\mathbf{x}\right| \leqslant 2\sup_{\mathbf{x}\in U^2}|D(\mathbf{x})|.$$

It remains to find a suitable lower bound for $|\int_{U^2}F(\mathbf{x})D(\mathbf{x})\,d\mathbf{x}|$.

It is not difficult to prove

Lemma 2.6. *Suppose, for* $i=1,\ldots,j$, *that* $\mathbf{r}_i=(r_{i1},r_{i2})$ *satisfies* $|\mathbf{r}_i|=n$. *Suppose further that* $\mathbf{r}_1,\ldots,\mathbf{r}_j$ *are all different. Then if* $\mathbf{s}=(s_1,s_2)$, *where*

$$s_1=\max_{1\leqslant i\leqslant j} r_{i1} \quad \text{and} \quad s_2=\max_{1\leqslant i\leqslant j} r_{i2},$$

then for any $\mathbf{s}$*-box* B, *exactly one of the following three conditions holds:*

(*i*) $g_{\mathbf{r}_1}\cdots g_{\mathbf{r}_j}=R_{\mathbf{s}}$; *or*
(*ii*) $g_{\mathbf{r}_1}\cdots g_{\mathbf{r}_j}=-R_{\mathbf{s}}$; *or*
(*iii*) $g_{\mathbf{r}_1}\cdots g_{\mathbf{r}_j}=0$.

Furthermore, (*iii*) *holds in any* $\mathbf{s}$*-box* B *where* $B\cap\mathscr{P}\neq\varnothing$.

Note that

$$F(\mathbf{x})=\alpha F_1(\mathbf{x})+\sum_{j=2}^{n+1}\alpha^j F_j(\mathbf{x}), \tag{17}$$

where

$$F_1(\mathbf{x})=\sum_{|\mathbf{r}|=n} g_{\mathbf{r}}(\mathbf{x}),$$

and where, for $j=2,\ldots,n+1$,

$$F_j(\mathbf{x})=\sum_{\substack{|\mathbf{r}_1|=\cdots=|\mathbf{r}_j|=n\\ \mathbf{r}_{i_1}\neq\mathbf{r}_{i_2}\text{ if } i_1\neq i_2}} g_{\mathbf{r}_1}(\mathbf{x})\cdots g_{\mathbf{r}_j}(\mathbf{x}).$$

From the proof of Lemma 2.5, it is clear that for every $\mathbf{r}$ satisfying $|\mathbf{r}|=n$,

$$\int_{U^2} g_{\mathbf{r}}(\mathbf{x})D(\mathbf{x})\,d\mathbf{x}\geqslant(2^n-N)2^{-2n-4}N,$$

so that

$$\int_{U^2} F_1(\mathbf{x})D(\mathbf{x})\,d\mathbf{x}\geqslant(n+1)(2^n-N)2^{-2n-4}N. \tag{18}$$

Lemma 2.7. *For* $j=2,\ldots,n+1$,

$$\left|\int_{U^2} F_j(\mathbf{x})D(\mathbf{x})\,d\mathbf{x}\right|\leqslant\sum_{r=0}^{n-j+1}\sum_{h=1}^{n-r}2^{-n-h-4}N\binom{h-1}{j-2}. \tag{19}$$

Before we prove Lemma 2.7, we shall first deduce Theorem 3B. From

(19), we have that

$$\left|\sum_{j=2}^{n+1}\alpha^j\int_{U^2}F_j(\mathbf{x})D(\mathbf{x})\,d\mathbf{x}\right|$$

$$\leqslant\sum_{j=2}^{n+1}\sum_{r=0}^{n-j+1}\sum_{h=1}^{n-r}\alpha^j 2^{-n-h-4}N\binom{h-1}{j-2}$$

$$=\sum_{r=0}^{n-1}\sum_{h=1}^{n-r}\sum_{j=2}^{h+1}\alpha^2 2^{-n-h-4}N\binom{h-1}{j-2}\alpha^{j-2}$$

$$\leqslant N\sum_{r=0}^{n-1}\sum_{h=1}^{\infty}2^{-n-h-4}\alpha^2(1+\alpha)^h$$

$$\leqslant Nn2^{-n-4}\alpha^2\sum_{h=0}^{\infty}\left(\frac{1+\alpha}{2}\right)^h\leqslant Nn2^{-n-2}\alpha^2, \tag{20}$$

since $0<\alpha<\frac{1}{2}$. We now choose n to satisfy

$$2N\leqslant 2^n<4N. \tag{21}$$

Then combining (17), (18), (20) and (21) and choosing

$$\alpha=2^{-6},$$

we have

$$\int_{U^2}F(\mathbf{x})D(\mathbf{x})\,d\mathbf{x}\geqslant\alpha n(2^n-N)2^{-2n-4}N-Nn2^{-n-2}\alpha^2$$

$$>\frac{\alpha n}{256}-\frac{\alpha^2 n}{8}=2^{-15}n,$$

proving Theorem 3B.

Proof of Lemma 2.7. It can be shown, as in the proof of Lemma 2.5, that for any $\mathbf{s}$-box B, where $\mathbf{s}$ is defined in Lemma 2.6, we have

$$\left|\int_B g_{\mathbf{r}_1}(\mathbf{x})\cdots g_{\mathbf{r}_j}(\mathbf{x})Nx_1x_2\,d\mathbf{x}\right|\leqslant 2^{-2|\mathbf{s}|-4}N.$$

Since there are $2^{|\mathbf{s}|}$ $\mathbf{s}$-boxes in U^2, it follows that

$$\left|\int_{U^2} g_{\mathbf{r}_1}(\mathbf{x})\cdots g_{\mathbf{r}_j}(\mathbf{x})Nx_1x_2\,d\mathbf{x}\right|\leqslant 2^{-|\mathbf{s}|-4}N.$$

If we use the convention $\mathbf{r}_i=(r_{i1},r_{i2})$ and $r_{i_1 1}<r_{i_2 1}$ if $i_1<i_2$, we have, writing $h=r_{j1}-r_{11}$, that

$$\left|\int_{U^2} g_{\mathbf{r}_1}(\mathbf{x})\cdots g_{\mathbf{r}_j}(\mathbf{x})Nx_1x_2\,d\mathbf{x}\right|\leqslant 2^{-n-h-4}N.$$

The lemma follows on noting that $\binom{h-1}{j-2}$ is the number of choices for $r_{21},\ldots,r_{(j-1)1}$, fixing $r_{11}=r$ and $r_{j1}=r+h$. ■

The method can also be used to prove

Theorem 1C. *There exists a constant c_8 such that for any distribution $\mathscr{P}$ of N points in U_0^2, we have $\|D(\mathscr{P})\|_1 > c_8 (\log N)^{\frac{1}{2}}$.*

To prove Theorem 1C, Halász considered the auxiliary function

$$F(\mathbf{x}) = \prod_{|\mathbf{r}|=n} (1 + \mathrm{i}n^{-\frac{1}{2}} g_{\mathbf{r}}(\mathbf{x})) - 1,$$

where $\mathrm{i} = \sqrt{-1}$. Then $|F(\mathbf{x})| \leqslant (1 + 1/n)^{\frac{1}{2}(n+1)} + 1 \leqslant c_9$, so that

$$\left| \int_{U^2} F(\mathbf{x}) D(\mathbf{x})\,\mathrm{d}\mathbf{x} \right| \leqslant c_9 \int_{U^2} |D(\mathbf{x})|\,\mathrm{d}\mathbf{x}.$$

One now handles the left-hand side in a similar way.

We conclude this section by discussing Halász's recent new ideas for proving Theorem 3C, which can be viewed as a stronger form of Theorem 3B.

Suppose that $\mathscr{P} = \{\mathbf{p}_1, \ldots, \mathbf{p}_N\}$ is a distribution of N points in U_0^2. For technical reasons, we consider open squares of the form $\mathbf{x} + Q_\rho$, where $Q_\rho = (-\frac{1}{2}\rho, \frac{1}{2}\rho) \times (-\frac{1}{2}\rho, \frac{1}{2}\rho)$ and $\mathbf{x} + Q_\rho$ denotes the translation of Q_ρ by $\mathbf{x}$. Write

$$D_\rho(\mathbf{x}) = D[\mathscr{P}; \mathbf{x} + Q_\rho] = \sum_{i=1}^{N} \chi_\rho(\mathbf{p}_i - \mathbf{x}) - N\rho^2, \tag{22}$$

where χ_ρ denotes the characteristic function of the square Q_ρ.

The vertices of the square $\mathbf{p}_i + Q_\rho$ are given by

$$\mathbf{p}_i + ((-1)^s \tfrac{1}{2}\rho, (-1)^t \tfrac{1}{2}\rho),$$

where $s, t \in \{0, 1\}$.

Definition. A vertex of the square $\mathbf{p}_i + Q_\rho$ is said to be positive if $s = t$ and negative if $s \neq t$.

In other words, the 'lower left' and 'upper right' vertices are positive.

Halász considered, with suitably chosen $n \gg\ll \log N$, the auxiliary function

$$F(\mathbf{x}) = \prod_{|\mathbf{r}|=n}{}^{*} (1 + \alpha h_{\mathbf{r}}(\mathbf{x})) - 1, \tag{23}$$

where * denotes the restriction

$$\frac{n}{4} < r_1, r_2 < \frac{3n}{4} \quad \text{if} \quad \mathbf{r} = (r_1, r_2), \tag{24}$$

and where $\alpha \in (0, \frac{1}{2})$ is to be determined suitably. The functions $h_{\mathbf{r}}(\mathbf{x})$ are defined as follows: For any $\mathbf{r}$-box B,

$$h_{\mathbf{r}}(\mathbf{x}) = \begin{cases} R_{\mathbf{r}}(\mathbf{x}) & (B \subset \mathscr{R} \text{ and contains positive but no negative vertices}), \\ -R_{\mathbf{r}}(\mathbf{x}) & (B \subset \mathscr{R} \text{ and contains negative but no positive vertices}), \\ 0 & (\text{otherwise}); \end{cases} \tag{25}$$

where

$$\mathscr{R} = [\tfrac{1}{4}, \tfrac{3}{4}) \times [\tfrac{1}{4}, \tfrac{3}{4}). \tag{26}$$

Then

$$\int_{U^2} |F(\mathbf{x})| \, d\mathbf{x} \leqslant 2,$$

so that

$$\int_{\mathscr{R}} D_\rho(\mathbf{x}) F(\mathbf{x}) \, d\mathbf{x} \leqslant 2 \sup_{\mathbf{x} \in \mathscr{R}} |D_\rho(\mathbf{x})|. \tag{27}$$

It remains to find a suitable lower bound for the left-hand side of (27).

Note that

$$\begin{aligned} \int_{\mathscr{R}} D_\rho(\mathbf{x}) F(\mathbf{x}) \, d\mathbf{x} &= \int_{\mathscr{R}} \sum_{i=1}^{N} \chi_\rho(\mathbf{p}_i - \mathbf{x}) F(\mathbf{x}) \, d\mathbf{x} = \sum_{i=1}^{N} \int_{\mathbf{p}_i + Q_\rho} F(\mathbf{x}) \, d\mathbf{x} \\ &= \alpha \sum_{i=1}^{N} \int_{\mathbf{p}_i + Q_\rho} F_1(\mathbf{x}) \, d\mathbf{x} + \sum_{i=1}^{N} \sum_{j=2}^{n+1} \alpha^j \int_{\mathbf{p}_i + Q_\rho} F_j(\mathbf{x}) \, d\mathbf{x} \\ &= I_1(\rho) + I_2(\rho), \end{aligned} \tag{28}$$

say, where

$$F_1(\mathbf{x}) = \sum_{|\mathbf{r}| = n}^{*} h_{\mathbf{r}}(\mathbf{x}) \tag{29}$$

and, for $j = 2, \ldots, n+1$,

$$F_j(\mathbf{x}) = \sum_{\substack{|\mathbf{r}_1| = \cdots = |\mathbf{r}_j| = n \\ \mathbf{r}_{i_1} \neq \mathbf{r}_{i_2} \text{ if } i_1 \neq i_2}}^{*} h_{\mathbf{r}_1}(\mathbf{x}) \cdots h_{\mathbf{r}_j}(\mathbf{x}).$$

Here * again denotes that the restriction (24) applies.

Firstly, we need to show that

$$|I_2(\rho)| \leqslant c_{10} N 2^{-n} \alpha^2 n, \tag{30}$$

where c_{10} is a positive absolute constant. We therefore have to evaluate integrals of the type

$$\int_{\mathbf{p}_i+Q_\rho} h_{\mathbf{r}_1}(\mathbf{x})\cdots h_{\mathbf{r}_j}(\mathbf{x})\,d\mathbf{x}. \tag{31}$$

To do this, we can divide the unit square into **s**-boxes, where **s** is defined as in Lemma 2.6. Then in view of Lemma 2.6, at most four such **s**-boxes (i.e. those containing vertices of $\mathbf{p}_i + Q_\rho$) will give non-zero contribution to the integral (31). An argument analogous to Lemma 2.7 and (20) will give (30).

Our main task is, therefore, to show that for some values of ρ, we have

$$I_1(\rho) \geqslant c_{11}\alpha \log N. \tag{32}$$

The proof of Theorem 3C will be complete if we choose α sufficiently small independent of N, in view of (22), (27) and (28).

We therefore have to investigate integrals of the form

$$\int_{\mathbf{p}_i+Q_\rho} h_{\mathbf{r}}(\mathbf{x})\,d\mathbf{x}.$$

Note that if an **r**-box B does not contain a vertex of $\mathbf{p}_i + Q_\rho$, then

$$\int_{(\mathbf{p}_i+Q_\rho)\cap B} h_{\mathbf{r}}(\mathbf{x})\,d\mathbf{x} = 0.$$

Also, if the **r**-box B contains a vertex of $\mathbf{p}_i + Q_\rho$, then

$$\int_{(\mathbf{p}_i+Q_\rho)\cap B} R_{\mathbf{r}}(\mathbf{x})\,d\mathbf{x}$$

is non-negative or non-positive depending on whether the vertex is positive or negative. Hence, for any **r**-box B, in view of (25),

$$\int_{(\mathbf{p}_i+Q_\rho)\cap B} h_{\mathbf{r}}(\mathbf{x})\,d\mathbf{x} \geqslant 0 \tag{33}$$

always.

This trivial estimate is clearly insufficient. Halász's new idea is that for certain values of ρ and for certain **r**-boxes B, one can get better estimates than (33).

Definition. Let B be an **r**-box. A vertex $\mathbf{v}$ of a square $\mathbf{p}_i + Q_\rho$ is said to be well inside B if $\mathbf{v}$ lies in a rectangle with the same centre as B but shrunk by a factor $\frac{15}{16}$ on each side.

Here the choice $\frac{15}{16}$ is arbitrary.

Definition. A vertex $\mathbf{v}$ of a square $\mathbf{p}_i + Q_\rho$ is said to be $\mathbf{r}$-good if $\mathbf{v} \in \mathcal{R}$ and $\mathbf{v}$ lies well inside an $\mathbf{r}$-box which does not contain vertices of the opposite sign.

It is not difficult to see that

Lemma 2.8. *Suppose that an $\mathbf{r}$-box B contains an $\mathbf{r}$-good vertex of the square $\mathbf{p}_i + Q_\rho$. Then*

$$\int_{(\mathbf{p}_i + Q_\rho) \cap B} h_{\mathbf{r}}(\mathbf{x})\, d\mathbf{x} > c_{12} 2^{-n},$$

where $|\mathbf{r}| = n$.

It follows from Lemma 2.8 that

$$I_1(\rho) \geqslant c_{12} 2^{-n} \alpha \sum_{i=1}^{N} \sum_{|\mathbf{r}|=n}^{*} \psi(i, \mathbf{r}, \rho),$$

where

$$\psi(i, \mathbf{r}, \rho) = \begin{cases} 1 & (\mathbf{p}_i + Q_\rho \text{ has an } \mathbf{r}\text{-good vertex}), \\ 0 & (\text{otherwise}). \end{cases}$$

It therefore remains to prove

Lemma 2.9. *There exists $\rho \in (0, \frac{1}{16})$ such that*

$$\sum_{i=1}^{N} \sum_{|\mathbf{r}|=n}^{*} \psi(i, \mathbf{r}, \rho) > c_{13} N n. \tag{34}$$

We shall prove that there exists a subset $\mathcal{P}'$ of $\mathcal{P}$ of at least $N/5$ points such that for every $\mathbf{p}_i \in \mathcal{P}'$ and for every $\mathbf{r}$ satisfying $|\mathbf{r}| = n$ and (24), we have

$$\int_0^{\frac{1}{16}} \psi(i, \mathbf{r}, \rho)\, d\rho \geqslant \tfrac{1}{64}. \tag{35}$$

Lemma 2.9 will follow immediately from (35).

Let $\mathcal{Q} = (\frac{1}{8}, \frac{7}{8})^2$. We may clearly assume that for every $\rho \in (0, \frac{1}{8}]$ and every $\mathbf{x} \in \mathcal{Q}$,

$$|D_\rho(\mathbf{x})| = \left| \sum_{i=1}^{N} \chi_\rho(\mathbf{p}_i - \mathbf{x}) - N\rho^2 \right| < c_{14} \log N; \tag{36}$$

for otherwise Theorem 3C follows immediately.

Consider the square $\mathcal{R}$. Dividing $\mathcal{R}$ into squares of side $\frac{1}{8}$, it follows from (36) that the number of points of $\mathcal{P}$ in $\mathcal{R}$ is at least

$$N/4 - 16 c_{14} \log N \geqslant N/5$$

for sufficiently large N. Let

$$\mathcal{P}' = \mathcal{P} \cap \mathcal{R}.$$

Lemma 2.10. *Any aligned rectangle of sides $a, b < \frac{1}{8}$ in $\mathcal{Q}$ contains at most $4abN$ points of $\mathcal{P}$ provided that*

$$abN > c_{14} \log N. \tag{37}$$

Proof. Let A be a rectangle in $\mathcal{Q}$ with sides $a < b < \frac{1}{8}$. Then A is contained in a square A_1 in $\mathcal{Q}$ of side b such that $A_1 \backslash A$ contains a square A_2 of side $b-a$. Then $A \subset A_1 \backslash A_2$. By (36), the number of points of $\mathcal{P}$ in A_1 does not exceed $Nb^2 + c_{14} \log N$, and the number of points of $\mathcal{P}$ in A_2 is at least $N(b-a)^2 - c_{14} \log N$. Hence the number of points of $\mathcal{P}$ in A does not exceed

$$N(b^2 - (b-a)^2) + 2c_{14} \log N < 2abN + 2c_{14} \log N < 4abN. \qquad \blacksquare$$

Let $\mathbf{p}_i \in \mathcal{P}'$. Then $\mathbf{p}_i \in \mathcal{R}$. Suppose that $\mathbf{p}_i \in [\frac{1}{4}, \frac{1}{2}]^2$ (the other possibilities are treated *mutatis mutandis*). Then the positive vertex $\mathbf{p}_i + (\frac{1}{2}\rho, \frac{1}{2}\rho)$ of the square $\mathbf{p}_i + Q_\rho$ clearly lies in $\mathcal{R}$. (35) will follow from the following two lemmas and their analogues.

Lemma 2.11. *Let* $\mathbf{r}$ *satisfy* $|\mathbf{r}| = n$ *and* (24). *Then the set*

$$S_1(i, \mathbf{r}) = \{\rho \in (0, \tfrac{1}{16}) : \mathbf{p}_i + (\tfrac{1}{2}\rho, \tfrac{1}{2}\rho) \textit{ lies well inside an } \mathbf{r}\textit{-box}\}$$

has measure at least $\frac{1}{32}$.

Lemma 2.12. *Let* $\mathbf{r}$ *satisfy* $|\mathbf{r}| = n$ *and* (24). *Then the set*

$$S_2(i, \mathbf{r}) = \{\rho \in (0, \tfrac{1}{16}) : \textit{the } \mathbf{r}\textit{-box containing } \mathbf{p}_i + (\tfrac{1}{2}\rho, \tfrac{1}{2}\rho) \textit{ contains no negative vertices}\}$$

has measure at least $\frac{3}{64}$.

Proof of Lemma 2.11. Suppose that the vertex $\mathbf{p}_i + (\frac{1}{2}\rho, \frac{1}{2}\rho)$ does not lie well inside an $\mathbf{r}$-box, where $\mathbf{r} = (r_1, r_2)$, then, writing $\mathbf{p}_i = (p_i^{(1)}, p_i^{(2)})$, either $p_i^{(1)} + \frac{1}{2}\rho$ is 'very near' an integer multiple of 2^{-r_1} or $p_i^{(2)} + \frac{1}{2}\rho$ is 'very near' an integer multiple of 2^{-r_2}. More precisely, writing $\|u\|$ to be the distance of u from the nearest integer, we have either

$$\|2^{r_1}(p_i^{(1)} + \tfrac{1}{2}\rho)\| < \tfrac{1}{32} \tag{38}$$

or

$$\|2^{r_2}(p_i^{(2)} + \tfrac{1}{2}\rho)\| < \tfrac{1}{32}. \tag{39}$$

Clearly, for sufficiently large n, each of the sets $\{\rho \in (0, \frac{1}{16}) : (38) \text{ holds}\}$ and $\{\rho \in (0, \frac{1}{16}) : (39) \text{ holds}\}$ has measure at most $\frac{1}{64}$. $\blacksquare$

Proof of Lemma 2.12. Suppose that an $\mathbf{r}$-box B, where $\mathbf{r} = (r_1, r_2)$, contains $\mathbf{p}_i + (\frac{1}{2}\rho, \frac{1}{2}\rho)$ and a negative vertex of a square $\mathbf{p}_j + Q_\rho$. There are two possibilities. Either (I) B contains the vertex $\mathbf{p}_j + (-\frac{1}{2}\rho, \frac{1}{2}\rho)$; or (II) B contains the vertex $\mathbf{p}_j + (\frac{1}{2}\rho, -\frac{1}{2}\rho)$. Consider first of all case (I). Writing $\mathbf{p}_i = (p_i^{(1)}, p_i^{(2)})$ and $\mathbf{p}_j = (p_j^{(1)}, p_j^{(2)})$, we must have

$$|(p_i^{(1)} + \tfrac{1}{2}\rho) - (p_j^{(1)} - \tfrac{1}{2}\rho)| < 2^{-r_1} \tag{40}$$

and

$$|(p_i^{(2)} + \tfrac{1}{2}\rho) - (p_j^{(2)} + \tfrac{1}{2}\rho)| < 2^{-r_2}.$$

It follows that

$$\mathbf{p}_j \in (p_i^{(1)} - 2^{-r_1}, p_i^{(1)} + \tfrac{1}{16} + 2^{-r_1}) \times (p_i^{(2)} - 2^{-r_2}, p_i^{(2)} + 2^{-r_2}), \tag{41}$$

a rectangle of sides less than $\frac{1}{8}$. It follows from Lemma 2.10 that the number of such $\mathbf{p}_j$ is less than

$$4(\tfrac{1}{16} + 2\cdot 2^{-r_1})(2\cdot 2^{-r_2})N < 2^{-r_2}N \tag{42}$$

provided that

$$(\tfrac{1}{16} + 2\cdot 2^{-r_1})(2\cdot 2^{-r_2})N > c_{14} \log N,$$

a condition clearly satisfied for large N in view of (24) and the choice $n \gg \ll \log N$. By (40), each such $\mathbf{p}_j$ in (41) determines an interval of length $2\cdot 2^{-r_1}$ for ρ. Combining this with (42), the total measure of these values of ρ does not exceed $2\cdot 2^{-n}N$. An analogous argument applies for case (II). Hence the set $(0, \frac{1}{16}) \backslash S_2(i, \mathbf{r})$ has measure at most $4\cdot 2^{-n}N < \frac{1}{64}$ if $n \gg \ll \log N$ is chosen suitably. ■

2.3 Application to approximate evaluation of certain functions

In this section, we prove Theorem 6A and indicate how we may prove Theorem 6C.

The proof of Theorem 6A is basically similar to that of Theorem 1B by Schmidt (see §2.1), and involves establishing the existence of positive constants $c_{15} = c_{15}(g)$ and $c_{16} = c_{16}(g, m)$ for $m = 1, 2, \ldots$ and an auxiliary function $F(\mathbf{x}) = F[\mathscr{P}; h; g; \mathbf{x}]$ such that, writing $D(\mathbf{x}) = D[\mathscr{P}; h; g; B(\mathbf{x})]$, we have

$$\int_{U^K} F(\mathbf{x}) D(\mathbf{x})\, d\mu > c_{15}(g)(\log N)^{K-1} \tag{43}$$

and, for $m = 1, 2, \ldots,$

$$\int_{U^K} F^{2m}(\mathbf{x})\, d\mu < c_{16}(g, m)(\log N)^{m(K-1)}. \tag{44}$$

As before, the auxiliary function F chosen is again of the type (8), so that (9) still holds. Again if

$$n \gg \ll \log N \tag{45}$$

is a suitably chosen integer, then (44) holds for suitable $c_{16}(g, m)$. It therefore remains to prove (43).

The main difficulty in establishing (43) is that the function g may take different signs in any **r**-box. To overcome this, we have to look for regions in U^K where g is 'predominantly positive' or 'predominantly negative', and deal with the remaining regions in a trivial way.

Suppose that g is a Lebesgue-integrable function in U^K. Let S be a measurable subset of U^K such that $\mu(S) > 0$ and $g(\mathbf{y}) \neq 0$ for every $\mathbf{y} \in S$. Then, replacing g by $-g$ if necessary, we may assume that there exist two positive constants $c_{17} = c_{17}(g)$ and $c_{18} = c_{18}(g)$ and a subset $S_1 \subset S$ such that

$$\mu(S_1) = c_{17}(g) \tag{46}$$

and

$$g(\mathbf{y}) \geqslant c_{18}(g) \quad \text{for every} \quad \mathbf{y} \in S_1. \tag{47}$$

Consider the function g^- defined for all $\mathbf{y} \in U^K$ by

$$g^-(\mathbf{y}) = \max\{-g(\mathbf{y}), 0\}. \tag{48}$$

Clearly g^- is Lebesgue-integrable in U^K. We write

$$c_{19}(g) = 2^{-K-3} c_{17}(g) c_{18}(g). \tag{49}$$

Then there is a positive constant $c_{20} = c_{20}(g)$ satisfying

$$c_{20}(g) \leqslant 2^{-K-4} c_{17}(g) \tag{50}$$

such that for every measurable set $E \subset U^K$,

$$\int_E g^-(\mathbf{y})\,d\mu \leqslant c_{19}(g) \quad \text{if} \quad \mu(E) \leqslant c_{20}(g). \tag{51}$$

We now try to approximate the region S_1. We shall call a set in U^K an elementary box in U^K if it is a **t**-box, where $\mathbf{t} = (t_1, \ldots, t_K) \in \mathbb{N}^K$. Since S_1 is measurable, there is a finite union T^* of elementary boxes in U^K such that

$$\mu(T^* \Delta S_1) \leqslant c_{20}(g). \tag{52}$$

Here $T^* \Delta S_1$ denotes the symmetric difference of T^* and S_1. Writing

$$E = T^* \backslash S_1, \tag{53}$$

we have, in view of (52), that

$$\mu(E) \leqslant c_{20}(g). \tag{54}$$

Also it follows easily from (50) that

$$\mu(T^*) \geqslant \tfrac{1}{2}c_{17}(g). \tag{55}$$

T^* is by definition a finite union of elementary boxes in U^K. Hence, for every $i = 1, \ldots, K$, there is one such elementary box with maximal t_i. Let T_i denote this maximal value of t_i, and write

$$T = T_1 + \cdots + T_K. \tag{56}$$

We shall only prove (43) and (44) when

$$N > c_{17}(g)2^{2T-3}. \tag{57}$$

Suppose that N is given such that (57) holds. Let $n \in \mathbb{N}$ be defined by

$$c_{17}(g)2^{n-3} < N \leqslant c_{17}(g)2^{n-2}. \tag{58}$$

Then

$$n \geqslant 2T \tag{59}$$

and (45) holds. We shall consider a function of the type (8).

Lemma 2.13. *If (58) holds, then for every* $\mathbf{r}$ *satisfying* $|\mathbf{r}| = n$ *and* $r_i \geqslant T_i$ *for every* $i = 1, \ldots, K$, *there is an* $\mathbf{r}$*-function* $f_\mathbf{r}$ *satisfying*

$$\int_{U^K} f_\mathbf{r}(\mathbf{x})D(\mathbf{x})\,\mathrm{d}\mu \geqslant 2^{-n-2K-4}c_{17}(g)c_{18}(g)N. \tag{60}$$

Lemma 2.14. *If (58) holds, then for every* $\mathbf{r}$ *satisfying* $|\mathbf{r}| = n$, *there is an* $\mathbf{r}$*-function* $f_\mathbf{r}$ *satisfying*

$$\int_{U^K} f_\mathbf{r}(\mathbf{x})D(\mathbf{x})\,\mathrm{d}\mu \geqslant 0. \tag{61}$$

The inequality (43) follows easily from Lemmas 2.13 and 2.14. Let F be defined by (8), where, for all $\mathbf{r}$ satisfying the hypotheses of Lemma 2.13, $f_\mathbf{r}$ is chosen to satisfy (60), and where, for the remaining $\mathbf{r}$, $f_\mathbf{r}$ is chosen to satisfy (61). By (56), the number of $\mathbf{r}$ satisfying the hypotheses of Lemma 2.13 is at least $c_{21}(K)(n-T)^{K-1}$, where $c_{21}(K)$ is a positive constant. Combining this with (59)–(61) and (8), it follows easily that

$$\int_{U^K} F(\mathbf{x})D(\mathbf{x})\,\mathrm{d}\mu \geqslant c_{22}(g)2^{-n}n^{K-1}N, \tag{62}$$

where $c_{22}(g)$ is a positive constant. (43) follows from (58) and (62).

Lemma 2.14 is trivial, so we only indicate the proof of Lemma 2.13 here. To prove Lemma 2.13, we first of all split the integral (60) into integrals over

r-boxes as in the proof of Lemma 2.5. We say that an **r**-box is 'good' if it is contained in T^* and does not contain any point of the distribution $\mathscr{P}$. For any **r**-box which is not 'good', we simply choose $f_{\mathbf{r}}(\mathbf{x}) = R_{\mathbf{r}}(\mathbf{x})$ or $f_{\mathbf{r}}(\mathbf{x}) = -R_{\mathbf{r}}(\mathbf{x})$ so that the integral $\int f_{\mathbf{r}}(\mathbf{x})D(\mathbf{x})\,\mathrm{d}\mu$ over the **r**-box is non-negative. Let B be a 'good' **r**-box given by

$$B = [m_1 2^{-r_1}, (m_1+1)2^{-r_1}) \times \cdots \times [m_K 2^{-r_K}, (m_K+1)2^{-r_K}),$$

and let B' be the box

$$B' = [m_1 2^{-r_1}, (m_1+\tfrac{1}{2})2^{-r_1}) \times \cdots \times [m_K 2^{-r_K}, (m_K+\tfrac{1}{2})2^{-r_K}).$$

Then it is not too difficult to see that $\int_B R_{\mathbf{r}}(\mathbf{x})D(\mathbf{x})\,\mathrm{d}\mu$ is equal to

$$\int_{B'} \sum_{\alpha_1=0}^{1} \cdots \sum_{\alpha_K=0}^{1} (-1)^{\alpha_1+\cdots+\alpha_K} D((x_1+\alpha_1 2^{-r_1-1}, \ldots, x_K+\alpha_K 2^{-r_K-1}))\,\mathrm{d}\mu. \tag{63}$$

By the definition of $Z[\mathscr{P}; h; B(\mathbf{x})]$, the sum

$$\left| \sum_{\alpha_1=0}^{1} \cdots \sum_{\alpha_K=0}^{1} (-1)^{\alpha_1+\cdots+\alpha_K} Z[\mathscr{P}; h; B((x_1+\alpha_1 2^{-r_1-1}, \ldots, x_K+\alpha_K 2^{-r_K-1}))] \right| = \left| \sum_{\mathbf{y}\in\mathscr{P}\cap B^*(\mathbf{x})} h(\mathbf{y}) \right|, \tag{64}$$

where $B^*(\mathbf{x}) = [x_1, x_1+2^{-r_1-1}) \times \cdots \times [x_K, x_K+2^{-r_K-1}) \subset B$ for every $\mathbf{x}\in B'$. Since B is 'good', $\mathscr{P}\cap B^*(\mathbf{x}) = \varnothing$ for every $\mathbf{x}\in B'$, and so (64) vanishes. Hence, by the definition of $D(\mathbf{x})$, (63) is equal to

$$-N\int_{B'} \left(\sum_{\alpha_1=0}^{1} \cdots \sum_{\alpha_K=0}^{1} (-1)^{\alpha_1+\cdots+\alpha_K} \int_0^{x_1+\alpha_1 2^{-r_1-1}} \cdots \int_0^{x_K+\alpha_K 2^{-r_K-1}} g(\mathbf{y})\,\mathrm{d}\mu \right) \mathrm{d}\mu$$
$$= (-1)^{K+1} N \int_{B'} \left(\int_{x_1}^{x_1+2^{-r_1-1}} \cdots \int_{x_K}^{x_K+2^{-r_K-1}} g(\mathbf{y})\,\mathrm{d}\mu \right) \mathrm{d}\mu$$
$$= (-1)^{K+1} N \int_B K_B(\mathbf{y}) g(\mathbf{y})\,\mathrm{d}\mu,$$

where

$$K_B(\mathbf{y}) = (2^{-r_1-1} - |y_1 - (m_1+\tfrac{1}{2})2^{-r_1}|) \cdots (2^{-r_K-1} - |y_K - (m_K+\tfrac{1}{2})2^{-r_K}|).$$

For $\mathbf{x}\in B$, let $f_{\mathbf{r}}(\mathbf{x}) = (-1)^{K+1} R_{\mathbf{r}}(\mathbf{x})$. Then for any 'good' **r**-box B,

$$\int_B f_{\mathbf{r}}(\mathbf{x}) D(\mathbf{x})\,\mathrm{d}\mu = N \int_B K_B(\mathbf{y}) g(\mathbf{y})\,\mathrm{d}\mu.$$

It follows that

$$\int_{U^K} f_{\mathbf{r}}(\mathbf{x})D(\mathbf{x})\,\mathrm{d}\mu \geqslant N \sum_{B\text{`good'}} \int_B K_B(\mathbf{y})g(\mathbf{y})\,\mathrm{d}\mu.$$

By (55), for any $\mathbf{r}$ satisfying the hypotheses of Lemma 2.13, we have at least $(\frac{1}{2}c_{17}(g)2^n - N)$ 'good' $\mathbf{r}$-boxes in U^K. It follows from (47), (53), (48), (51), (58), (49), (54) and (50) that

$$\begin{aligned}
&\int_{U^K} f_{\mathbf{r}}(\mathbf{x})D(\mathbf{x})\,\mathrm{d}\mu \\
&\geqslant Nc_{18}(g) \sum_{B\text{`good'}} \int_B K_B(\mathbf{y})\,\mathrm{d}\mu - N2^{-n-K}c_{18}(g)\mu(E) - N2^{-n-K}\int_E g^-(\mathbf{y})\,\mathrm{d}\mu \\
&\geqslant N_{18}(g)2^{-2n-2K}(\tfrac{1}{2}c_{17}(g)2^n - N) - N2^{-n-K}c_{18}(g)\mu(E) - N2^{-n-K}c_{19}(g) \\
&\geqslant N2^{-n-K}(2^{-K-2}c_{17}(g)c_{18}(g) - c_{18}(g)\mu(E) - c_{19}(g)) \\
&= N2^{-n-K}(2^{-K-3}c_{17}(g)c_{18}(g) - c_{18}(g)\mu(E)) \\
&\geqslant N2^{-n-2K-4}c_{17}(g)c_{18}(g).
\end{aligned}$$

This completes the proof of Lemma 2.13.

We briefly indicate how we can prove Theorem 6C, using Halász's method with ideas in this section. For Halász's method to work, we need analogues of (18) and (19). An analogue of (18) can easily be obtained in the same way as in the proof of (43). To prove an analogue of (19), we need to make sure that in the regions we have chosen, the value of $|g|$ is not 'too large' 'too often'. So instead of considering a set like S_1, we look at the following. Replacing g by $-g$ if necessary, we may assume that there exist three positive constants $c_{23} = c_{23}(g)$, $c_{24} = c_{24}(g)$ and $c_{25} = c_{25}(g)$ and a subset $S_2 \subset S$ such that

$$\mu(S_2) = c_{23}(g)$$

and

$$c_{24}(g) \leqslant g(\mathbf{y}) \leqslant c_{25}(g) \quad \text{for every} \quad \mathbf{y} \in S_2.$$

This extra upper bound on the value $g(\mathbf{y})$ in S_2 is sufficient to enable us to obtain upper bounds of the type (19). For a detailed account, see Chen (TA).

2.4 An application of Halász's method

Let $\mathscr{P}$ be a distribution of N points in U_0^2. For every $\mathbf{p} \in \mathscr{P}$, let $\lambda(\mathbf{p}) = \pm 1$. Let B denote any rectangle in U^2 with sides parallel to the coordinate axes.

Consider the function

$$T(N) = \sup_{\mathscr{P}} \inf_{\lambda} \sup_{B} \left| \sum_{\mathbf{p} \in \mathscr{P} \cap B} \lambda(\mathbf{p}) \right|,$$

where the supremum is taken over all distributions $\mathscr{P}$ of N points in U_0^2 and all aligned rectangles B in U^2, and the infimum is taken over all '2-colourings' λ of $\mathscr{P}$.

Theorem 10A (Beck (1981*b*)). *We have, for* $N \geq 2$,

$$\log N \ll T(N) \ll (\log N)^4. \tag{65}$$

Note that the assertion $T(N) \gg \log N$ implies the existence of a distribution $\mathscr{P}$ of N points in U_0^2 such that if 'weights' $\lambda(\mathbf{p}) = \pm 1$ are freely assigned to the points of $\mathscr{P}$, then there is an aligned rectangle B in U^2 such that

$$\left| \sum_{\mathbf{p} \in \mathscr{P} \cap B} \lambda(\mathbf{p}) \right| \gg (\log N) N^{-1} \sum_{\mathbf{p} \in \mathscr{P}} |\lambda(\mathbf{p})|. \tag{66}$$

Using a variant of Halász's method in §2.2, Roth recently generalized the lower bound result in Theorem 10A to arbitrary real weights $\lambda(\mathbf{p})$.

Theorem 10B (Roth (1985)). *For every natural number* $N \geq 1$, *there exists a distribution* $\mathscr{P}$ *of* N *points in* U_0^2 *such that for every real-valued function* λ *defined on* U_0^2, *there exists an aligned rectangle* B *in* U^2 *such that* (66) *holds.*

The rest of this section is devoted to the proof of Theorem 10B. For a proof of the upper bound in Theorem 10A, see §8.2.

For $\mathbf{v} = (v_1, v_2)$, let $\delta(\mathbf{v}) = \|v_1\| \, \|v_2\|$, where $\|\cdot\|$ denotes the distance to the nearest integer. Let n be a non-negative integer.

Definition. A distribution $\mathscr{P}$ of a finite number of points in U_0^2 is said to be well-separated with respect to the non-negative integer n if

$$\inf_{\mathbf{p}', \mathbf{p}'' \in \mathscr{P}} \delta(\mathbf{p}' - \mathbf{p}'') \geq 2^{-n}. \tag{67}$$

Let $\mathscr{P}$ be a distribution of a finite number of points in U_0^2, and let λ be a real-valued function defined on U_0^2. Let

$$Z(\boldsymbol{\alpha}; \mathbf{x}) = \sum_{\mathbf{p} \in \mathscr{P} \cap (B(\mathbf{x}) + \boldsymbol{\alpha})} \lambda(\mathbf{p}), \tag{68}$$

where for every $\mathbf{x} \in U^2$, $B(\mathbf{x}) = [0, x_1) \times [0, x_2)$. Note that here we use the convention of vector addition modulo U_0^2.

Lemma 2.15. *Suppose that the distribution $\mathscr{P}$ is well-separated with respect to the non-negative integer n. Then there exist $\boldsymbol{\alpha}$ and $\mathbf{x}$ such that*

$$|Z(\boldsymbol{\alpha};\mathbf{x})| \geqslant (n+1)2^{-n-4} \sum_{\mathbf{p}\in\mathscr{P}} |\lambda(\mathbf{p})|. \tag{69}$$

Theorem 10B will follow from Lemma 2.15 if we can construct a distribution $\mathscr{P}$ of N points in U_0^2 which satisfies (67) for some non-negative integer $n \gg \ll \log N$. To do this, we can, for example, use the following well-known result in diophantine approximation: If θ is an irrational number whose continued fraction has bounded partial quotients, then $\nu\|\nu\theta\| > c_{26} > 0$ for all $\nu = 1, 2, 3, \ldots$, where $c_{26} = c_{26}(\theta)$ depends only on θ. The set

$$\{(\{\nu\theta\}, \nu N^{-1}) : 0 \leqslant \nu < N\}$$

clearly has the desired property. In fact, it is quite easy to construct other sets with the desired well-separation property.

We shall follow the notation of §2.1.

Let r be a non-negative integer.

Definition. By a modified r-function, we mean a function defined in U_0 of the type $C_r(x)R_r(x)$, where $R_r(x)$ is the Rademacher function defined in §2.1 and where $C_r(x)$ is constant on any r-interval.

It is clear that if $M_r(x)$ is a modified r-function, then

$$\int_U M_r(x)\,\mathrm{d}x = 0. \tag{70}$$

Let $\mathbf{r} = (r_1, r_2)$, where r_1 and r_2 are non-negative integers.

Definition. By a modified $\mathbf{r}$-function, we mean a function defined in U_0^2 of the type $C_{\mathbf{r}}(\mathbf{x})R_{\mathbf{r}}(\mathbf{x})$, where $R_{\mathbf{r}}(\mathbf{x}) = R_{r_1}(x_1)R_{r_2}(x_2)$ for every $\mathbf{x} = (x_1, x_2)$ and where $C_{\mathbf{r}}(\mathbf{x})$ is constant in any $\mathbf{r}$-box.

It is trivial that for a fixed x_1, a modified $\mathbf{r}$-function is a modified r_2-function in the variable x_2, and that for a fixed x_2, a modified $\mathbf{r}$-function is a modified r_1-function in the variable x_1.

Let $\mathscr{P}$ be a fixed distribution, well separated with respect to the non-negative integer n. Let λ be a real-valued function defined on U_0^2. We may assume, without loss of generality, that

$$\sum_{\mathbf{p}\in\mathscr{P}} \lambda(\mathbf{p}) \geqslant 0. \tag{71}$$

Let $\beta(\mathbf{p}) = \pm 1$ such that

$$\beta(\mathbf{p})\lambda(\mathbf{p}) = |\lambda(\mathbf{p})|. \tag{72}$$

For every $\mathbf{r}=(r_1,r_2)$, let $|\mathbf{r}|=r_1+r_2$ and let $\mathscr{A}_{\mathbf{r}}$ denote the set of all $\mathbf{r}$-boxes in U_0^2. For every $B\in\mathscr{A}_{\mathbf{r}}$, let χ_B denote the characteristic function of B. We consider

$$g_{\mathbf{r}}(\boldsymbol{\alpha};\mathbf{x})=R_{\mathbf{r}}(\mathbf{x})\sum_{\mathbf{p}\in\mathscr{P}}\beta(\mathbf{p})\sum_{B\in\mathscr{A}_{\mathbf{r}}}\chi_B(\mathbf{p}-\boldsymbol{\alpha})\chi_B(\mathbf{x}), \tag{73}$$

and, using the key idea of Halász, we let

$$F(\boldsymbol{\alpha};\mathbf{x})=\prod_{|\mathbf{r}|=n}(1+g_{\mathbf{r}}(\boldsymbol{\alpha};\mathbf{x})). \tag{74}$$

Note that in view of (67), any $\mathbf{r}$-box with $|\mathbf{r}|=n$ contains at most one point of the type $(\mathbf{p}-\boldsymbol{\alpha})$, where $\mathbf{p}\in\mathscr{P}$, and so

$$1+g_{\mathbf{r}}(\boldsymbol{\alpha};\mathbf{x})\geqslant 0. \tag{75}$$

In analogy to (17), we can write

$$F(\boldsymbol{\alpha};\mathbf{x})=1+F_1(\boldsymbol{\alpha};\mathbf{x})+\sum_{j=2}^{n+1}F_j(\boldsymbol{\alpha};\mathbf{x}), \tag{76}$$

where

$$F_1(\boldsymbol{\alpha};\mathbf{x})=\sum_{|\mathbf{r}|=n}g_{\mathbf{r}}(\boldsymbol{\alpha};\mathbf{x}), \tag{77}$$

and where, for $j=2,\ldots,n+1$,

$$F_j(\boldsymbol{\alpha};\mathbf{x})=\sum_{\substack{|\mathbf{r}_1|=\cdots=|\mathbf{r}_j|=n\\ \mathbf{r}_{i_1}\neq\mathbf{r}_{i_2}\text{ if } i_1\neq i_2}}g_{\mathbf{r}_1}(\boldsymbol{\alpha};\mathbf{x})\cdots g_{\mathbf{r}_j}(\boldsymbol{\alpha};\mathbf{x}). \tag{78}$$

We shall deduce Lemma 2.15 from the following two lemmas.

Lemma 2.16. *For every* $\boldsymbol{\alpha}$ *and every* $j=1,\ldots,n+1$,

$$\int_{U^2}F_j(\boldsymbol{\alpha};\mathbf{x})\,d\mathbf{x}=0. \tag{79}$$

Lemma 2.17.

(i) *We have*

$$\int_{U^2}\int_{U^2}F_1(\boldsymbol{\alpha};\mathbf{x})Z(\boldsymbol{\alpha};\mathbf{x})\,d\mathbf{x}\,d\boldsymbol{\alpha}=(n+1)2^{-n-4}\sum_{\mathbf{p}\in\mathscr{P}}|\lambda(\mathbf{p})|. \tag{80}$$

(ii) *For every* $j=2,\ldots,n+1$,

$$\int_{U^2}\int_{U^2}F_j(\boldsymbol{\alpha};\mathbf{x})Z(\boldsymbol{\alpha};\mathbf{x})\,d\mathbf{x}\,d\boldsymbol{\alpha}=0. \tag{81}$$

Before we prove Lemmas 2.16 and 2.17, we shall deduce Lemma 2.15. In view of (75), it follows from (76) and Lemma 2.16 that

$$\int_{U^2}\int_{U^2} |F(\boldsymbol{\alpha};\mathbf{x})|\,d\mathbf{x}\,d\boldsymbol{\alpha} = \int_{U^2}\int_{U^2} F(\boldsymbol{\alpha};\mathbf{x})\,d\mathbf{x}\,d\boldsymbol{\alpha} = 1. \tag{82}$$

On the other hand, combining (76) and Lemma 2.17, we have

$$\begin{aligned}
&\int_{U^2}\int_{U^2} F(\boldsymbol{\alpha};\mathbf{x})Z(\boldsymbol{\alpha};\mathbf{x})\,d\mathbf{x}\,d\boldsymbol{\alpha}\\
&= \int_{U^2}\int_{U^2} Z(\boldsymbol{\alpha};\mathbf{x})\,d\mathbf{x}\,d\boldsymbol{\alpha} + (n+1)2^{-n-4}\sum_{\mathbf{p}\in\mathscr{P}}|\lambda(\mathbf{p})|\\
&= \int_{U^2} x_1x_2\sum_{\mathbf{p}\in\mathscr{P}}\lambda(\mathbf{p})\,d\mathbf{x} + (n+1)2^{-n-4}\sum_{\mathbf{p}\in\mathscr{P}}|\lambda(\mathbf{p})|\\
&\geqslant (n+1)2^{-n-4}\sum_{\mathbf{p}\in\mathscr{P}}|\lambda(\mathbf{p})|, \qquad (83)
\end{aligned}$$

in view of the assumption (71). Lemma 2.15 follows on combining (82), (83) and

$$\sup_{\boldsymbol{\alpha},\mathbf{x}}|Z(\boldsymbol{\alpha};\mathbf{x})| = \sup_{\boldsymbol{\alpha},\mathbf{x}}|Z(\boldsymbol{\alpha};\mathbf{x})|\int_{U^2}\int_{U^2}|F(\boldsymbol{\alpha};\mathbf{x})|\,d\mathbf{x}\,d\boldsymbol{\alpha}.$$

It remains to prove Lemmas 2.16 and 2.17.

By (78), we see that to prove Lemma 2.16, it clearly suffices to prove that for distinct $\mathbf{r}_1,\ldots,\mathbf{r}_j$ satisfying $|\mathbf{r}_1| = \cdots = |\mathbf{r}_j| = n$, we have

$$\int_{U^2} g_{\mathbf{r}_1}(\mathbf{x})\cdots g_{\mathbf{r}_j}(\mathbf{x})\,d\mathbf{x} = 0. \tag{84}$$

For $i = 1,\ldots,j$, let $\mathbf{r}_i = (r_{i1}, r_{i2})$, and let $\mathbf{s} = (s_1, s_2)$ be defined by

$$s_1 = \max_{1\leqslant i\leqslant j} r_{i1} \quad\text{and}\quad s_2 = \max_{1\leqslant i\leqslant j} r_{i2}. \tag{85}$$

Let B be an $\mathbf{s}$-box. In view of (67), it is clear that if $\chi_B(\mathbf{p}-\boldsymbol{\alpha}) = 1$, then for each $i = 1,\ldots,j$,

$$\chi_B(\mathbf{x})g_{\mathbf{r}_i}(\boldsymbol{\alpha};\mathbf{x}) = R_{\mathbf{r}_i}(\mathbf{x})\beta(\mathbf{p})\chi_B(\mathbf{x}), \tag{86}$$

so that

$$\prod_{i=1}^{j} g_{\mathbf{r}_i}(\boldsymbol{\alpha};\mathbf{x}) = C_{\mathbf{s}}(\boldsymbol{\alpha};\mathbf{x})\prod_{i=1}^{j} R_{\mathbf{r}_i}(\mathbf{x}), \tag{87}$$

where

$$\chi_B(\mathbf{x})C_{\mathbf{s}}(\boldsymbol{\alpha};\mathbf{x}) = (\beta(\mathbf{p}))^j\chi_B(\mathbf{x}) \tag{88}$$

is constant in the $\mathbf{s}$-box B. Assume now that

$$r_{11} < r_{21} < \cdots < r_{j1}.$$

Clearly

$$\prod_{i=1}^{j} R_{\mathbf{r}_i}(\mathbf{x}) = Q(\mathbf{x})R_{\mathbf{s}}(\mathbf{x}), \tag{89}$$

where

$$Q(\mathbf{x}) = \left(\prod_{i=1}^{j-1} R_{r_{i1}}(x_1)\right)\left(\prod_{i=2}^{j} R_{r_{i2}}(x_2)\right) \tag{90}$$

is constant in any s-box. Note also that $Q(\mathbf{x}) = 1$ if $j = 1$. It easily follows from (90) that

$$\int_{U^2} Q(\mathbf{x})\,d\mathbf{x} = 0 \quad \text{if} \quad j \geqslant 2. \tag{91}$$

Combining (87)–(89), we have that if B is an s-box and $\chi_B(\mathbf{p} - \boldsymbol{\alpha}) = 1$, then

$$\chi_B(\mathbf{x}) \prod_{i=1}^{j} g_{\mathbf{r}_i}(\boldsymbol{\alpha}; \mathbf{x}) = (\beta(\mathbf{p}))^j \chi_B(\mathbf{x}) Q(\mathbf{x}) R_{\mathbf{s}}(\mathbf{x}). \tag{92}$$

Note that (89) is a modified s-function, and therefore so is (87) for every value of $\boldsymbol{\alpha}$. (84) follows on noting (70). This proves Lemma 2.16.

To prove Lemma 2.17, note that for each s, defined above by (85), we have

$$\int_{U^2}\left(\prod_{i=1}^{j} g_{\mathbf{r}_i}(\boldsymbol{\alpha}; \mathbf{x})\right) Z(\boldsymbol{\alpha}; \mathbf{x})\,d\mathbf{x} = \sum_{B \in \mathscr{A}_{\mathbf{s}}} \int_B \left(\prod_{i=1}^{j} g_{\mathbf{r}_i}(\boldsymbol{\alpha}; \mathbf{x})\right) Z(\boldsymbol{\alpha}; \mathbf{x})\,d\mathbf{x}. \tag{93}$$

We therefore investigate

$$\int_B \left(\prod_{i=1}^{j} g_{\mathbf{r}_i}(\boldsymbol{\alpha}; \mathbf{x})\right) Z(\boldsymbol{\alpha}; \mathbf{x})\,d\mathbf{x} \tag{94}$$

for a fixed s-box B. In view of (87) and (89), we consider first

$$\int_B R_{\mathbf{s}}(\mathbf{x}) Z(\boldsymbol{\alpha}; \mathbf{x})\,d\mathbf{x}. \tag{95}$$

Suppose that

$$B = [m_1 2^{-s_1}, (m_1 + 1)2^{-s_1}) \times [m_2 2^{-s_2}, (m_2 + 1)2^{-s_2}).$$

Let

$$B' = [m_1 2^{-s_1}, (m_1 + \tfrac{1}{2})2^{-s_1}) \times [m_2 2^{-s_2}, (m_2 + \tfrac{1}{2})2^{-s_2}).$$

Then the integral (95) is equal to

$$\int_{B'} \sum_{\varepsilon_1 = 0}^{1} \sum_{\varepsilon_2 = 0}^{1} (-1)^{\varepsilon_1 + \varepsilon_2} Z(\boldsymbol{\alpha}; \mathbf{y} + (\varepsilon_1 2^{-s_1 - 1}, \varepsilon_2 2^{-s_2 - 1}))\,d\mathbf{y}$$

$$= \int_{B'} \sum_{\mathbf{p} \in \mathscr{P}} \lambda(\mathbf{p}) \chi_{B'}(\mathbf{p} - \boldsymbol{\alpha} - \mathbf{y} + (m_1 2^{-s_1}, m_2 2^{-s_2}))\,d\mathbf{y}$$

$$= \sum_{\mathbf{p} \in \mathscr{P}} \lambda(\mathbf{p}) \int_{B'} \chi_{B'}(\mathbf{p} - \boldsymbol{\alpha} - \mathbf{y} + (m_1 2^{-s_1}, m_2 2^{-s_2}))\,d\mathbf{y}. \tag{96}$$

Note that $B' + \mathbf{y} - (m_1 2^{-s_1}, m_2 2^{-s_2}) \subset B$ for every $\mathbf{y} \in B'$. It follows from (67) that at most one $\mathbf{p} \in \mathscr{P}$ contributes to the sum on the right-hand side of (96). If there is no $\mathbf{p} \in \mathscr{P}$ such that $\mathbf{p} - \boldsymbol{\alpha} \in B$, then the integral (95) vanishes. If $\mathbf{p}^* \in \mathscr{P}$ satisfies $\mathbf{p}^* - \boldsymbol{\alpha} \in B$, then the integral (94) is equal to

$$(\beta(\mathbf{p}^*))^j q(B) \int_B R_{\mathbf{s}}(\mathbf{x}) Z(\boldsymbol{\alpha}; \mathbf{x})\, d\mathbf{x},$$

where $q(B)$ is the constant value of $Q(\mathbf{x})$ in B. Note that

$$q(B) = (\mu(B))^{-1} \int_B Q(\mathbf{x})\, d\mathbf{x}. \tag{97}$$

In either case, the integral (94) is equal to

$$q(B) \sum_{\mathbf{p} \in \mathscr{P}} |\lambda(\mathbf{p})| (\beta(\mathbf{p}))^{j-1} \int_{B'} \chi_{B'}(\mathbf{p} - \boldsymbol{\alpha} - \mathbf{y} + (m_1 2^{-s_1}, m_2 2^{-s_2}))\, d\mathbf{y}.$$

Now

$$\int_{U^2} \chi_{B'}(\mathbf{p} - \boldsymbol{\alpha} - \mathbf{y} + (m_1 2^{-s_1}, m_2 2^{-s_2}))\, d\boldsymbol{\alpha} = \mu(B') = \frac{\mu(B)}{4}.$$

It follows from (97) that

$$\int_{U^2} \int_B \left(\prod_{i=1}^{j} g_{\mathbf{r}_i}(\mathbf{x}) \right) Z(\boldsymbol{\alpha}; \mathbf{x})\, d\mathbf{x}\, d\boldsymbol{\alpha}$$

$$= \frac{\mu(B)}{16} \sum_{\mathbf{p} \in \mathscr{P}} |\lambda(\mathbf{p})| (\beta(\mathbf{p}))^{j-1} \int_B Q(\mathbf{x})\, d\mathbf{x}. \tag{98}$$

If $j = 1$, then $Q(\mathbf{x}) = 1$, $|\mathbf{s}| = n$ and $\mu(B) = 2^{-n}$. Summing over all $B \in \mathscr{A}_{\mathbf{s}}$, we see that the integral (93) is equal to

$$2^{-n-4} \sum_{\mathbf{p} \in \mathscr{P}} |\lambda(\mathbf{p})|.$$

(80) follows on noting that there are $(n+1)$ values of $\mathbf{r}$ satisfying $|\mathbf{r}| = n$. If $j \geqslant 2$, then by (98) and (91),

$$\int_{U^2} \int_{U^2} \left(\prod_{i=1}^{j} g_{\mathbf{r}_i}(\mathbf{x}) \right) Z(\boldsymbol{\alpha}; \mathbf{x})\, d\mathbf{x}\, d\boldsymbol{\alpha}$$

$$= 2^{-|\mathbf{s}|-4} \sum_{\mathbf{p} \in \mathscr{P}} |\lambda(\mathbf{p})| (\beta(\mathbf{p}))^{j-1} \int_{U^2} Q(\mathbf{x})\, d\mathbf{x} = 0.$$

(81) follows immediately, and the proof of Lemma 2.17 is now complete.

3
Upper bounds

3.1 Davenport's theorem and Roth's variation

Not long after Roth proved Theorem 1A, Davenport was able to show that a special case of Roth's theorem is best possible. We state Davenport's theorem as follows.

Theorem 2A. *There exists a constant c_1 such that for every even natural number N, there exists a distribution $\mathscr{P}$ of N points in U_0^2 such that*

$$\int_{U^2} |D[\mathscr{P}; B(\mathbf{x})]|^2 \,d\mathbf{x} < c_1 \log N.$$

More than 20 years later, Roth was able to extend Davenport's ideas to prove an analogue in U_0^3. Roth's theorem below is perhaps the most difficult of all the upper bound theorems known so far.

Theorem 2B. *There exists a constant c_2 such that for every natural number $N \geqslant 2$, there exists a distribution $\mathscr{P}$ of N points in U_0^3 such that*

$$\int_{U^3} |D[\mathscr{P}; B(\mathbf{x})]|^2 \,d\mathbf{x} < c_2 (\log N)^2.$$

The key to the proofs of the above theorems is the following result in diophantine approximation: Let θ be any irrational number having a continued fraction with bounded partial quotients. Then there exists a positive constant $c_3 = c_3(\theta)$, depending on θ, such that

$$\nu \| \nu\theta \| > c_3 > 0 \tag{1}$$

for all positive integers ν, where $\|\cdot\|$ denotes the distance from the nearest integer.

Lemma 3.1. *Let W_1 be any integer and W a natural number. Then*

$$\sum_{\nu=1}^{\infty} \frac{1}{\nu^2} \left| \sum_{n=W_1}^{W_1+W-1} e(\theta n \nu) \right|^2 \ll \log(2W).$$

Proof. It is well-known that

$$\left|\sum_{n=W_1}^{W_1+W-1} e(\theta n\nu)\right| \ll \min\{W, \|\nu\theta\|^{-1}\},$$

so that

$$S=\sum_{\nu=1}^{\infty}\frac{1}{\nu^2}\left|\sum_{n=W_1}^{W_1+W-1} e(\theta n\nu)\right|^2 \ll \sum_{m=1}^{\infty} 2^{-2m} \sum_{2^{m-1}\leqslant \nu<2^m} \min\{W^2, \|\nu\theta\|^{-2}\}.$$

For any pair m, p of natural numbers, there are at most two values of ν in the interval $2^{m-1}\leqslant \nu < 2^m$ for which

$$pc_3 2^{-m} \leqslant \|\nu\theta\| < (p+1)c_3 2^{-m};$$

for otherwise the difference $(\nu_1 - \nu_2)$ of two of them would contradict (1). Hence

$$\begin{aligned} S &\ll \sum_{m=1}^{\infty}\sum_{p=1}^{\infty} \min\{2^{-2m}W^2, p^{-2}\} \\ &= \sum_{2^m\leqslant W}\sum_{p=1}^{\infty} \min\{2^{-2m}W^2, p^{-2}\} + \sum_{2^m> W}\sum_{p=1}^{\infty} \min\{2^{-2m}W^2, p^{-2}\} \\ &\ll \sum_{2^m\leqslant W}\sum_{p=1}^{\infty} p^{-2} + \sum_{2^m>W}\left(2^{-2m}W^2 2^m W^{-1} + \sum_{p>2^m W^{-1}} p^{-2}\right) \\ &\ll \sum_{2^m\leqslant W} 1 + \sum_{2^m>W} 2^{-m}W \ll \log(2W). \end{aligned}$$

■

We shall be concerned with (2- and) 3-dimensional euclidean space, and denote a typical point by (x, y, z). The vectors $\mathbf{i}, \mathbf{j}, \mathbf{k}$ denote, respectively, $(1,0,0), (0,1,0), (0,0,1)$. The symbol Λ is reserved for non-degenerate lattices in the xy-plane.

Let θ be an irrational number having a continued fraction with bounded partial quotients. We keep θ fixed throughout, and constants in the subsequent argument may depend on this choice of θ. Let

$$\Lambda_0 = \Lambda_0(\theta\mathbf{i}+\mathbf{j}, \mathbf{i})$$

be the lattice generated by $\theta\mathbf{i}+\mathbf{j}$ and $\mathbf{i}$.

For any lattice Λ and any rectangle of the form $R=[0, X]\times[Y_1, Y_2)$, let $Z[\Lambda; R]$ denote the number of points of Λ falling into R, and write

$$E[\Lambda; R] = Z[\Lambda; R] - |d(\Lambda)|^{-1}A(R),$$

where $d(\Lambda)$ is the determinant of the lattice Λ and $A(R)$ is the area of R.

It is clear that $|d(\Lambda_0)| = 1$.

Let M be a natural number. We are interested in the M points of Λ_0 falling into $[0, 1)\times[0, M)$. Let $R^* = [0, X)\times[Y_1, Y_2)$, where $0 < X \leqslant 1$ and where the integers Y_1, Y_2 satisfy $0 \leqslant Y_1 < Y_2 \leqslant M$.

Let $\psi(x) = x - [x] - \frac{1}{2}$ when $x \notin \mathbb{Z}$ and $\psi(x) = 0$ when $x \in \mathbb{Z}$. Then since $0 < X \leqslant 1$, we have

$$\psi(x - X) - \psi(x) = \begin{cases} 1 - X & (0 < \{x\} < X), \\ -X & (\{x\} > X); \end{cases}$$

so that

$$Z[\Lambda_0; R^*] = \sum_{n=Y_1}^{Y_2-1} (X + \psi(\theta n - X) - \psi(\theta n))$$

for all but a finite number of X in the interval $0 < X \leqslant 1$. We comment here that the use of the function ψ is a technical device. One really wants to study the characteristic function.

It follows that

$$E[\Lambda_0; R^*] = \sum_{n=Y_1}^{Y_2-1} (\psi(\theta n - X) - \psi(\theta n)) \tag{2}$$

for all but a finite number of X in the interval $0 < X \leqslant 1$.

Now $\psi(x)$ has the Fourier expansion

$$\psi(x) = -\sum_{\nu \neq 0} \frac{e(x\nu)}{2\pi i\nu},$$

so that the right-hand side of (2) has the expansion

$$\sum_{\nu \neq 0} \left(\frac{1 - e(-\nu X)}{2\pi i\nu}\right)\left(\sum_{n=Y_1}^{Y_2-1} e(\theta n\nu)\right). \tag{3}$$

We would like to square expression (3) and integrate with respect to X along the interval $(0, 1]$. Unfortunately, the term 1 in $(1 - e(-\nu X))$ proves to be a nuisance.

In order to overcome this difficulty, Davenport introduced another lattice $\Lambda_0' = \Lambda_0'(-\theta\mathbf{i} + \mathbf{j}, \mathbf{i})$ and considered the $2M$ points of $\Lambda_0 \cup \Lambda_0'$ in $[0, 1) \times [0, M)$. Then, since $\psi(x)$ is an odd function,

$$\begin{aligned} &Z[\Lambda_0 \cup \Lambda_0'; R^*] - 2A(R^*) \\ &= \sum_{n=Y_1}^{Y_2-1} (\psi(\theta n - X) - \psi(\theta n) + \psi(-\theta n - X) - \psi(-\theta n)) \\ &= \sum_{n=Y_1}^{Y_2-1} (\psi(\theta n - X) - \psi(\theta n + X)) \end{aligned} \tag{4}$$

for all but a finite number of X in the interval $0 < X \leqslant 1$.

Now the right-hand side of (4) has the expansion

$$\sum_{\nu \neq 0} \left(\frac{e(\nu X) - e(-\nu X)}{2\pi i\nu}\right)\left(\sum_{n=Y_1}^{Y_2-1} e(\theta n\nu)\right),$$

so that by Parseval's theorem and Lemma 3.1,

$$\int_0^1 |Z[\Lambda_0 \cup \Lambda_0'; R^*] - 2A(R^*)|^2 \, dX \ll \sum_{\nu=1}^{\infty} \frac{1}{\nu^2} \left| \sum_{n=Y_1}^{Y_2-1} e(\theta n \nu) \right|^2$$

$$\ll \log(2(Y_2 - Y_1)) \ll \log(2M).$$

If Y is any real number satisfying $0 < Y \leqslant M$, then for $R = [0, X) \times [0, Y)$, we take $R^* = [0, X) \times [0, -[-Y])$, where $-[-Y]$ is the least integer not less than Y. Then

$$Z[\Lambda_0 \cup \Lambda_0'; R] = Z[\Lambda_0 \cup \Lambda_0'; R^*]$$

and

$$A(R) - A(R^*) \ll 1,$$

so that

$$\int_0^1 |Z[\Lambda_0 \cup \Lambda_0'; R] - 2A(R)|^2 \, dX \ll \log(2M).$$

It follows that

$$\int_0^M \int_0^1 |Z[\Lambda_0 \cup \Lambda_0'; R] - 2A(R)|^2 \, dX \, dY \ll M \log(2M).$$

Rescaling in the Y-direction by a factor $1/M$, we see that the set

$$\mathscr{P} = \{(\{\pm \theta n\}, (n-1)/M) : 1 \leqslant n \leqslant M\}$$

of $2M$ points in U_0^2 satisfies

$$\int_{U^2} |D[\mathscr{P}; B(\mathbf{x})]|^2 \, d\mathbf{x} \ll_\theta \log(2M),$$

completing the proof of Davenport's theorem.

Instead of introducing the extra lattice Λ_0', Roth devised an ingenious variation of the argument. This new idea proved to be extremely important in later work on upper bound theorems.

For any real number t, let $t\mathbf{i} + \Lambda_0$ be the lattice given by

$$t\mathbf{i} + \Lambda_0 = \{t\mathbf{i} + \mathbf{u} : \mathbf{u} \in \Lambda_0\},$$

i.e. $t\mathbf{i} + \Lambda_0$ is a translation in the x-direction of the lattice Λ_0. Then

$$E[t\mathbf{i} + \Lambda_0; R^*] = \sum_{n=Y_1}^{Y_2-1} (\psi(t + \theta n - X) - \psi(t + \theta n))$$

has the expansion

$$\sum_{\nu \neq 0} \left(\frac{1 - e(-\nu X)}{2\pi i \nu} \right) \left(\sum_{n=Y_1}^{Y_2-1} e(\theta n \nu) \right) e(\nu t),$$

so that by Parseval's theorem,

$$\int_0^1 |E[t\mathbf{i}+\Lambda_0; R^*]|^2\,dt \ll \sum_{\nu=1}^{\infty}\frac{1}{\nu^2}\left|\sum_{n=Y_1}^{Y_2-1} e(\theta n\nu)\right|^2.$$

Hence Roth was able to prove

Lemma 3.2. *Let M be a natural number and suppose that $0 < X_2' - X_1' \leqslant 1$ and $0 < Y_2' - Y_1' \leqslant M$. If $R = [X_1', X_2') \times [Y_1', Y_2')$, then*

$$\int_0^1 |E[t\mathbf{i}+\Lambda_0; R]|^2\,dt \ll \log(2M).$$

Note that since $E[t\mathbf{i}+\Lambda_0; R]$ is periodic in t with period 1, we can assume that $X_1' = 0$ when proving Lemma 3.2. On the other hand, the following form of Lemma 3.2 is better-suited for the proof of Roth's theorem.

Lemma 3.2′. *Let M be a natural number and suppose that $0 < X_2'' - X_1'' \leqslant M^{-1}$ and $0 < Y_2'' - Y_1'' \leqslant 1$. If $R = [X_1'', X_2'') \times [Y_1'', Y_2'')$, then*

$$\int_0^1 |E[M^{-1}t\mathbf{i}+M^{-1}\Lambda_0; R]|^2\,dt \ll \log(2M).$$

In other words,

$$\int_0^1 |E[t\mathbf{i}+M^{-1}\Lambda_0; R]|^2\,dt \ll \log(2M).$$

Note that the second inequality of Lemma 3.2′ follows from the first, as we have, in view of periodicity,

$$\int_0^M |E[M^{-1}t\mathbf{i}+M^{-1}\Lambda_0; R]|^2\,dt \ll M\log(2M).$$

The idea of Roth is to consider layers of 2-dimensional lattices in 3-dimensional euclidean space.

If $\mathscr{S}$ is any subset of the 3-dimensional euclidean space, we define, for any vector $\mathbf{v}^*$,

$$\mathbf{v}^* + \mathscr{S} = \{\mathbf{v}^* + \mathbf{v} : \mathbf{v} \in \mathscr{S}\}.$$

In other words, $\mathbf{v}^* + \mathscr{S}$ is a translation by $\mathbf{v}^*$ of the set $\mathscr{S}$.

Roth considered sets Ω of the type

$$\bigcup_{\nu=p_1}^{p_2} (\nu\mathbf{k} + \mathbf{w}_\nu + \Lambda), \tag{5}$$

where p_1, p_2 are non-negative integers, $\mathbf{w}_\nu = (x_\nu, y_\nu, 0)$ is a vector in the xy-plane for each ν, and Λ is a lattice in the xy-plane. He also considered boxes B of the type

$$[X', X'') \times [Y', Y'') \times [Z', Z''). \tag{6}$$

If Ω is a set of the type (5) and B a box of the type (6) with $p_1 \leqslant Z' < Z'' \leqslant p_2 + 1$, we write $Z[\Omega; B]$ for the number of points of Ω falling into B, and

$$E[\Omega; B] = Z[\Omega; B] - |d(\Lambda)|^{-1} V(B),$$

where $V(B)$ is the volume of B.

The sets Ω that Roth constructed are obtained from Λ_0 as follows: Recall that $\Lambda_0 = \Lambda(\mathbf{u}, \mathbf{i})$, where $\mathbf{u} = \theta\mathbf{i} + \mathbf{j}$. For any non-negative integer m, write

$$\Lambda_m = 2^{-m}\Lambda_0 = \Lambda(2^{-m}\mathbf{u}, 2^{-m}\mathbf{i}).$$

Define

$$\mathbf{q}_0 = \mathbf{0}, \mathbf{q}_1 = \tfrac{1}{2}\mathbf{u}, \mathbf{q}_2 = \tfrac{1}{2}\mathbf{i}, \mathbf{q}_3 = \tfrac{1}{2}\mathbf{u} + \tfrac{1}{2}\mathbf{i}.$$

Then for every m,

$$\Lambda_{m+1} = \bigcup_{\tau=0}^{3} (2^{-m}\mathbf{q}_\tau + \Lambda_m).$$

Roth defined $\Omega_0, \Omega_1, \ldots$ successively by

$$\Omega_0 = \Lambda_0, \Omega_{m+1} = \bigcup_{\tau=0}^{3} (\tau 4^m \mathbf{k} + 2^{-m}\mathbf{q}_\tau + \Omega_m).$$

Then it is easily seen that

Lemma 3.3. *Ω_m has a representation of the type (5) with $p_1 = 0$, $p_2 = 4^m - 1$ and $\Lambda = \Lambda_0$. Furthermore, the projection of Ω_m onto the xy-plane is Λ_m.*

The special boxes that Roth considered are defined as follows.

Definition. A box of the type $[0, X) \times [0, Y) \times [0, Z)$ is said to be admissible with respect to m if

$$0 < X \leqslant 2^{-m}, \quad 0 < Y \leqslant 1, \quad 0 < Z \leqslant 4^m.$$

Theorem 2B can then be easily deduced from the following lemma.

Lemma 3.4. *There exists a number $c_4 = c_4(\theta)$ such that for any m,*

$$\int_0^1 \int_0^1 |E[s\mathbf{u} + t\mathbf{i} + \Omega_m; B]|^2 \, ds\, dt \leqslant c_4(m+1)^2 \tag{7}$$

for every box B that is admissible with respect to m.

For let the natural number $N \geqslant 2$ be given. Choose m so that $2^{m-1} < N \leqslant 2^m$. Writing $B = B(X, Y, Z) = [0, X) \times [0, Y) \times [0, Z)$, we have

$$\int_0^1 \int_0^1 \int_0^{4^m} \int_0^{N2^{-m}} \int_0^{2^{-m}} |E[s\mathbf{u} + t\mathbf{i} + \Omega_m; B(X, Y, Z)]|^2 \, \mathrm{d}X \, \mathrm{d}Y \, \mathrm{d}Z \, \mathrm{d}s \, \mathrm{d}t$$
$$\leqslant c_4(m+1)^2 N.$$

Hence there exist s^*, t^* satisfying $0 \leqslant s^*, t^* < 1$ such that

$$\int_0^{4^m} \int_0^{N2^{-m}} \int_0^{2^{-m}} |E[s^*\mathbf{u} + t^*\mathbf{i} + \Omega_m; B(X, Y, Z)]|^2 \, \mathrm{d}X \, \mathrm{d}Y \, \mathrm{d}Z$$
$$\leqslant c_4(m+1)^2 N.$$

By Lemma 3.3, there are exactly N points of $s^*\mathbf{u} + t^*\mathbf{i} + \Omega_m$ in the region $[0, 2^{-m}) \times [0, N2^{-m}) \times [0, 4^m)$. If these are the points

$$(2^{-m}x_\nu, N2^{-m}y_\nu, 4^m z_\nu) \quad (\nu = 1, \ldots, N),$$

then the set

$$\mathscr{P} = \{(x_\nu, y_\nu, z_\nu) : \nu = 1, \ldots, N\}$$

of N points in U_0^3 satisfies

$$\int_{U^3} |D[\mathscr{P}; B(\mathbf{x})]|^2 \, \mathrm{d}\mathbf{x} \ll_\theta (\log N)^2,$$

proving Roth's theorem.

It remains to prove Lemma 3.4. We do this by induction on m. The result is trivial when $m = 0$ if c_4 is chosen large enough. Suppose that $m \geqslant 0$ and that (7) holds for all boxes admissible with respect to m. Suppose now that the box $B^* = [0, X^*) \times [0, Y^*) \times [0, Z^*)$ is admissible with respect to $(m+1)$. Let the integer μ satisfy $\mu 4^m < Z^* \leqslant (\mu+1)4^m$. Then $0 \leqslant \mu \leqslant 3$. We may assume that $0 < \mu \leqslant 3$, for the case $\mu = 0$ is trivial. We have

$$B^* = \left(\bigcup_{\tau=0}^{\mu-1} B^{(\tau)} \right) \cup B^{**},$$

where, for $0 \leqslant \tau \leqslant \mu - 1$,

$$B^{(\tau)} = [0, X^*) \times [0, Y^*) \times [\tau 4^m, (\tau+1)4^m)$$

and

$$B^{**} = [0, X^*) \times [0, Y^*) \times [\mu 4^m, Z^*).$$

The idea is to use Lemma 3.2′ on the projection of $B^{(\tau)}$ onto the xy-plane, with $M = 2^m$, and to note that $-\mu 4^m\mathbf{k} + B^{**}$ is admissible with respect to m.

Writing

$$E_1(s,t)=\sum_{\tau=0}^{\mu-1} E[s\mathbf{u}+t\mathbf{i}+\Omega_{m+1}; B^{(\tau)}]$$

and

$$E_2(s,t)=E[s\mathbf{u}+t\mathbf{i}+\Omega_{m+1}; B^{**}],$$

we have

$$\int_0^1\int_0^1 |E[s\mathbf{u}+t\mathbf{i}+\Omega_{m+1}; B^*]|^2\, ds\, dt = I_1+I_2+2J,$$

where, for $\beta=1,2$,

$$I_\beta=\int_0^1\int_0^1 |E_\beta(s,t)|^2\, ds\, dt,$$

and where

$$J=\int_0^1\int_0^1 E_1(s,t)E_2(s,t)\, ds\, dt.$$

Consider first

$$I_1\leqslant \mu\sum_{\tau=0}^{\mu-1}\int_0^1\int_0^1 |E[s\mathbf{u}+t\mathbf{i}+\Omega_{m+1}; B^{(\tau)}]|^2\, ds\, dt.$$

Then writing $R_0=[0,X^*)\times[0,Y^*)$, we see that

$$E[s\mathbf{u}+t\mathbf{i}+\Omega_{m+1}; B^{(\tau)}]=E[s\mathbf{u}+t\mathbf{i}+2^{-m}\mathbf{q}_\tau+\Lambda_m; R_0]. \tag{8}$$

In view of periodicity in s and t, we have

$$I_1\leqslant \mu^2\int_0^1\int_0^1 |E[s\mathbf{u}+t\mathbf{i}+\Lambda_m; R_0]|^2\, ds\, dt \ll m+1 \tag{9}$$

by Lemma 3.2′ with $M=2^m$. On the other hand,

$$\begin{aligned}
I_2&=\int_0^1\int_0^1 |E[s\mathbf{u}+t\mathbf{i}+\Omega_{m+1}; B^{**}]|^2\, ds\, dt\\
&=\int_0^1\int_0^1 |E[s\mathbf{u}+t\mathbf{i}+2^{-m}\mathbf{q}_\mu+\Omega_m; -\mu 4^m\mathbf{k}+B^{**}]|^2\, ds\, dt\\
&=\int_0^1\int_0^1 |E[s\mathbf{u}+t\mathbf{i}+\Omega_m; -\mu 4^m\mathbf{k}+B^{**}]|^2\, ds\, dt,
\end{aligned}$$

in view of periodicity. Since $-\mu 4^m\mathbf{k}+B^{**}$ is admissible with respect to m, we have, by the induction hypothesis, that

$$I_2\leqslant c_4(m+1)^2.$$

To complete the proof of Lemma 3.4, we need to show that $J \ll m+1$. However, by applying Schwarz's inequality to J and using the estimates for I_1 and I_2, we only get $J \ll (m+1)^{\frac{3}{2}}$. We therefore need extra ideas (we comment here that this new idea, in a slightly different form, was first used in Roth (1976*b*); variations of this idea were subsequently used in Roth (1980) and Chen (1980, 1983)).

Note that by periodicity in s and t,

$$J = \int_0^1 \int_0^1 E_1(s + a2^{-m}, t + b2^{-m}) E_2(s + a2^{-m}, t + b2^{-m}) \, ds \, dt \qquad (10)$$

for every pair of integers a, b. Furthermore, by (8), $E_1(s, t)$ is connected with the lattice Λ_m, and so it is in fact periodic in both s and t with period 2^{-m}. Hence, summing (10) over the ranges $0 \leqslant a, b \leqslant 2^m - 1$, we have

$$4^m J = \int_0^1 \int_0^1 E_1(s, t) D(s, t) \, ds \, dt,$$

where

$$D(s, t) = \sum_{a=0}^{2^m-1} \sum_{b=0}^{2^m-1} E_2(s + a2^{-m}, t + b2^{-m}).$$

By Schwarz's inequality and (9),

$$(4^m J)^2 \ll (m+1) \int_0^1 \int_0^1 |D(s, t)|^2 \, ds \, dt. \qquad (11)$$

We now show that $D(s, t)$ is connected with a very special Ω, where 'the sheets Λ are exactly on top of each other'. More precisely,

$$D(s, t) = Z[s\mathbf{u} + t\mathbf{i} + \Omega'; B^{**}] - 4^m V(B^{**}),$$

where Ω' is obtained from

$$\bigcup_{a=0}^{2^m-1} \bigcup_{b=0}^{2^m-1} (a2^{-m}\mathbf{u} + b2^{-m}\mathbf{i} + \Omega_{m+1})$$

by restricting the last coordinate to the interval $[\mu 4^m, (\mu+1)4^m)$, i.e.

$$\Omega' = \mu 4^m \mathbf{k} + 2^{-m}\mathbf{q}_\mu + \bigcup_{a=0}^{2^m-1} \bigcup_{b=0}^{2^m-1} (a2^{-m}\mathbf{u} + b2^{-m}\mathbf{i} + \Omega_m)$$

$$= \mu 4^m \mathbf{k} + 2^{-m}\mathbf{q}_\mu + \Omega'',$$

say. In view of Lemma 3.3, it is not difficult to see that Ω'' is of the form

$$\Omega'' = \bigcup_{\nu=0}^{4^m-1} (\nu\mathbf{k} + \Lambda_m).$$

Let

$$B^{***} = [0, X^*) \times [0, Y^*) \times [\mu 4^m, -[-Z^*]).$$

Then

$$Z[\mathbf{su} + t\mathbf{i} + \Omega'; B^{**}] = Z[\mathbf{su} + t\mathbf{i} + \Omega'; B^{***}]$$

and

$$|4^m V(B^{**}) - 4^m V(B^{***})| \leqslant 2^m.$$

Then using the projection onto the xy-plane, we have

$$D(s, t) = (-[-Z^*] - \mu 4^m)E[\mathbf{su} + t\mathbf{i} + 2^{-m}\mathbf{q}_\mu + \Lambda_m; R_0] + O(2^m).$$

Since $0 \leqslant -[-Z^*] - \mu 4^m \leqslant 4^m$, we have, by (9), that

$$\int_0^1 \int_0^1 |D(s, t)|^2 \, ds \, dt \ll 4^{2m}(m + 1). \tag{12}$$

Combining (11) and (12), we have $J \ll m + 1$. Lemma 3.4 follows.

3.2 Halton's theorem. Faure's theorem

In 1960, Halton proved the following theorem.

Theorem 4. *There exists a constant $c_5(k)$ such that for every natural number $N \geqslant 2$, there exists a distribution $\mathscr{P}$ of N points in U_0^{k+1} such that*

$$\sup_{\mathbf{x} \in U_1^{k+1}} |D[\mathscr{P}; B(\mathbf{x})]| < c_5(k)(\log N)^k.$$

To prove Theorem 4, Halton considered sets of the following type. In order to avoid repeating definitions in §§3.3–3.5, the definitions in this section are more general than necessary. A slight variation of sets of this general type was first introduced in Chen (1980) to prove Theorem 2D.

The first idea is to consider, instead of sets of N points in U_0^{k+1}, infinite sets of points in $U_0^k \times [0, \infty)$ such that there is an average of one point per unit volume. We then consider those N points contained in $U_0^k \times [0, N)$ and multiply the last coordinate of each of these N points by the factor $1/N$ to obtain a set of N points in U_0^{k+1}.

A box in $U_0^k \times [0, \infty)$ is a set of the form

$$I_1 \times I_2 \times \cdots \times I_k \times I_0, \tag{13}$$

where, for each $j = 1, \ldots, k$, I_j is an interval of the form $[\alpha_j, \beta_j)$ and contained in U_0, while I_0 is an interval of the form $[\alpha_0, \beta_0)$ satisfying $0 \leqslant \alpha_0 < \beta_0$.

The second idea is to look for distributions such that many boxes contain the right number of points (i.e. equal to the volume of the boxes), while making sure that other boxes can be approximated to one of these special boxes, or a union of these special boxes. Our special boxes are defined as follows.

Definition. Let p be a prime and s be a non-negative integer. By an elementary p-type interval of order s, we mean an interval of the form $[\alpha, \beta)$, contained in U_0, where α and β are consecutive integer multiples of p^{-s}.

Let h be a non-negative integer, and let $p_1, \ldots, p_k$ be primes, not necessarily distinct. Let q be a natural number (we only need $q = 1$ in §§3.2–3.3).

Definition. By an elementary q-box of order h with respect to the primes $p_1, \ldots, p_k$ in $U_0^k \times [0, \infty)$, we mean a set of the form $I_1 \times \cdots \times I_k \times I_0$, where for each $j = 1, \ldots, k, I_j$ is an elementary p_j-type interval of order s_j $(0 \leqslant s_j \leqslant h)$, and where I_0 is of the form $[\alpha_0, \beta_0)$, with α_0, β_0 being consecutive non-negative integer multiples of $qp_1^{s_1} \cdots p_k^{s_k}$.

It is clear that any elementary q-box in $U_0^k \times [0, \infty)$ has 'volume' q.

Definition. By a q-set of class h with respect to the primes $p_1, \ldots, p_k$ in $U_0^k \times [0, \infty)$, we mean an infinite set $\mathcal{Q}$ of points in $U_0^k \times [0, \infty)$ which has the property that every elementary q-box of order h with respect to the primes $p_1, \ldots, p_k$ in $U_0^k \times [0, \infty)$ contains exactly one point of $\mathcal{Q}$.

The existence of 1-sets (and hence q-sets in general) in $U_0^k \times [0, \infty)$ has been established in certain cases. The more general notion of q-sets is required in order that we may prove Theorem 2D later.

Lemma 3.5 (Hammersley (1960), Halton (1960)). *Suppose that $p_1, \ldots, p_k$ are distinct primes. For each $n \in \mathbb{N}^*$ and each $j = 1, \ldots, k$, write*

$$n = \sum_{\nu=1}^{\infty} a_{j,\nu} p_j^{\nu-1} \quad (0 \leqslant a_{j,\nu} < p_j),$$

where the integers $a_{j,\nu}$ are uniquely determined by n, and write

$$x_j(n) = \sum_{\nu=1}^{\infty} a_{j,\nu} p_j^{-\nu}.$$

Then the set

$$\{(x_1(n), \ldots, x_k(n), n) : n \in \mathbb{N}^*\} \tag{14}$$

is a 1-set of class h with respect to the primes $p_1, \ldots, p_k$ in $U_0^k \times [0, \infty)$ for any non-negative integer h.

Suppose that

$$x = \sum_{\nu=1}^{\infty} a_\nu p^{-\nu} \quad (0 \leqslant a_\nu < p),$$

where all but a finite number of the integers a_ν are zero. We define

$$\mathscr{C}x = \sum_{\nu=1}^{\infty} b_\nu p^{-\nu},$$

where $0 \leqslant b_\nu < p$ and

$$b_\nu \equiv \sum_{\mu \geqslant \nu} \binom{\mu-1}{\nu-1} a_\mu \pmod{p}.$$

For any integer $m > 1$, let $\mathscr{C}^m x = \mathscr{C}\mathscr{C}^{m-1}x$.

Lemma 3.6 (Faure (1982)). *Suppose that p is a prime and $p \geqslant k$. For each $n \in \mathbb{N}^*$, write*

$$n = \sum_{\nu=1}^{\infty} a_\nu p^{\nu-1} \quad \textit{and} \quad x_n = \sum_{\nu=1}^{\infty} a_\nu p^{-\nu}.$$

Then the set

$$\{(x_n, \mathscr{C}x_n, \ldots, \mathscr{C}^{k-1}x_n, n) : n \in \mathbb{N}^*\} \tag{15}$$

is a 1-set of class h with respect to the primes $p, \ldots, p$ in $U_0^k \times [0, \infty)$ for any non-negative integer h.

To deduce Theorem 4, we begin by defining an appropriate discrepancy function for q-sets in $U_0^k \times [0, \infty)$.

For any q-set $\mathscr{Q}$ and any box B of the type (13) in $U_0^k \times [0, \infty)$, we let $Z[\mathscr{Q}; B]$ denote the number of points of $\mathscr{Q}$ falling into B, and write

$$E[\mathscr{Q}; B] = Z[\mathscr{Q}; B] - q^{-1}\mu(B), \tag{16}$$

where $\mu(B)$ is the volume of B. We note that $q^{-1}\mu(B)$ is the 'expected number' of points of $\mathscr{Q}$ falling into B. Note also that E is 'additive', i.e. if $B = B_1 \cup B_2$, where $B_1 \cap B_2 = \varnothing$, then

$$E[\mathscr{Q}; B] = E[\mathscr{Q}; B_1] + E[\mathscr{Q}; B_2]. \tag{17}$$

We are interested in boxes B where for each $j = 1, \ldots, k$, I_j is a union of elementary p_j-type intervals, and where I_0 is of special type. More precisely,

Definition. Let s be a non-negative integer. Suppose that $B^* = I_1 \times \cdots \times I_k \times I^*$, where for each $j = 1, \ldots, k$, $I_j = [0, \eta_j)$, where $0 < \eta_j \leqslant 1$ and η_j is an integer multiple of p_j^{-s}. Suppose further that $I^* = [0, Y)$, where Y is positive but otherwise unrestricted. Then we say that B^* is a box of class s with respect to the primes $p_1, \ldots, p_k$ in $U_0^k \times [0, \infty)$.

Halton's theorem can be deduced from Lemma 3.5 and the following lemma (note that Halton only proved Lemma 3.7 in the case where $\mathscr{Q}$ is the set (14)).

Lemma 3.7 (Halton (1960)). *For any* 1*-set* $\mathscr{Q}$ *of class* h *with respect to the primes* $p_1, \ldots, p_k$ *in* $U_0^k \times [0, \infty)$ *and for any box* B^* *of class* h *with respect to the same primes in* $U_0^k \times [0, \infty)$, *we have*

$$|E[\mathscr{Q}; B^*]| < (p_1 \cdots p_k)h^k. \tag{18}$$

Proof. Note that since each of the $I_j = [0, \eta_j)$ is a disjoint union of at most $p_j h$ elementary p_j-type intervals of order at most h, B^* is a disjoint union of at most $(p_1 \cdots p_k)h^k$ boxes of the type

$$I_1 \times \cdots \times I_k \times [0, Y), \tag{19}$$

where, for each $j = 1, \ldots, k$, I_j is an elementary p_j-type interval of order s_j where $0 \leqslant s_j \leqslant h$. To prove (18), it clearly suffices, in view of (17), to prove that for each box B of the type (19), $|E[\mathscr{Q}; B]| < 1$. Let Y_0 denote the greatest integer multiple of $p_1^{s_1} \cdots p_k^{s_k}$ not exceeding Y. Then $[0, Y) = [0, Y_0) \cup [Y_0, Y)$. The box $I_1 \times \cdots \times I_k \times [0, Y_0)$ is a union of a finite number of elementary 1-boxes of order h with respect to the primes $p_1, \ldots, p_k$ in $U_0^k \times [0, \infty)$, and so contains the expected number of points of $\mathscr{Q}$. Hence $E[\mathscr{Q}; I_1 \times \cdots \times I_k \times [0, Y_0)] = 0$. On the other hand, the remainder $I_1 \times \cdots \times I_k \times [Y_0, Y)$ is contained in an elementary 1-box of order h with respect to the primes $p_1, \ldots, p_k$ in $U_0^k \times [0, \infty)$, and so contains at most one point of $\mathscr{Q}$. Furthermore, $\mu(I_1 \times \cdots \times I_k \times [Y_0, Y)) < 1$. It follows that $|E[\mathscr{Q}; I_1 \times \cdots \times I_k \times [Y_0, Y)]| < 1$, and the proof of the lemma is complete. ■

To prove Lemma 3.5, simply note that for every $j = 1, \ldots, k$, if I_j is an elementary p_j-type interval of order s_j, then the relation $x_j(n) \in I_j$ is satisfied by precisely all the non-negative integers n of a residue class modulo $p_j^{s_j}$. The result follows easily from the Chinese remainder theorem.

We now deduce Theorem 4. Let the natural number $N \geqslant 2$ be given, and choose integer h to satisfy

$$2^{h-1} < N \leqslant 2^h. \tag{20}$$

For any $\mathbf{x} = (x_1, \ldots, x_k) \in U_1^k$ and for any Y satisfying

$$0 < Y \leqslant N, \tag{21}$$

let $B(\mathbf{x}, Y)$ denote the box

$$B(\mathbf{x}, Y) = [0, x_1) \times \cdots \times [0, x_k) \times [0, Y). \tag{22}$$

Let $\boldsymbol{\eta} = \boldsymbol{\eta}(\mathbf{x}) = (\eta_1, \ldots, \eta_k)$ be defined such that, for every $j = 1, \ldots, k$,

$$\eta_j = \eta_j(x_j) = -p_j^{-h}[-p_j^h x_j], \tag{23}$$

i.e. η_j is the least integer multiple of p_j^{-h} not less than x_j. Suppose that $\mathcal{Q}$ is the set (14).

Lemma 3.8. *For any* $\mathbf{x} \in U_1^k$ *and any* Y *satisfying* (21),

$$|E[\mathcal{Q}; B(\mathbf{x}, Y)] - E[\mathcal{Q}; B^*(\boldsymbol{\eta}(\mathbf{x}), Y)]| \leqslant k.$$

Remark. In fact, Lemma 3.8 works for every 1-set of class h with respect to the primes $p_1, \ldots, p_k$ in $U_0^k \times [0, \infty)$.

Proof of Lemma 3.8. For $j = 0, \ldots, k$, for any fixed $\mathbf{x} \in U_1^k$ and for any fixed Y satisfying (21), let

$$B^{(j)} = [0, \eta_1) \times \cdots \times [0, \eta_j) \times [0, x_{j+1}) \times \cdots \times [0, x_k) \times [0, Y).$$

Then clearly $B^{(0)} = B(\mathbf{x}, Y)$ and $B^{(k)} = B^*(\boldsymbol{\eta}(\mathbf{x}), Y)$. To prove Lemma 3.8, it suffices to show that for every $j = 1, \ldots, k$,

$$|E[\mathcal{Q}; B^{(j)}] - E[\mathcal{Q}; B^{(j-1)}]| \leqslant 1.$$

This follows from observing that the complement of $B^{(j-1)}$ in $B^{(j)}$ (note that $B^{(j-1)} \subset B^{(j)}$) is contained in an elementary 1-box of order h with respect to the primes $p_1, \ldots, p_k$ in $U_0^k \times [0, \infty)$. ■

It now follows from Lemmas 3.7 and 3.8 that for any box $B(\mathbf{x}, Y)$ with $\mathbf{x} \in U_1^k$ and $0 < Y \leqslant N$,

$$|E[\mathcal{Q}; B(\mathbf{x}, Y)]| < (p_1 \cdots p_k)h^k + k. \tag{24}$$

The box $[0, 1)^k \times [0, N)$ contains precisely the points

$$(x_1(n), \ldots, x_k(n), n) \quad (n = 0, \ldots, N-1).$$

Choosing $p_1, \ldots, p_k$ to be the first k primes, we see that the set

$$\mathcal{P} = \{(x_1(n), \ldots, x_k(n), n/N) : 0 \leqslant n < N\}$$

satisfies, in view of (20) and (24),

$$\sup_{\mathbf{x} \in U_1^{k+1}} |D[\mathcal{P}; B(\mathbf{x})]| \ll_k (\log N)^k.$$

This completes the proof of Halton's theorem.

In fact, we have also shown the following: Let $\mathbf{s}_1, \mathbf{s}_2, \mathbf{s}_3, \ldots$ be an infinite sequence in U_0^k. For any $n \in \mathbb{N}$ and $\boldsymbol{\alpha} = (\alpha_1, \ldots, \alpha_k) \in U_1^k$, let $Z(n, \boldsymbol{\alpha})$ denote the number of $\mathbf{s}_\nu = (s_{1\nu}, \ldots, s_{k\nu})$ $(\nu = 1, \ldots, n)$ that satisfy $0 \leqslant s_{j\nu} < \alpha_j$

for every $j = 1, \ldots, k$. Then the Halton sequence $\mathbf{s}_n = \mathbf{x}(n-1) = (x_1(n-1), \ldots, x_k(n-1))$ as defined in Lemma 3.5 satisfies

$$\sup_{\boldsymbol{\alpha} \in U_1^k} |Z(n, \boldsymbol{\alpha}) - n\alpha_1 \cdots \alpha_k| \ll (\log n)^k,$$

where the implicit constant in the inequality depends only on k.

Before we prove Lemma 3.6, it is perhaps appropriate to look at the background first. Much work was done on constructing 1-sets with respect to the primes 2, 2 in $U_0^2 \times [0, \infty)$, notably by Sobol. In Sobol (1967), it was shown that if for each $n \in \mathbb{N}^*$,

(i) $x_n = \sum_{v=1}^{\infty} a_v 2^{-v}$, where $n = \sum_{v=1}^{\infty} a_v 2^{v-1}$ $(0 \leqslant a_v < 2)$, and
(ii) $y_n = \sum_{v=1}^{\infty} b_v 2^{-v}$, where $0 \leqslant b_v < 2$ and

$$\begin{bmatrix} b_1 \\ b_2 \\ b_3 \\ b_4 \\ b_5 \\ \vdots \end{bmatrix} \equiv \begin{bmatrix} 1 & 1 & 1 & 1 & 1 & \cdots \\ 0 & 1 & 0 & 1 & 0 & \\ 0 & 0 & 1 & 1 & 0 & \\ 0 & 0 & 0 & 1 & 0 & \\ 0 & 0 & 0 & 0 & 1 & \\ \vdots & & & & & \ddots \end{bmatrix} \begin{bmatrix} a_1 \\ a_2 \\ a_3 \\ a_4 \\ a_5 \\ \vdots \end{bmatrix} \pmod 2,$$

then the set $\{(x_n, y_n, n) : n \in \mathbb{N}^*\}$ is a 1-set of any class with respect to the primes 2, 2 in $U_0^2 \times [0, \infty)$. Note that the matrix consists of binomial coefficients modulo 2.

We now prove Lemma 3.6. Suppose that B is an elementary 1-box of order h with respect to the primes $p, \ldots, p$ in $U_0^k \times [0, \infty)$ (in fact, we do not have to consider any restriction imposed by h since the conclusion of Lemma 3.6 is valid for every non-negative integer h; we therefore omit further reference to h). Then

$$\begin{aligned} B = [z_1 p^{-s_1}, (z_1+1)p^{-s_1}) \times \cdots \times [z_k p^{-s_k}, (z_k+1)p^{-s_k}) \\ \times [z_0 p^{s_1+\cdots+s_k}, (z_0+1)p^{s_1+\cdots+s_k}), \end{aligned}$$

where $z_1, \ldots, z_k, z_0$ are non-negative integers. Writing, for each $n \in \mathbb{N}^*$, $n = \sum_{v=1}^{\infty} a_v(n) p^{v-1}$ and $x_n = \sum_{v=1}^{\infty} a_v(n) p^{-v}$, we have that if

$$(x_n, \mathscr{C}x_n, \ldots, \mathscr{C}^{k-1}x_n, n) \in B, \tag{25}$$

we must have

$$z_0 p^{s_1+\cdots+s_k} \leqslant n < (z_0+1)p^{s_1+\cdots+s_k}. \tag{26}$$

Then the number $a_v(n)$ for $v > s_1 + \cdots + s_k$ are determined uniquely by the given $s_1, \ldots, s_k$ and z_0. On the other hand,

Lemma 3.9. *If $\mathscr{C}$ is the matrix $\left(\binom{\mu-1}{\nu-1}\right)$, then for any $t \geqslant 1$, $\mathscr{C}^t$ is the matrix $\left(t^{\mu-\nu}\binom{\mu-1}{\nu-1}\right)$.*

It follows from Lemma 3.9 that for $j = 2, \ldots, k$,

$$\mathscr{C}^{j-1}x_n = \sum_{\nu=1}^{\infty} b_\nu^{(j)}(n)p^{-\nu},$$

where

$$b_\nu^{(j)}(n) = \sum_{\mu \geqslant \nu} \binom{\mu-1}{\nu-1}(j-1)^{\mu-\nu}a_\mu(n). \tag{27}$$

Suppose that for each $j = 1, \ldots, k$,

$$z_j p^{-s_j} = y_{j,1}p^{-1} + \cdots + y_{j,s_j}p^{-s_j},$$

then adopting the convention $b_\nu^{(j)}(n) = a_\nu(n)$ and $(j-1)^{\mu-\nu} = 1$ if $j = 1$, we have that for each $j = 1, \ldots, k$,

$$\mathscr{C}^{j-1}x_n \in [z_j p^{-s_j}, (z_j+1)p^{-s_j})$$

if and only if $b_\nu^{(j)}(n) = y_{j,\nu}$ for every $\nu = 1, \ldots, s_j$, i.e.

$$\sum_{\mu \geqslant \nu} \binom{\mu-1}{\nu-1}(j-1)^{\mu-\nu}a_\mu(n) \equiv y_{j,\nu} \pmod p \quad (\nu = 1, \ldots, s_j). \tag{28}$$

Hence the number of points of the set $\{(x_n, \mathscr{C}x_n, \ldots, \mathscr{C}^{k-1}x_n, n) : n \in \mathbb{N}^*\}$ that fall into B is the number of solutions of the $s_1 + \cdots + s_k$ simultaneous linear congruences in the unknowns $a_1(n), \ldots, a_m(n)$, where $m = s_1 + \cdots + s_k$. However, the simultaneous linear congruences (28) ($j = 1, \ldots, k$) have unique solution if the determinant of the associated matrix is non-zero modulo p. The determinant in question is

$$D = \begin{vmatrix} A_1 \\ \vdots \\ A_k \end{vmatrix},$$

where, for each $j = 1, \ldots, k$,

$$A_j = \begin{bmatrix} \binom{0}{0} & \binom{1}{0}(j-1) & \binom{2}{0}(j-1)^2 & \cdot & \cdot & \cdot & \binom{m-1}{0}(j-1)^{m-1} \\ & \binom{1}{1} & \binom{2}{1}(j-1) & \cdot & \cdot & \cdot & \binom{m-1}{1}(j-1)^{m-2} \\ & & \ddots & & & & \\ \bigcirc & & \binom{s_j-1}{s_j-1} & \binom{s_j}{s_j-1}(j-1) & \cdots & & \binom{m-1}{s_j-1}(j-1)^{m-s_j} \end{bmatrix},$$

so that D is a generalized Vandermonde determinant with

$$D = \prod_{1 \leqslant j < J \leqslant k} (J - j)^{s_j s_J}.$$

If $k \leqslant p$, then $1 \leqslant J - j \leqslant p - 1$, so that clearly $D \not\equiv 0 \pmod p$.

We mention here a generalization of Halton's theorem. Instead of considering distributions in U_0^K, we consider distributions in a Lebesgue measurable set in $\mathbb{R}^K$ with finite measure. In Chapter 8, we shall prove

Theorem 11A (Beck). *Let $A \subset \mathbb{R}^K$ be a Lebesgue measurable set with $\mu(A) = 1$. For $\mathbf{x} = (x_1, \ldots, x_K) \in \mathbb{R}^K$, let*

$$B^{\infty}(\mathbf{x}) = \prod_{j=1}^{k} (-\infty, x_j).$$

For every natural number $N \geqslant 2$, there exists a distribution $\mathscr{P} = \{\mathbf{p}_1, \ldots, \mathbf{p}_N\}$ of N points in A such that

$$\sup_{\mathbf{x} \in \mathbb{R}^K} \left| \sum_{\mathbf{p}_i \in B^{\infty}(\mathbf{x})} 1 - N\mu(A \cap B^{\infty}(\mathbf{x})) \right| \ll (\log N)^{K + \frac{3}{2} + \varepsilon},$$

where the implicit constant depends only on K and ε.

3.3 Roth's probabilistic method

In this section, we prove the following theorem of Roth.

Theorem 2C. *There exists a constant $c_6(k)$ such that for every natural number $N \geqslant 2$, there exists a distribution $\mathscr{P}$ of N points in U_0^{k+1} such that*

$$\int_{U^{k+1}} |D[\mathscr{P}; B(\mathbf{x})]|^2 \, d\mathbf{x} < c_6(k)(\log N)^k.$$

Roth proved Theorem 2C by considering 'translations' of a set of a similar type to (14), and taking the average of the discrepancy over a 'long' interval. More precisely, let $p_1, \ldots, p_k$ be distinct primes and suppose that h is a non-negative integer. Consider the subset

$$\{(x_1(n), \ldots, x_k(n), n) : 0 \leqslant n < (p_1 \cdots p_k)^h\} \tag{29}$$

of the set (14) (it is possible to consider the set (14) without modification; we then have to modify (31) below). The set (29) is then extended by 'periodicity'. Writing $\mathbf{X}(n) = (X_1(n), \ldots, X_k(n))$, we define $X_j(n) = x_j(n)$ for all $j = 1, \ldots, k$ if $0 \leqslant n < (p_1 \cdots p_k)^h$; and let

$$\mathbf{X}(n) = \mathbf{X}(n + (p_1 \cdots p_k)^h)$$

otherwise. Let

$$\Omega = \{(\mathbf{X}(n), n): n \in \mathbb{Z}\}. \tag{30}$$

For any real number t, we define the translation $\Omega(t)$ by

$$\Omega(t) = \{(\mathbf{X}(n), n+t): n \in \mathbb{Z}\}. \tag{31}$$

Clearly $\Omega(0) = \Omega$. Also note that the subset $\{(\mathbf{X}(n), n+t): n+t \geqslant 0\}$ of $\Omega(t)$ is a 1-set of class h with respect to the primes $p_1, \ldots, p_k$ in $U_0^k \times [0, \infty)$.

We define, for any box B in $U_0^k \times [0, \infty)$,

$$E[\Omega(t); B] = Z[\Omega(t); B] - \mu(B),$$

where, as before, $Z[\Omega(t); B]$ denotes the number of points of $\Omega(t)$ falling into B, and $\mu(B)$ denotes the volume of B.

Theorem 2C can easily be deduced from the following lemma. Let

$$M = p_1 \cdots p_k. \tag{32}$$

Lemma 3.10. *For any box B^* of class h with respect to the primes $p_1, \ldots, p_k$ in $U_0^k \times [0, \infty)$, we have*

$$\int_0^{M^h} |E[\Omega(t); B^*]|^2 \, \mathrm{d}t < (4h)^k (p_1 \cdots p_k)^2 M^h.$$

It is convenient to express $E[\Omega(t); B^*]$ as a sum of a finite number of 1-dimensional discrepancy functions.

We use R to denote a residue class. In particular, $R(m, q)$ denotes the residue class of integers congruent to m modulo q. For any real number t, we denote by $t + R$ the set $\{t + n: n \in R\}$, and write

$$F[t + R; I^*] = Z[t + R; I^*] - q^{-1}\ell(I^*), \tag{33}$$

where $Z[t + R; I^*]$ denotes the number of elements of $t + R$ falling into the interval I^*, q is the modulus of the residue class R, and $\ell(I^*)$ is the length of I^*. It is obvious that

$$|F[t + R; I^*]| \leqslant 1 \tag{34}$$

always.

Let $B^* = [0, \eta_1) \times \cdots \times [0, \eta_k) \times I^*$, where $I^* = [0, Y)$ and where, for each $j = 1, \ldots, k$, $[0, \eta_j)$ is a union of $L_j < p_j h$ elementary p_j-type intervals I_j of order at most h and such that there are at most $(p_j - 1)$ elementary p_j-type intervals of any one order in the union.

Lemma 3.11. *Suppose that for each $j = 1, \ldots, k$, I_j is of order s_j, where*

$0 \leqslant s_j \leqslant h$. Then there is precisely one residue class R modulo $p_1^{s_1} \cdots p_k^{s_k}$ such that

$$E[\Omega(t); I_1 \times \cdots \times I_k \times I^*] = F[t+R; I^*].$$

Proof. Suppose that $0 \leqslant n < (p_1 \cdots p_k)^h$ and $X_j(n) \in I_j$. Then since $X_j(n) = \sum_{v=1}^{\infty} a_{j,v} p_j^{-v}$ where $n = \sum_{v=1}^{\infty} a_{j,v} p_j^{v-1}$, the numbers $a_{j,1}, \ldots, a_{j,s_j}$ are determined uniquely, but the remaining $a_{j,v}$ are left arbitrary. Since $p_j^{s_j}$ divides $(p_1 \cdots p_k)^h$, it follows that those integers n satisfying $X_j(n) \in I_j$ constitute precisely a residue class modulo $p_j^{s_j}$. By the Chinese remainder theorem, there exists a unique residue class R modulo $p_1^{s_1} \cdots p_k^{s_k}$ such that

$$\mathbf{X}(n) \in I_1 \times \cdots \times I_k \Leftrightarrow n \in R.$$

The lemma follows immediately. ∎

Suppose that for each $j = 1, \ldots, k$, $[0, \eta_j]$ is the union of the elementary p_j-type intervals

$$I_{j,1}, \ldots, I_{j,L_j}.$$

Then

$$B^* = \bigcup_{l_1=1}^{L_1} \cdots \bigcup_{l_k=1}^{L_k} (I_{1,l_1} \times \cdots \times I_{k,l_k} \times I^*).$$

Writing $\mathbf{l} = (l_1, \ldots, l_k)$ and writing $R(\mathbf{l})$ for the residue class such that

$$E[\Omega(t); I_{1,l_1} \times \cdots \times I_{k,l_k} \times I^*] = F[t + R(\mathbf{l}); I^*],$$

we have, since E is additive,

$$E[\Omega(t); B^*] = \sum_{l_1=1}^{L_1} \cdots \sum_{l_k=1}^{L_k} F[t + R(\mathbf{l}); I^*],$$

so that, omitting reference to I^*,

$$\int_0^{M^h} |E[\Omega(t); B^*]|^2 \, dt$$

$$= \sum_{l_1'=1}^{L_1} \cdots \sum_{l_k'=1}^{L_k} \sum_{l_1''=1}^{L_1} \cdots \sum_{l_k''=1}^{L_k} \int_0^{M^h} F[t + R(\mathbf{l}')] F[t + R(\mathbf{l}'')] \, dt. \qquad (35)$$

Clearly, application of (34) alone is not sufficient to give the desired result. Roth proved the following

Lemma 3.12. *Suppose that for each $j = 1, \ldots, k$, $0 \leqslant s_j', s_j'' \leqslant h$ and suppose that $m', m'' \in \mathbb{Z}$. Then, writing, for each $j = 1, \ldots, k$, $u_j = \min\{s_j', s_j''\}$ and*

$d_j = |s'_j - s''_j|$, *we have*

$$\int_0^{M^h} F[t + R(m', p_1^{s'_1} \cdots p_k^{s'_k})] F[t + R(m'', p_1^{s''_1} \cdots p_k^{s''_k}]\, dt$$

$$= p_1^{-d_1} \cdots p_k^{-d_k} \int_0^{M^h} F[t + R(m', p_1^{u_1} \cdots p_k^{u_k})] F[t + R(m'', p_1^{u_1} \cdots p_k^{u_k})]\, dt. \quad (36)$$

Proof. We shall only show that

$$I = \int_0^{M^h} F[t + R(m', p_1^{s'_1} P')] F[t + R(m'', p_1^{s''_1} P'')]\, dt$$

$$= p^{-d_1} \int_0^{M^h} F[t + R(m', p_1^{u_1} P')] F[t + R(m'', p_1^{u_1} P'')]\, dt, \qquad (37)$$

where $P' = p_2^{s'_2} \cdots p_k^{s'_k}$ and $P'' = p_2^{s''_2} \cdots p_k^{s''_k}$. (36) then follows by repeating the argument on the other primes. To prove (37), we may assume, without loss of generality, that $s'_1 \leqslant s''_1$, so that $u_1 = s'_1$. Then the function $F[t + R(m', p_1^{s'_1} P')]$ is periodic in t with period $p_1^{u_1}(p_2 \cdots p_k)^h$ and therefore

$$F[t + ap_1^{u_1}(p_2 \cdots p_k)^h + R(m', p_1^{s'_1} P')]$$

is independent of the choice of the integer a. Furthermore, since $p_1^{u_1}(p_2 \cdots p_k)^h$ divides M^h, the period of the integrand in (37), we have

$$I = \int_0^{M^h} F[t + R(m', p_1^{u_1} P')] F[t + ap_1^{u_1}(p_2 \cdots p_k)^h + R(m'', p_1^{s''_1} P'')]\, dt$$

for every integer a. Hence

$$p^{d_1} I = \sum_{a=1}^{p^{d_1}} \int_0^{M^h} F[t + R(m', p_1^{u_1} P')] F[t + ap_1^{u_1}(p_2 \cdots p_k)^h + R(m'', p_1^{s''_1} P'')]\, dt$$

$$= \int_0^{M^h} F[t + R(m', p_1^{u_1} P')] F[t + R(m'', p_1^{u_1} P'')]\, dt. \qquad \blacksquare$$

It now follows from (35) and Lemma 3.12 that if for each $j = 1, \ldots, k$ and $l_j = 1, \ldots, L_j$, the interval I_{j,l_j} is of order $s(j, l_j)$, then, in view of (34),

$$M^{-h} \int_0^{M^h} |E[\Omega(t); B^*]|^2\, dt \leqslant \sigma_1 \cdots \sigma_k,$$

where, for $j = 1, \ldots, k$,

$$\sigma_j = \sum_{l'_j=1}^{L_j} \sum_{l''_j=1}^{L_j} p_j^{-|s(j,l'_j) - s(j,l''_j)|}.$$

We can write

$$\sigma_j = \sum_{b=0}^{h} \sum_{\substack{\ell'_j=1 \\ \min\{s(j,\ell'_j),s(j,\ell''_j)\}=b}}^{L_j} \sum_{\ell''_j=1}^{L_j} p_j^{-|s(j,\ell'_j)-s(j,\ell''_j)|}.$$

If $L_j = 1$, we have $\sigma_j = 1$. If $L_j > 1$, then since there are at most $(p_j - 1)$ elementary p_j-type intervals of any fixed order, we have

$$\sigma_j < 2\sum_{b=1}^{h} \sum_{\substack{\ell'_j=1 \\ b=s(j,\ell'_j)\leqslant s(j,\ell''_j)}}^{L_j} \sum_{\ell''_j=1}^{L_j} p_j^{-|s(j,\ell'_j)-s(j,\ell''_j)|}$$

$$\leqslant 2h(p_j-1)^2 \sum_{d=0}^{\infty} p_j^{-d} < 4hp_j^2.$$

This completes the proof of Lemma 3.10.

We now deduce Theorem 2C. Let $p_1, \ldots, p_k$ be the first k primes. For any natural number $N \geqslant 2$, let the integer h satisfy $2^{h-1} < N \leqslant 2^h$. For any $\mathbf{x} = (x_1, \ldots, x_k) \in U_1^k$ and for any Y satisfying $0 < Y \leqslant N$, let $B(\mathbf{x}, Y) = [0, x_1) \times \cdots \times [0, x_k) \times [0, Y)$. Then in view of Lemmas 3.10 and 3.8, we have that

$$\int_0^{M^h} |E[\Omega(t); B(\mathbf{x}, Y)]|^2 \, dt \ll_k M^h h^k$$

for every $B(\mathbf{x}, Y)$. Hence

$$\int_0^{M^h} \int_0^N \int_{U^k} |E[\Omega(t); B(\mathbf{x}, Y)]|^2 \, d\mathbf{x} \, dY \, dt \ll_k N M^h h^k,$$

so that there exists a real number t^*, satisfying $0 \leqslant t^* < M^h$, such that

$$\int_0^N \int_{U^k} |E[\Omega(t^*); B(\mathbf{x}, Y)]|^2 \, d\mathbf{x} \, dY \ll_k N(\log N)^k.$$

It follows that the set $\mathscr{P} = \{(X_1(n), \ldots, X_k(n), N^{-1}(n + t^*)) : 0 \leqslant n + t^* < N\}$ of N points in U_0^{k+1} gives the desired result.

3.4 An inductive argument using Roth's probabilistic method

In this section, we show how we can modify Roth's approach to Theorem 2C in order to prove Theorem 2D. We prove the following theorem of Chen.

Theorem 2D. *Let $W > 0$. There exists a constant $c_7(k, W)$ such that for every natural number $N \geqslant 2$, there exists a distribution $\mathscr{P}$ of N points in U_0^{k+1} such* *that*

$$\int_{U^{k+1}} |D[\mathscr{P}; B(\mathbf{x})]|^W \, d\mathbf{x} < c_7(k, W)(\log N)^{\frac{1}{2}kW}.$$

Note first of all that it is sufficient to prove Theorem 2D for even positive integers W, since for any given distribution $\mathscr{P}$, $\|D(\mathscr{P})\|_W$ is an increasing function of W.

Let Ω be the set (30). For any real number t, let $\Omega(t)$ be defined by (31) as in the previous section. It is easily seen that to prove Theorem 2D, it suffices to prove the following generalization of Lemma 3.10. Let $M = p_1 \cdots p_k$ as in (32).

Lemma 3.13. *For any box B^* of class h with respect to the primes $p_1, \ldots, p_k$ in $U_0^k \times [0, \infty)$ and for any even positive integer W, there exists a constant $c_8(p_1, \ldots, p_k, W)$ such that*

$$\int_0^{M^h} |E[\Omega(t); B^*]|^W \, dt < c_8(p_1, \ldots, p_k, W) M^h h^{\frac{1}{2}kW}.$$

A direct proof of Lemma 3.13 using Roth's method does not seem to work, as it does not seem possible to generalize Lemma 3.12 to the case when the integrand in (36) is a product of more than two discrepancy functions. On the other hand, if one uses induction on h, one can prove Lemma 3.13 for $k = 1$ using (34). For $k \geq 2$, our induction hypotheses are too weak.

The new idea is to strengthen the induction hypotheses and prove a more general form of Lemma 3.13. This is the motivation for introducing the following variation of q-sets (in $U_0^k \times \mathbb{R}$ instead of in $U_0^k \times [0, \infty)$). We comment here that the definition in §3.2 of q-sets in $U_0^k \times [0, \infty)$ was first given in Chen (1983) after the first proof of Theorem 2D and the result of Faure (1982), whereas the idea of periodic q-sets in this section was introduced in Chen (1980) from ideas in Roth (1980).

The important property of the set Ω is that Lemma 3.11 holds. However, there are other sets of points besides Ω for which the corresponding Lemma 3.11 also holds.

As in Roth's proof of Theorem 2C, we let $p_1, \ldots, p_k$ be distinct primes.

Definition. By a periodic 1-set of class h with respect to the primes $p_1, \ldots, p_k$ in $U_0^k \times \mathbb{R}$, we mean a set of the form

$$\omega^{(1)} = \{(x_1(n), \ldots, x_k(n), n) : n \in \mathbb{Z}\}$$

having the following property: For every $j = 1, \ldots, k$, if I_j is an elementary

p_j-type interval of order s, satisfying $0 \leqslant s \leqslant h$, then the set

$$\{n \in \mathbb{Z} : x_j(n) \in I_j\}$$

is a residue class modulo p_j^s.

In other words, Lemma 3.11 holds with Ω replaced by $\omega^{(1)}$.

It is convenient to introduce the following more general type of sets.

Definition. Suppose that $(q, p_1 \cdots p_k) = 1$. By a periodic q-set of class h with respect to the primes $p_1, \ldots, p_k$ in $U_0^k \times \mathbb{R}$, we mean a set of the form

$$\omega^{(q)} = \{(x_1(n), \ldots, x_k(n), n) : n \in Q\}, \tag{38}$$

where Q is a residue class modulo q, and having the following property: For every $j = 1, \ldots, k$, if I_j is an elementary p_j-type interval of order s, satisfying $0 \leqslant s \leqslant h$, then the set

$$\{n \in Q : x_j(n) \in I_j\}$$

is a residue class modulo qp_j^s.

Note that the set $\omega^{(q)}$ has the property that if I'_j and I''_j are two elementary p_j-type intervals of order at most h, then

$$I'_j \subset I''_j \Leftrightarrow \{n \in Q : x_j(n) \in I'_j\} \subset \{n \in Q : x_j(n) \in I''_j\}. \tag{39}$$

Definition. We shall call a set of the form

$$\omega^{(q)}(t) = \{(\mathbf{x}(n), n + t) : (\mathbf{x}(n), n) \in \omega^{(q)}\}$$

a translated periodic q-set of class h with respect to the primes $p_1, \ldots, p_k$ in $U_0^k \times \mathbb{R}$ if the set $\omega^{(q)}$ is a periodic q-set of class h with respect to the same primes in $U_0^k \times \mathbb{R}$.

The following lemmas are obvious.

Lemma 3.14. *Suppose that $\omega^{(q)}(t)$ is a translated periodic q-set of class h with respect to the primes $p_1, \ldots, p_k$ in $U_0^k \times \mathbb{R}$. Then, for every s satisfying $0 \leqslant s \leqslant h$, $\omega^{(q)}(t)$ is also a translated periodic q-set of class s with respect to the same primes in $U_0^k \times \mathbb{R}$.*

Lemma 3.15. *Suppose that $\omega^{(1)}(t) = \{(\mathbf{x}(n), n + t) : n \in \mathbb{Z}\}$ is a translated periodic 1-set of class h with respect to the primes $p_1, \ldots, p_k$ in $U_0^k \times \mathbb{R}$. Then, for every $j = 1, \ldots, k$, the set*

$$\{(x_1(n), \ldots, x_{j-1}(n), x_{j+1}(n), \ldots, x_k(n), n + t) : n \in \mathbb{Z}\}$$

is a translated periodic 1*-set of class* h *with respect to the primes* $p_1,\ldots,p_{j-1}, p_{j+1},\ldots,p_k$ *in* $U_0^{k-1}\times\mathbb{R}$. *On the other hand, if* R_0 *is a residue class modulo* q, *where* $(q,p_1\cdots p_k)=1$, *then the set* $\{(\mathbf{x}(n),n+t):n\in R_0\}$ *is a translated periodic* q*-set of class* h *with respect to the primes* $p_1,\ldots,p_k$ *in* $U_0^k\times\mathbb{R}$.

For any box B in $U_0^k\times[0,\infty)$, we let $Z[\omega^{(q)}(t);B]$ denote the number of points of $\omega^{(q)}(t)$ falling into B, and write

$$E[\omega^{(q)}(t);B]=Z[\omega^{(q)}(t);B]-q^{-1}\mu(B). \tag{40}$$

Then clearly the function E is additive.

Lemma 3.16. *Suppose that* $\omega^{(q)}$ *is a periodic* q*-set of class* h *with respect to the primes* $p_1,\ldots,p_k$ *in* $U_0^k\times\mathbb{R}$, *where* $(q,p_1\cdots p_k)=1$; *and suppose that* B^* *is a box of class* s *with respect to the same primes in* $U_0^k\times[0,\infty)$, *where* $0\leqslant s\leqslant h$. *Then the function* $E[\omega^{(q)}(t);B^*]$ *is periodic in* t *with period* $q(p_1\cdots p_k)^s$.

Proof. Note that B^* is a finite union of mutually disjoint boxes of the type $B=I_1\times\cdots\times I_k\times I^*$, where, for $j=1,\ldots,k$, I_j is an elementary p_j-type interval of order s. By additivity, it suffices to show that Lemma 3.16 holds with B^* replaced by B. If $\omega^{(q)}$ is the set (38), we shall show that

$$E[\omega^{(q)}(t);I_1\times\cdots\times I_k\times I^*]=F[t+(R\cap Q);I^*],$$

where R is a residue class modulo $(p_1\cdots p_k)^s$. By the Chinese remainder theorem, we have that for each $j=1,\ldots,k$,

$$\{n\in Q:x_j(n)\in I_j\}=Q\cap R_j,$$

where R_j is a residue class modulo p_j^s. Then

$$\bigcap_{j=1}^{k}\{n\in Q:x_j(n)\in I_j\}=Q\cap(R_1\cap\cdots\cap R_k).$$

The result follows easily. ∎

We can now state a more general form of Lemma 3.13.

Lemma 3.17. *Let* W *be an even positive integer. For a suitable constant* $c_9=c_9(p_1,\ldots,p_k,W)$, *we have, for any periodic* 1*-set* $\omega^{(1)}$ *of class* h *with respect to the primes* $p_1,\ldots,p_k$ *in* $U_0^k\times\mathbb{R}$ *and for any box* B^* *of class* h *with respect to the same primes in* $U_0^k\times[0,\infty)$, *that*

$$\int_0^{M^h}|E[\omega^{(1)}(t);B^*]|^W\,dt<c_9(p_1,\ldots,p_k,W)M^h h^{\frac{1}{2}kW}.$$

Lemma 3.17 is in fact equivalent to the following superficially more general form.

Lemma 3.17′. *For the same constant $c_9(p_1,\dots,p_k,W)$ as in Lemma 3.17, we have, for any periodic q-set $\omega^{(q)}$ of class h with respect to the primes $p_1,\dots,p_k$ in $U_0^k\times\mathbb{R}$ and for any box B^* of class h with respect to the same primes in $U_0^k\times[0,\infty)$, that*

$$\int_0^{qM^h}|E[\omega^{(q)}(t);B^*]|^W\,dt<c_9(p_1,\dots,p_k,W)qM^h h^{\frac{1}{2}kW}.$$

To show this equivalence, simply note that if $\omega^{(q)}$ is the set (38), where Q is the residue class n_0 modulo q, then

$$\omega^{(q)}(-n_0)=\{(\mathbf{x}(n),n-n_0):n\in Q\}$$

is of the form

$$\{(\mathbf{x}(n_0+qn'),qn'):n'\in\mathbb{Z}\}.$$

It is not difficult to show that the set

$$\{(\mathbf{x}(n_0+qn'),n'):n'\in\mathbb{Z}\}$$

is a periodic 1-set of class h with respect to the primes $p_1,\dots,p_k$ in $U_0^k\times\mathbb{R}$. The equivalence follows from the periodicity of E.

Let $\omega=\omega^{(1)}$ be a periodic 1-set of class h with respect to the primes $p_1,\dots,p_k$ in $U_0^k\times\mathbb{R}$. We keep this set ω fixed. To prove Lemma 3.17, we use induction on h. For $k\geqslant 2$, we also make use of corresponding results in lower-dimensional cases. We therefore begin by investigating some properties of ω and the box B^*.

Let

$$B^*=B^*(\boldsymbol{\eta},Y)=[0,\eta_1)\times\cdots\times[0,\eta_k)\times[0,Y) \tag{41}$$

be a box of class h with respect to the primes $p_1,\dots,p_k$ in $U_0^k\times[0,\infty)$. For $j=1,\dots,k$ and $s=0,\dots,h$, let $\xi_{j,s}$ denote the greatest integer multiple of p_j^{-s} not exceeding η_j. Furthermore, for $s\neq 0$, write

$$v_{j,s}=p_j^s(\xi_{j,s}-\xi_{j,s-1}).$$

Clearly $v_{j,s}$ is an integer and $0\leqslant v_{j,s}<p_j$ for $j=1,\dots,k$ and $s=1,\dots,h$. Let

$$I^*=[0,Y).$$

For $s=0,\dots,h$, let

$$B_s^*=[0,\xi_{1,s})\times\cdots\times[0,\xi_{k,s})\times I^*, \tag{42}$$

i.e. B_s^* is the largest box of class s with respect to the primes $p_1,\dots,p_k$ in $U_0^k\times[0,\infty)$ which is contained in B^*. We now consider the complement of B_{s-1}^* in B_s^*. We let $B_{1,s}$ denote the part of this which is contained in

$$[\xi_{1,s-1}, \xi_{1,s}) \times [0, 1)^{k-1} \times I^*,$$

$B_{2,s}$ denote the part of the remainder which is contained in

$$[0, 1) \times [\xi_{2,s-1}, \xi_{2,s}) \times [0, 1)^{k-2} \times I^*,$$

and so on. In other words, for $j = 1, \ldots, k$ and $s = 1, \ldots, h$, we let

$$B_{j,s} = [0, \xi_{1,s-1}) \times \cdots \times [0, \xi_{j-1,s-1}) \times [\xi_{j,s-1}, \xi_{j,s}) \times [0, \xi_{j+1,s}) \times \cdots \times [0, \xi_{k,s}) \times I^*. \tag{43}$$

Then for $s = 1, \ldots, h$,

$$B_s^* = B_{s-1}^* \cup B_{1,s} \cup \cdots \cup B_{k,s}. \tag{44}$$

Note that the union is pairwise disjoint, so it follows by additivity that for $s = 1, \ldots, h$,

$$E[\omega(t); B_s^*] = E[\omega(t); B_{s-1}^*] + \sum_{j=1}^{k} E[\omega(t); B_{j,s}]. \tag{45}$$

For $j = 1, \ldots, k$ and $s = 1, \ldots, h$, we may suppose that $B_{j,s} \neq \varnothing$; for otherwise $B_{j,s} = \varnothing$, and so obviously $E[\omega(t); B_{j,s}] = 0$, so that we can omit reference to $E[\omega(t); B_{j,s}]$ in (45). Then the interval $[\xi_{j,s-1}, \xi_{j,s})$ in (43) is a union of exactly $v_{j,s}$ mutually disjoint elementary p_j-type intervals of order s. We denote them by $J_{j,s,\alpha}$ $(\alpha = 1, \ldots, v_{j,s})$. For $\alpha = 1, \ldots, v_{j,s}$, let

$$B_{j,s,\alpha} = [0, \xi_{1,s-1}) \times \cdots \times [0, \xi_{j-1,s-1}) \times J_{j,s,\alpha} \times [0, \xi_{j+1,s}) \times \cdots \times [0, \xi_{k,s}) \times I^*. \tag{46}$$

Then for $j = 1, \ldots, k$ and $s = 1, \ldots, h$,

$$B_{j,s} = \bigcup_{\alpha=1}^{v_{j,s}} B_{j,s,\alpha},$$

and the union is pairwise disjoint, so that by additivity,

$$E[\omega(t); B_{j,s}] = \sum_{\alpha=1}^{v_{j,s}} E[\omega(t); B_{j,s,\alpha}]. \tag{47}$$

For $j = 1, \ldots, k$ and $s = 1, \ldots, h-1$, let

$$\bar{J}_{j,s} = [\xi_{j,s}, \xi_{j,s} + p_j^{-s}), \tag{48}$$

i.e. $\bar{J}_{j,s}$ is the elementary p_j-type interval of order s containing the complement of $[0, \xi_{j,s})$ in $[0, \eta_j)$. We modify $B_{j,h}$ ((43) with $s = h$) by replacing $[\xi_{j,h-1}, \xi_{j,h})$ by $\bar{J}_{j,s}$, which is the elementary p_j-type interval of order s

which contains $[\xi_{j,h-1}, \xi_{j,h})$. Accordingly, for $j = 1, \ldots, k$ and $s = 1, \ldots, h-1$, we let the modification $\bar{B}_{j,s}$ be defined by

$$\bar{B}_{j,s} = [0, \xi_{1,h-1}) \times \cdots \times [0, \xi_{j-1,h-1}) \times \bar{J}_{j,s} \times [0, \xi_{j+1,h}) \times \cdots \times [0, \xi_{k,h}) \times I^*. \tag{49}$$

Note that $B_{j,h}$ has $(h-1)$ modifications, namely $\bar{B}_{j,1}, \ldots, \bar{B}_{j,h-1}$.

The next lemma contains the key idea underlying Roth's proof of Lemma 3.10. This idea is contained in the proof of Lemma 3.12.

For $j = 1, \ldots, k$, write

$$M_j = Mp_j^{-1} = p_1 \cdots p_{j-1} p_{j+1} \cdots p_k.$$

Lemma 3.18. *For $j = 1, \ldots, k$ and for $\alpha = 1, \ldots, v_{j,h}$, we have that*

$$\sum_{a=0}^{p_j-1} E[\omega(t + aM_j^h p_j^{h-1}); B_{j,h,\alpha}] = E[\omega(t); \bar{B}_{j,h-1}]. \tag{50}$$

Also, for $j = 1, \ldots, k$ and $s = 2, \ldots, h-1$, we have that

$$\sum_{a=0}^{p_j-1} E[\omega(t + aM_j^h p_j^{s-1}); \bar{B}_{j,s}] = E[\omega(t); \bar{B}_{j,s-1}]. \tag{51}$$

Proof. We first prove (50). Recall the definitions of $B_{j,h,\alpha}$ ((46) with $s = h$) and $\bar{B}_{j,h-1}$ ((49) with $s = h-1$). We note that the only difference is that $J_{j,h,\alpha}$ in the former is replaced by $\bar{J}_{j,h-1}$ in the latter. By (48), we see that $\bar{J}_{j,h-1}$ is the elementary p_j-type interval of order $(h-1)$ containing $J_{j,h,\alpha}$. If $\omega = \{(\mathbf{x}(n), n) : n \in \mathbb{Z}\}$, let $R_{j,h,\alpha} = \{n \in \mathbb{Z} : x_j(n) \in J_{j,h,\alpha}\}$ and $R_0 = \{n \in \mathbb{Z} : x_j(n) \in \bar{J}_{j,h-1}\}$. Then by (39), since $J_{j,h,\alpha} \subset \bar{J}_{j,h-1}$, we have that $R_{j,h,\alpha} \subset R_0$, so R_0 is the residue class modulo p_j^{h-1} containing $R_{j,h,\alpha}$. For $i = 1, \ldots, j-1, j+1, \ldots, k$, let J_i denote any elementary p_i-type interval of order h. Then we see that since $[0, \xi_{i,h-1})$ and $[0, \xi_{i,h})$ are the union of a finite number of mutually disjoint intervals of the type J_i, it follows by additivity that to prove (50), it suffices to prove that

$$\sum_{a=0}^{p_j-1} E[\omega(t + aM_j^h p_j^{h-1}); J_1 \times \cdots \times J_{j-1} \times J_{j,h,\alpha} \times J_{j+1} \times \cdots \times J_k \times I^*]$$
$$= E[\omega(t); J_1 \times \cdots \times J_{j-1} \times \bar{J}_{j,h-1} \times J_{j+1} \times \cdots \times J_k \times I^*]. \tag{52}$$

For $i = 1, \ldots, j-1, j+1, \ldots, k$, let $R_i = \{n \in \mathbb{Z} : x_i(n) \in J_i\}$. Then the summand on the left-hand side of (52) is, by (33) and (40), equal to

$$F[t + aM_j^h p_j^{h-1} + (R_1 \cap \cdots \cap R_{j-1} \cap R_{j,h,\alpha} \cap R_{j+1} \cap \cdots \cap R_k); I^*].$$

Now, for each $i \neq j$, we have $aM_j^h p_j^{h-1} + R_i = R_i$; so it follows that the summand on the left-hand side of (52) is equal to

$$F[t+(R_1\cap\cdots\cap R_{j-1}\cap(aM_j^hp_j^{h-1}+R_{j,h,a})\cap R_{j+1}\cap\cdots\cap R_k); I^*]. \quad (53)$$

On the other hand,

$$\bigcup_{a=0}^{p_j-1}(aM_j^hp_j^{h-1}+R_{j,h,a})=R_0. \quad (54)$$

It follows from (53) and (54) that the left-hand side of (52) is equal to

$$F[t+(R_1\cap\cdots\cap R_{j-1}\cap R_0\cap R_{j+1}\cap\cdots\cap R_k); I^*],$$

which is equal to the right-hand side of (52). This completes the proof of (50). A similar argument gives (51). ■

For $s=1,\ldots,h$, write, for the sake of simplicity,

$$G_s(t)=\sum_{j=1}^{k}E[\omega(t); B_{j,s}]. \quad (55)$$

Let

$$T_0=\int_0^{M^h}E[\omega(t); B_{h-1}^*]^W\,\mathrm{d}t; \quad (56)$$

and for $w=2,\ldots,W$,

$$T_w=\int_0^{M^h}E[\omega(t); B_{h-1}^*]^{W-w}G_h^w(t)\,\mathrm{d}t. \quad (57)$$

For $j=1,\ldots,k$, we write

$$S_j=\int_0^{M^h}E[\omega(t); B_0^*]^{W-1}E[\omega(t); \bar{B}_{j,1}]\,\mathrm{d}t; \quad (58)$$

further, for $s=1,\ldots,h-1$ and $w=1,\ldots,W-1$, we write

$$S_{j,s,w}=\int_0^{M^h}E[\omega(t); B_{s-1}^*]^{W-w-1}G_s^w(t)E[\omega(t); \bar{B}_{j,s}]\,\mathrm{d}t. \quad (59)$$

The proof of Lemma 3.17 is based on the following identity.

Lemma 3.19. *We have, for $h\geqslant 2$, that*

$$\int_0^{M^h}E[\omega(t); B^*]^W\,\mathrm{d}t$$

$$=T_0+\sum_{w=2}^{W}\binom{W}{w}T_w+W\sum_{j=1}^{k}v_{j,h}p_j^{-(h-1)}S_j$$

$$+W\sum_{j=1}^{k}v_{j,h}\sum_{s=1}^{h-1}p_j^{-(h-s)}\sum_{w=1}^{W-1}\binom{W-1}{w}S_{j,s,w}. \quad (60)$$

Proof. By (41) and (42), we have $B^* = B_h^*$, and so by taking W-th powers on both sides of (45) with $s = h$ and using binomial expansion on the right-hand side, we have, in view of (55), that

$$E[\omega(t); B^*]^W = E[\omega(t); B_{h-1}^*]^W + \sum_{w=2}^{W} \binom{W}{w} E[\omega(t); B_{h-1}^*]^{W-w} G_h^w(t)$$
$$+ W \sum_{j=1}^{k} E[\omega(t); B_{h-1}^*]^{W-1} E[\omega(t); B_{j,h}].$$

It follows from (56) and (57) that to prove Lemma 3.19, it remains to show that for $j = 1, \ldots, k$,

$$\int_0^{M^h} E[\omega(t); B_{h-1}^*]^{W-1} E[\omega(t); B_{j,h}]\, dt$$
$$= v_{j,h} p_j^{-(h-1)} S_j + v_{j,h} \sum_{s=1}^{h-1} p_j^{-(h-s)} \sum_{w=1}^{W-1} \binom{W-1}{w} S_{j,s,w}. \tag{61}$$

Since B_{h-1}^* is a box of class $(h-1)$ with respect to the primes $p_1, \ldots, p_k$ in $U_0^k \times [0, \infty)$, we have, by Lemma 3.16, that $E[\omega(t); B_{h-1}^*]$ is periodic in t with period M^{h-1}, and therefore also periodic with period $M_j^h p_j^{h-1}$ for every $j = 1, \ldots, k$; so by (47) with $s = h$ and (50), we have, for $j = 1, \ldots, k$ and writing V_j for the left-hand side of (61), that

$$V_j = \sum_{\alpha=1}^{v_{j,s}} \int_0^{M^h} E[\omega(t); B_{h-1}^*]^{W-1} E[\omega(t); B_{j,h,\alpha}]\, dt$$
$$= v_{j,h} p_j^{-1} \int_0^{M^h} E[\omega(t); B_{h-1}^*]^{W-1} E[\omega(t); \bar{B}_{j,h-1}]\, dt. \tag{62}$$

By (45) and (55), we have, for $s = 1, \ldots, h-1$, that

$$E[\omega(t); B_s^*]^{W-1}$$
$$= E[\omega(t); B_{s-1}^*]^{W-1} + \sum_{w=1}^{W-1} \binom{W-1}{w} E[\omega(t); B_{s-1}^*]^{W-w-1} G_s^w(t). \tag{63}$$

On combining (62) and the case $s = h-1$ of (63) and (59), we have, for $j = 1, \ldots, k$, that

$$v_{j,h}^{-1} V_j = p_j^{-1} \int_0^{M^h} E[\omega(t); B_{h-2}^*]^{W-1} E[\omega(t); \bar{B}_{j,h-1}]\, dt$$
$$+ p_j^{-1} \sum_{w=1}^{W-1} \binom{W-1}{w} S_{j,h-1,w}. \tag{64}$$

For $s = 1, \ldots, h-2$, B_s^* is a box of class s with respect to the primes

$p_1, \ldots, p_k$ in $U_0^k \times [0, \infty)$. Hence, by Lemma 3.16, (51), (63) and (59), we have, for $j = 1, \ldots, k$ and $s = 1, \ldots, h-2$, that

$$\begin{aligned}\int_0^{M^h} & E[\omega(t); B_s^*]^{W-1} E[\omega(t); \bar{B}_{j,s+1}]\, dt \\ &= p_j^{-1} \int_0^{M^h} E[\omega(t); B_s^*]^{W-1} E[\omega(t); \bar{B}_{j,s}]\, dt \\ &= p_j^{-1} \int_0^{M^h} E[\omega(t); B_{s-1}^*]^{W-1} E[\omega(t); \bar{B}_{j,s}]\, dt \\ &\quad + p_j^{-1} \sum_{w=1}^{W-1} \binom{W-1}{w} S_{j,s,w}. \end{aligned} \tag{65}$$

(61) now follows, in view of (58), from (64) and repeated application of (65). This completes the proof of Lemma 3.19. ■

We first investigate the case $k = 1$, as the cases $k \geq 2$ depend on results in lower dimensions.

The following lemma is only applicable for the case $k = 1$.

Lemma 3.20. *We have, for $s = 1, \ldots, h$, that*

$$|G_s(t)| = |E[\omega(t); B_{1,s}]| < p_1; \tag{66}$$

also, for $s = 1, \ldots, h-1$, we have that

$$|E[\omega(t); \bar{B}_{1,s}]| \leq 1. \tag{67}$$

Proof. Recall (47) with $j = 1$. We have, for $s = 1, \ldots, h$, that

$$E[\omega(t); B_{1,s}] = \sum_{\alpha=1}^{\nu_{1,s}} E[\omega(t); B_{1,s,\alpha}].$$

Since $\nu_{1,s} < p_1$, it follows that to prove (66), it suffices to show that for each $\alpha = 1, \ldots, \nu_{1,s}$,

$$|E[\omega(t); B_{1,s,\alpha}]| \leq 1. \tag{68}$$

On recalling (46), we have, for $s = 1, \ldots, h$ and $\alpha = 1, \ldots, \nu_{1,s}$, that $B_{1,s,\alpha} = J_{1,s,\alpha} \times I^*$, where $J_{1,s,\alpha}$ is an elementary p_1-type interval of order s. It follows from (33), (40) and the remark before (46) that

$$E[\omega(t); B_{1,s,\alpha}] = F[t + R_{1,s,\alpha}; I^*]. \tag{69}$$

(68) now follows from (69) and (34). (67) follows similarly, since for $s = 1, \ldots, h-1$, we have, by (49) and (48), that $\bar{B}_{1,s} = \bar{J}_{1,s} \times I^*$, and $\bar{J}_{1,s}$ is an elementary p_1-type interval of order s. The proof of Lemma 3.20 is now complete. ■

We shall prove by induction on h that Lemma 3.17 for $k=1$ holds for a constant C satisfying

$$C = C(p_1, W) = (2^{W+1}p_1)^W. \tag{70}$$

Note that in particular,

$$C > p_1^W. \tag{71}$$

Suppose that $B^* = I_1 \times I^*$, where $I_1 = [0, 1)$. Then by (33), (40) and (34),

$$|E[\omega(t); B^*]| = |F[t+\mathbb{Z}; I^*]| \leqslant 1,$$

and Lemma 3.17 for $k=1$ is proved. We may therefore suppose that $I_1 \neq [0, 1)$, and hence, by (42), that $B_0^* = \varnothing$. For $h=1$, we have, by (44) and (66), that

$$|E[\omega(t); B^*]| = |E[\omega(t); B_1^*]| = |E[\omega(t); B_{1,1}]| < p_1,$$

and so in view of (71), Lemma 3.17 for $k=1$ holds for $h=1$. Suppose now that $h>1$, and suppose that Lemma 3.17 for $k=1$ holds when h is replaced by any smaller positive integer, so that in particular, for any $s=2,\dots,h$, we have, in view of Lemma 3.16, that

$$\int_0^{M^h} E[\omega(t); B_{s-1}^*]^W \,\mathrm{d}t \leqslant CM^h(s-1)^{\frac{1}{2}W}. \tag{72}$$

On applying Hölder's inequality, (66) and (72) to (57), we have, for $w=2,\dots,W$, that

$$T_w \leqslant C^{(W-w)/W} p_1^w M^h (h-1)^{\frac{1}{2}(W-w)} < C^{1-1/W} p_1 M^h (h-1)^{\frac{1}{2}W-1}$$

in view of (71), so that

$$\sum_{w=2}^{W} \binom{W}{w} T_w \leqslant C^{1-1/W} 2^W p_1 M^h (h-1)^{\frac{1}{2}W-1}. \tag{73}$$

On the other hand, since $B_0^* = \varnothing$, we have that $E[\omega(t); B_0^*] = 0$, and so by (58),

$$S_1 = 0. \tag{74}$$

On applying Hölder's inequality, (66), (67) and (72) to (59), we have, for $s=2,\dots,h-1$ and $w=1,\dots,W-1$, that, in view of (71),

$$S_{1,s,w} \leqslant C^{(W-w-1)/W} p_1^w M^h (s-1)^{\frac{1}{2}(W-w-1)} < C^{1-1/W} M^h (h-1)^{\frac{1}{2}W-1};$$

furthermore

$$S_{1,1,W-1} \leqslant p_1^{W-1} M^h < C^{1-1/W} M^h (h-1)^{\frac{1}{2}W-1}.$$

For $w = 1, \ldots, W-2$, we have $S_{1,1,w} = 0$. It follows that

$$Wv_{1,h}\sum_{s=1}^{h-1} p_1^{-(h-s)}\sum_{w=1}^{W-1}\binom{W-1}{w}S_{1,s,w} \leqslant C^{1-1/W}2^{W-1}Wp_1M^h(h-1)^{\frac{1}{2}W-1}. \tag{75}$$

On the other hand, we have, by (56) and (72), that

$$T_0 \leqslant CM^h(h-1)^{\frac{1}{2}W}. \tag{76}$$

On combining (60), (73)–(76), we have that

$$\begin{aligned}\int_0^{M^h} & E[\omega(t); B^*]^W \, dt \\ & \leqslant CM^h(h-1)^{\frac{1}{2}W} + C^{1-1/W}2^{W-1}(2+W)p_1M^h(h-1)^{\frac{1}{2}W-1} \\ & \leqslant CM^h\{(h-1)^{\frac{1}{2}W} + C^{-1/W}2^W Wp_1(h-1)^{\frac{1}{2}W-1}\} \\ & = CM^h\{(h-1)^{\frac{1}{2}W} + \tfrac{1}{2}W(h-1)^{\frac{1}{2}W-1}\} \leqslant CM^h h^{\frac{1}{2}W}\end{aligned}$$

in view of (70). The proof of Lemma 3.17 for $k = 1$ is now complete.

We can now consider the case $k \geqslant 2$.

Suppose, for every $j = 1, \ldots, k$, that the corresponding Lemma 3.17′ for periodic q-sets with respect to the primes $p_1, \ldots, p_{j-1}, p_{j+1}, \ldots, p_k$ in $U_0^{k-1} \times \mathbb{R}$ and boxes with respect to the same primes in $U_0^{k-1} \times [0, \infty)$ holds with $C_j(p_1, \ldots, p_{j-1}, p_{j+1}, \ldots, p_k, W)$ and M_j in place of c_9 and M respectively, and with the exponent of h replaced by $\frac{1}{2}(k-1)W$. We let $C_0 = C_0(p_1, \ldots, p_k, W)$ be defined by

$$C_0 = 1 + \max_{1 \leqslant j \leqslant k} C_j(p_1, \ldots, p_{j-1}, p_{j+1}, \ldots, p_k, W). \tag{77}$$

We shall prove by induction on h that Lemma 3.17 holds for a constant C satisfying

$$C = C(p_1, \ldots, p_k, W) = 2^{kW^2+W}(p_1 \cdots p_k)^{W+1}C_0(p_1, \ldots, p_k, W). \tag{78}$$

Clearly,

$$C > (p_1 \cdots p_k)^{W+1}C_0, \tag{79}$$

and

$$C > (p_1 \cdots p_k)^W. \tag{80}$$

First of all, we must establish a link which enables us to use the analogues of Lemma 3.17′ in lower-dimensional cases.

For $j = 1, \ldots, k$ and $s = 1, \ldots, h$ and $\alpha = 1, \ldots, v_{j,s}$, we define the set $\omega_{j,s,\alpha}(t)$ of points in k-dimensional space by

$$\omega_{j,s,\alpha}(t) = \{(x_1(n), \ldots, x_{j-1}(n), x_{j+1}(n), \ldots, x_k(n), n+t) : n \in R_{j,s,\alpha}\},$$

where $R_{j,s,\alpha}=\{n\in\mathbb{Z}:x_j(n)\in J_{j,s,\alpha}\}$ is a residue class modulo p_j^s. Then for every real number t, we have, by Lemma 3.15, that $\omega_{j,s,\alpha}(t)$ is a translated p_j^s-set of class h with respect to the primes $p_1,\ldots,p_{j-1},p_{j+1},\ldots,p_k$ in $U_0^{k-1}\times\mathbb{R}$. On the other hand, let $B_{j,s}^*$ be defined for the above values of j and s by

$$B_{j,s}^*=[0,\xi_{1,s-1})\times\cdots\times[0,\xi_{j-1,s-1})\times[0,\xi_{j+1,s})\times\cdots\times[0,\xi_{k,s})\times I^*.$$

Then $B_{j,s}^*$ is a box of class s with respect to the primes $p_1,\ldots,p_{j-1},p_{j+1},\ldots,p_k$ in $U_0^{k-1}\times[0,\infty)$. It is easy to see that for $j=1,\ldots,k$ and $s=1,\ldots,h$ and $\alpha=1,\ldots,v_{j,s}$,

$$E[\omega(t);B_{j,s,\alpha}]=E[\omega_{j,s,\alpha}(t);B_{j,s}^*]. \tag{81}$$

It follows that on our assumption of the lower-dimensional cases of Lemma 3.17, and hence Lemma 3.17′, we have, in view of (81), (77), Lemma 3.16 and the definitions of M and M_j, that for every $j=1,\ldots,k$ and $s=1,\ldots,h$ and $\alpha=1,\ldots,v_{j,s}$,

$$\begin{aligned}\int_0^{M^h}E[\omega(t);B_{j,s,\alpha}]^W\,dt&=M^{h-s}\int_0^{p_j^sM_j^s}E[\omega_{j,s,\alpha}(t);B_{j,s}^*]^W\,dt\\&\leqslant C_jM^hs^{\frac{1}{2}(k-1)W}<C_0M^hs^{\frac{1}{2}(k-1)W}.\end{aligned} \tag{82}$$

Similarly, for $j=1,\ldots,k$ and $s=1,\ldots,h-1$, we have, on relating $E[\omega(t);\bar{B}_{j,s}]$ to the discrepancy of certain translated p_j^s-sets with respect to the primes $p_1,\ldots,p_{j-1},p_{j+1},\ldots,p_k$ in $U_0^{k-1}\times\mathbb{R}$ in the box $B_{j,h}^*$, that

$$\int_0^{M^h}E[\omega(t);\bar{B}_{j,s}]^W\,dt<C_0M^hh^{\frac{1}{2}(k-1)W}. \tag{83}$$

By (55) and (47), we have, for $s=1,\ldots,h$, that

$$\begin{aligned}\int_0^{M^h}G_s^W(t)\,dt&=\int_0^{M^h}\left(\sum_{j=1}^{k}\sum_{\alpha=1}^{v_{j,s}}E[\omega(t);B_{j,s,\alpha}]\right)^W dt\\&\leqslant\left(\sum_{j=1}^{k}v_{j,s}\right)^W\sum_{j=1}^{k}\sum_{\alpha=1}^{v_{j,s}}\int_0^{M^h}E[\omega(t);B_{j,s,\alpha}]^W\,dt;\end{aligned} \tag{84}$$

so that on combining (82) and (84), we have, for $s=1,\ldots,h$, that

$$\begin{aligned}\int_0^{M^h}G_s^W(t)\,dt&\leqslant(p_1+\cdots+p_k)^{W+1}C_0M^hs^{\frac{1}{2}(k-1)W}\\&<(p_1\cdots p_k)^{W+1}C_0M^hs^{\frac{1}{2}(k-1)W}.\end{aligned} \tag{85}$$

We may assume that $B^*=I_1\times\cdots\times I_k\times I^*$, where for some $j=1,\ldots,k$, $I_j\neq[0,1)$; for otherwise, by (33), (40) and (34), we have that $|E[\omega(t);B^*]|=|F[t+\mathbb{Z};I^*]|\leqslant 1$, and Lemma 3.17 is proved. By (42), we see that $B_0^* = \varnothing$. On the other hand, it is easy to see that by the definition of $\omega(t)$, we have, for any fixed t and $\boldsymbol{\eta}$, that $E[\omega(t); B_1^*(\boldsymbol{\eta}, Y)]$ is periodic in Y with period $p_1 \cdots p_k$, so that we may assume that $0 \leqslant Y < p_1 \cdots p_k$ when estimating $E[\omega(t); B_1^*]$. With this restriction on $Y, Z[\omega(t); B_1^*] \leqslant p_1 \cdots p_k$ and $\mu(B_1^*) < p_1 \cdots p_k$, so that by (40), we have $|E[\omega(t); B_1^*]| \leqslant p_1 \cdots p_k$; by (80), Lemma 3.17 holds for $h = 1$ with C defined by (78).

Suppose now that $h \geqslant 2$, and that Lemma 3.17 holds when h is replaced by any smaller positive integer, so that in particular, for $s = 2, \ldots, h$, in view of Lemma 3.16,

$$\int_0^{M^h} E[\omega(t); B_{s-1}^*]^W \, dt \leqslant CM^h(s-1)^{\frac{1}{2}kW}. \tag{86}$$

On applying Hölder's inequality, (85) and (86) to (57), we have, for $w = 2, \ldots, W$, that

$$\begin{aligned} T_w &\leqslant C^{(W-w)/W} C_0^{w/W} (p_1 \cdots p_k)^{w(W+1)/W} M^h (h-1)^{\frac{1}{2}k(W-w)} h^{\frac{1}{2}(k-1)w} \\ &\leqslant C^{1-1/W} C_0^{1/W} (p_1 \cdots p_k)^{(W+1)/W} 2^{(k-1)W} M^h (h-1)^{\frac{1}{2}kW-1} \end{aligned}$$

by (79); hence

$$\sum_{w=2}^{W} \binom{W}{w} T_w \leqslant C^{1-1/W} C_0^{1/W} (p_1 \cdots p_k)^{(W+1)/W} 2^{kW} M^h (h-1)^{\frac{1}{2}kW-1}. \tag{87}$$

On the other hand, since $B_0^* = \varnothing$, we have that $E[\omega(t); B_0^*] = 0$, and so by (58), we have, for every $j = 1, \ldots, k$, that

$$S_j = 0. \tag{88}$$

On applying Hölder's inequality, (83), (85) and (86) to (59), we have, for $j = 1, \ldots, k$ and $s = 2, \ldots, h-1$ and $w = 1, \ldots, W-1$, that, in view of (79),

$$\begin{aligned} S_{j,s,w} &\leqslant C^{(W-w-1)/W} C_0^{(w+1)/W} (p_1 \cdots p_k)^{w(W+1)/W} \\ &\quad \cdot M^h (s-1)^{\frac{1}{2}k(W-w-1)} s^{\frac{1}{2}(k-1)w} h^{\frac{1}{2}(k-1)} \\ &\leqslant C^{1-1/W} C_0^{1/W} 2^{\frac{1}{2}(k-1)} M^h (h-1)^{\frac{1}{2}kW-1}; \end{aligned}$$

furthermore,

$$\begin{aligned} S_{j,1,W-1} &\leqslant C_0 (p_1 \cdots p_k)^{(W-1)(W+1)/W} M^h h^{\frac{1}{2}(k-1)} \\ &\leqslant C^{1-1/W} C_0^{1/W} 2^{\frac{1}{2}(k-1)} M^h (h-1)^{\frac{1}{2}kW-1}. \end{aligned}$$

For $j = 1, \ldots, k$ and $w = 1, \ldots, W-2$, we have that $S_{j,1,w} = 0$. It follows that

$$W \sum_{j=1}^{k} v_{j,h} \sum_{s=1}^{h-1} p_j^{-(h-s)} \sum_{w=1}^{W-1} \binom{W-1}{w} S_{j,s,w}$$

$$\leqslant C^{1-1/W}C_0^{1/W}(p_1+\cdots+p_k)2^{\frac{1}{2}(k-1)}2^{W-1}WM^h(h-1)^{\frac{1}{2}kW-1}$$
$$\leqslant C^{1-1/W}C_0^{1/W}(p_1\cdots p_k)^{(W+1)/W}2^{kW}M^h(h-1)^{\frac{1}{2}kW-1}, \tag{89}$$

since $2^{\frac{1}{2}(k-1)}2^{W-1}W \leqslant 2^{\frac{1}{2}(k-1)+2W-2} < 2^{kW}$. Also, by (56) and (86), we have that

$$T_0 \leqslant CM^h(h-1)^{\frac{1}{2}kW}. \tag{90}$$

On combining (60), (87)–(90), we have, in view of (78), that

$$\int_0^{M^h} E[\omega(t); B^*]^W \, dt$$
$$\leqslant CM^h\{(h-1)^{\frac{1}{2}kW} + 2C^{-1/W}C_0^{1/W}(p_1\cdots p_k)^{(W+1)/W}2^{kW}(h-1)^{\frac{1}{2}kW-1}\}$$
$$= CM^h\{(h-1)^{\frac{1}{2}kW} + (h-1)^{\frac{1}{2}kW-1}\} < CM^h h^{\frac{1}{2}kW}.$$

This completes the proof of Lemma 3.17.

3.5 A variation of Roth's probabilistic method

Roth's probabilistic method works mainly because the set Ω has a certain 'periodicity' property. More precisely, Lemma 3.11 holds. In Chen (1980), the (periodic) q-sets are constructed simply to satisfy the analogues of Lemma 3.11.

In Faure (1982), the set (15), even if extended to $U_0^k \times \mathbb{R}$, does not have a corresponding Lemma 3.11, so that Roth's probabilistic method using the idea of the translation variable t fails. We hope, therefore, to find a probabilistic method that works for every q-set that exists. In Chen (1983), it is shown that instead of considering translations in the $(k+1)$-th direction, one can consider modifications of q-sets in the other k directions.

Before we proceed, it is convenient to write down the analogues of Lemmas 3.14 and 3.15 for general q-sets as defined in §3.2. Let $p_1, \ldots, p_k$ be k primes, not necessarily distinct.

Lemma 3.21. *Suppose that $\mathcal{Q}$ is a q-set of class h with respect to the primes $p_1, \ldots, p_k$ in $U_0^k \times [0, \infty)$. Then, for every s satisfying $0 \leqslant s \leqslant h$, $\mathcal{Q}$ is also a q-set of class s with respect to the same primes in $U_0^k \times [0, \infty)$.*

Lemma 3.22. *Suppose that $\mathcal{Q}$ is a 1-set of class h with respect to the primes $p_1, \ldots, p_k$ in $U_0^k \times [0, \infty)$, and suppose that I is an elementary p_j-type interval of order s, with $1 \leqslant j \leqslant k$ and $0 \leqslant s \leqslant h$. Then the set*

$$\{(x_1, \ldots, x_{j-1}, x_{j+1}, \ldots, x_k, y) : x_j \in I \text{ and } (x_1, \ldots, x_k, y) \in \mathcal{Q}\}$$

is a p_j^s-set of class h with respect to the primes $p_1,\ldots,p_{j-1},p_{j+1},\ldots,p_k$ in $U_0^{k-1}\times[0,\infty)$.

To indicate how we modify a given q-set, we need some notation.

We denote by $\mathscr{A}_{k,h}$ the set of all matrices $A=(a_{i,t})$ with integer entries $a_{i,t}$ satisfying $0\leqslant a_{i,t}<p_i$ for each $i=1,\ldots,k$ and $t=1,\ldots,h$. It is clear that $\mathscr{A}_{k,h}$ has $(p_1\cdots p_k)^h$ elements.

Let $A=(a_{i,t})\in\mathscr{A}_{k,h}$, and let $\mathscr{Q}_q$ be a q-set of class h with respect to the primes $p_1,\ldots,p_k$ in $U_0^k\times[0,\infty)$. We construct the set $\mathscr{Q}_q(A)$ as follows: Suppose that $\mathbf{x}=(x_1,\ldots,x_k,y)\in\mathscr{Q}_q$. For $i=1,\ldots,k$, write

$$x_i=\sum_{t=1}^{h}b_{i,t}p_i^{-t}+c_i,$$

where $b_{i,t}$ are integers satisfying $0\leqslant b_{i,t}<p_i$, and where $0\leqslant c_i<p_i^{-h}$. For $i=1,\ldots,k$ and $t=1,\ldots,h$, let $b_{i,t}^A$ be defined by

$$0\leqslant b_{i,t}^A<p_i \quad\text{and}\quad b_{i,t}^A\equiv b_{i,t}+a_{i,t} \pmod{p_i},$$

and let

$$x_i^A=\sum_{t=1}^{h}b_{i,t}^Ap_i^{-t}+c_i.$$

Then it is clear that $x_i^A\in U_0$. Write

$$\mathbf{x}^A=(x_1^A,\ldots,x_k^A,y),$$

and let

$$\mathscr{Q}_q(A)=\{\mathbf{x}^A\in U_0^k\times[0,\infty):\mathbf{x}\in\mathscr{Q}_q\}.$$

By considering the effect of each entry of A separately, it is not difficult to show that

Lemma 3.23. *If $\mathscr{Q}_q$ is a q-set of class h with respect to the primes $p_1,\ldots,p_k$ in $U_0^k\times[0,\infty)$, then for every $A\in\mathscr{A}_{k,h}$, $\mathscr{Q}_q(A)$ is also a q-set of class h with respect to the same primes in $U_0^k\times[0,\infty)$.*

The analogue of Lemma 3.16 is the following

Lemma 3.24. *Suppose that $\mathscr{Q}_q$ is a q-set of class h with respect to the primes $p_1,\ldots,p_k$ in $U_0^k\times[0,\infty)$, $A=(a_{i,t})\in\mathscr{A}_{k,h}$ and $B=I_1\times\cdots I_k\times I^*$ is a box in $U_0^k\times[0,\infty)$, where for some fixed integer s satisfying $0\leqslant s\leqslant h$ and for each $i=1,\ldots,k$, I_i is a union of elementary p_i-type intervals of order s. Then the function $E[\mathscr{Q}_q(A);B]$ is independent of the entries $a_{i,t}$ of A for every $i=1,\ldots,k$ and $t=s+1,\ldots,h$.*

As before, we write

$$M=p_1\cdots p_k,$$

and, for each $j = 1, \ldots, k$,

$$M_j = Mp_j^{-1} = p_1 \cdots p_{j-1} p_{j+1} \cdots p_k.$$

It is convenient to introduce the following notation. Let $A \in \mathscr{A}_{k,h}$. For $j = 1, \ldots, k$ and $s = 0, \ldots, h$, we write $A_{j,s}$ for the matrix obtained from A by replacing $a_{j,s+1}, \ldots, a_{j,h}$ in A by 0, and write A_s for the matrix obtained from A by replacing $a_{i,s+1}, \ldots, a_{i,h}$ in A by 0 for every $i = 1, \ldots, k$. We see that there are $M_j^h p_j^s$ choices of $A_{j,s} \in \mathscr{A}_{k,h}$, and M^s choices of $A_s \in \mathscr{A}_{k,h}$. Throughout,

$$\sum_{A \in \mathscr{A}_{k,h}} f(A) \quad \text{and} \quad \sum_{A_s \in \mathscr{A}_{k,h}} f(A_s)$$

denote respectively a sum over the M^h choices of $A \in \mathscr{A}_{k,h}$ and a sum over the M^s choices of $A_s \in \mathscr{A}_{k,h}$, while

$$\sum_{A_{j,s} \in \mathscr{A}_{k,h}} f(A_{j,s})$$

denotes a sum over the $M_j^h p_j^s$ choices of $A_{j,s} \in \mathscr{A}_{k,h}$.

We can now state the key lemma for the proof of Theorem 2D. The following is the analogue of Lemma 3.17′.

Lemma 3.25. *Let W be an even positive integer. For the same constant $c_9(p_1, \ldots, p_k, W)$ as in Lemma* 3.17, *we have, for any q-set $\mathscr{Q}_q$ of class h with respect to the primes $p_1, \ldots, p_k$ in $U_0^k \times [0, \infty)$ and for any box B^* of class h with respect to the same primes in $U_0^k \times [0, \infty)$, that*

$$\sum_{A \in \mathscr{A}_{k,h}} E[\mathscr{Q}_q(A); B^*]^W < c_9(p_1, \ldots, p_k, W) M^h h^{\frac{1}{2}kW}.$$

It is not difficult to see that Theorem 2D follows easily from Lemma 3.25. Furthermore, it is sufficient to prove Lemma 3.25 for the case $q = 1$.

The proof of Lemma 3.25 is similar to the proof of Lemma 3.17. Suppose that $\mathscr{Q}$ is a 1-set of class h with respect to the primes $p_1, \ldots, p_k$ in $U_0^k \times [0, \infty)$ and suppose that B^*, as in (41), is a box of class h with respect to the same primes in $U_0^k \times [0, \infty)$. We partition the box B^* as in §3.4. By considering all modifications $\mathscr{Q}(A)$ $(A \in \mathscr{A}_{k,h})$ of $\mathscr{Q}$, we have the following lemma, the analogue of Lemma 3.19.

For $s = 1, \ldots, h$, write

$$G_s(A) = \sum_{j=1}^{k} E[\mathscr{Q}(A); B_{j,s}].$$

Let

$$T_0 = \sum_{A_{h-1} \in \mathscr{A}_{k,h}} E[\mathscr{Q}(A_{h-1}); B_{h-1}^*]^W,$$

and

$$T_1 = \sum_{A \in \mathscr{A}_{k,h}} E[\mathscr{Q}(A); B_0^*]^{W-1} G_h(A).$$

For $w = 2, \ldots, W$, let

$$T_w = \sum_{A \in \mathscr{A}_{k,h}} E[\mathscr{Q}(A_{h-1}); B_{h-1}^*]^{W-w} G_h^w(A).$$

Further, for $j = 1, \ldots, k$ and $s = 1, \ldots, h-1$ and $w = 1, \ldots, W-1$, write

$$S_{j,s,w} = \sum_{A_{j,s} \in \mathscr{A}_{k,h}} E[\mathscr{Q}(A_{s-1}); B_{s-1}^*]^{W-w-1} G_s^w(A_s) E[\mathscr{Q}(A_{j,s}); \bar{B}_{j,s}].$$

Lemma 3.26. *We have, for $h \geqslant 2$, that*

$$\sum_{A \in \mathscr{A}_{k,h}} E[\mathscr{Q}(A); B^*]^W = MT_0 + WT_1 + \sum_{w=2}^{W} \binom{W}{w} T_w + W \sum_{j=1}^{k} v_{j,h} \sum_{s=1}^{h-1} \sum_{w=1}^{W-1} \binom{W-1}{w} S_{j,s,w}.$$

The key to the proof of Lemma 3.26 is the following analogue of Lemma 3.18.

Lemma 3.27. *For $j = 1, \ldots, k$ and $\alpha = 1, \ldots, v_{j,h}$, we have*

$$\sum_{a_{j,h}=0}^{p_j-1} E[\mathscr{Q}(A); B_{j,h,\alpha}] = E[\mathscr{Q}(A_{j,h-1}); \bar{B}_{j,h-1}].$$

Also, for $j = 1, \ldots, k$ and $s = 1, \ldots, h-2$, we have

$$\sum_{a_{j,s+1}=0}^{p_j-1} E[\mathscr{Q}(A_{j,s+1}); \bar{B}_{j,s+1}] = E[\mathscr{Q}(A_{j,s}); \bar{B}_{j,s}].$$

We briefly return to the problem of existence of 1-sets with respect to certain primes in $U_0^k \times [0, \infty)$. It turns out that a slight elaboration of the ideas in this section enables one to prove the complement of Lemma 3.6. The key idea is the following generalization of Lemma 3.23.

Suppose that $\mathscr{Q}$ is a 1-set of class h with respect to the primes $p_1, \ldots, p_k$ in $U_0^k \times [0, \infty)$. For $i = 1, \ldots, k$ and $t = 1, \ldots, h$, let $\theta_{i,t}$ be a permutation of the numbers $0, 1, \ldots, p_i - 1$, and let $\Theta = (\theta_{i,t})$. Suppose that $\mathbf{x} = (x_1, \ldots, x_k, y) \in \mathscr{Q}$. For $i = 1, \ldots, k$, let

$$x_i = \sum_{t=1}^{h} b_{i,t} p_i^{-t} + c_i,$$

where $b_{i,t}$ are integers satisfying $0 \leqslant b_{i,t} < p_i$, and where $0 \leqslant c_i < p_i^{-h}$. For $i = 1, \ldots, k$ and $t = 1, \ldots, h$, let $b_{i,t}^{\Theta}$ be defined by

$$b_{i,t}^{\Theta} = \theta_{i,t}(b_{i,t}),$$

and write

$$x_i^{\Theta} = \sum_{t=1}^{h} b_{i,t}^{\Theta} p_i^{-t} + c_i.$$

Then $x_i^{\Theta} \in U_0$. Write

$$\mathbf{x}^{\Theta} = (x_1^{\Theta}, \ldots, x_k^{\Theta}, y),$$

and let

$$\mathscr{Q}(\Theta) = \{\mathbf{x}^{\Theta} \in U_0^k \times [0, \infty) : \mathbf{x} \in \mathscr{Q}\}.$$

Lemma 3.23′. *If $\mathscr{Q}$ is a 1-set of class h with respect to the primes $p_1, \ldots, p_k$ in $U_0^k \times [0, \infty)$, then $\mathscr{Q}(\Theta)$ is also a 1-set of class h with respect to the same primes in $U_0^k \times [0, \infty)$.*

We can now prove

Lemma 3.28. *Suppose that p is a prime and $p < k$. Then there are no 1-sets of class h with respect to the primes $p, \ldots, p$ in $U_0^k \times [0, \infty)$ unless $h = 0$.*

The special case $p = 2$ was proved by Sobol (1967).

Proof. By Lemma 3.21, it suffices to show that there are no 1-sets of class 1 with respect to the primes $p, \ldots, p$ in $U_0^k \times [0, \infty)$ if $p < k$. Suppose on the contrary that $\mathscr{Q}$ is a 1-set of class 1 with respect to the primes $p, \ldots, p$ in $U_0^k \times [0, \infty)$. We may assume, without loss of generality, that $\mathscr{Q}$ is of the form

$$\mathscr{Q} = \{\mathbf{x}(n) = (x_1(n), \ldots, x_k(n), n) : n \in \mathbb{N}^*\}. \tag{91}$$

For each $n \in \mathbb{N}^*$ and $i = 1, \ldots, k$, write

$$x_i(n) = b_i(n) p^{-1} + c_i(n),$$

where $b_i(n)$ are integers satisfying

$$0 \leqslant b_i(n) < p, \tag{92}$$

and where $0 \leqslant c_i(n) < p^{-1}$. It follows that each elementary 1-box of order 1 with respect to the primes $p, \ldots, p$ in $U_0^k \times [0, \infty)$ of the form

$$[0, 1)^{i-1} \times I \times [0, 1)^{k-i} \times [0, p),$$

where I is an elementary p-type interval of order 1, contains exactly one point of $\mathscr{Q}$, or, more precisely, exactly one point of

$$\{\mathbf{x}(n) : \mathbf{x}(n) \in \mathscr{Q} \text{ and } 0 \leqslant n \leqslant p - 1\}.$$

It follows from (92) that for each $i = 1, \ldots, k$, the integers $b_i(0), \ldots, b_i(p-1)$ are distinct and form a complete set of residues modulo p. Hence by

Lemma 3.23′, we may assume, without loss of generality, that for each $i = 1, \ldots, k$ and each $n = 0, \ldots, p-1$,

$$b_i(n) = n. \tag{93}$$

Consider now the k integers $b_1(p), \ldots, b_k(p)$. Since $p < k$, we have, by (92) and Dirichlet's box principle, that there exist j and J satisfying $1 \leqslant j < J \leqslant k$ such that

$$b_j(p) = b_J(p), \tag{94}$$

$= b$, say. It then follows from (93) and (94) that the box

$$[0, 1)^{j-1} \times [bp^{-1}, (b+1)p^{-1}) \times [0, 1)^{J-j-1} \times [bp^{-1}, (b+1)p^{-1}) \times [0, 1)^{k-J} \times [0, p^2),$$

an elementary 1-box of order 1 with respect to the primes $p, \ldots, p$ in $U_0^k \times [0, \infty)$, contains at least two points of $\mathcal{Q}$, a contradiction. The existence of 1-sets of class 0 with respect to the primes $p, \ldots, p$ in $U_0^k \times [0, \infty)$ is left as a simple exercise! ■

The non-existence of 1-sets has also been established in other cases. For example, it has been shown in Chen (1983) that

Lemma 3.29.

(i) *There are no 1-sets of class h with respect to the primes $2, 2, 3$ in $U_0^3 \times [0, \infty)$ unless $h \leqslant 1$.*
(ii) *There are no 1-sets of class h with respect to the primes $2, 3, 3$ in $U_0^3 \times [0, \infty)$ unless $h = 0$.*

Perhaps the following is true.

Conjecture. *Suppose that $p_1, \ldots, p_k$ are not all distinct but not all equal. Then there exists a positive integer h_0, depending at most on $p_1, \ldots, p_k$, such that for every $h \geqslant h_0$, there are no 1-sets of class h with respect to the primes $p_1, \ldots, p_k$ in $U_0^k \times [0, \infty)$.*

Any counterexample will be very interesting.

4

Lower bounds – a combinatorial method of Schmidt

4.1 The key inequality

Let $s_1, s_2, s_3, \ldots$ be a sequence in U_0, and suppose that $\alpha \in U_1$. We let $Z(n, \alpha)$ denote the number of s_ν $(1 \leqslant \nu \leqslant n)$ such that $s_\nu \in [0, \alpha)$, and write

$$D(n, \alpha) = Z(n, \alpha) - n\alpha.$$

We extend the definition of $D(n, \alpha)$ to all $\alpha \in \mathbb{R}$ by periodicity, i.e.

$$D(n, \alpha) = D(n, \{\alpha\}),$$

where $\{\alpha\}$ denotes the fractional part of α.

By intervals $I, J, \ldots$, we mean intervals of the type $(a, b]$, where $a, b \in \mathbb{Z}$ and $0 \leqslant a < b$. Then the number of integers in an interval I is equal to its length $l(I) = b - a$. For any interval I and any $\alpha \in \mathbb{R}$, write

$$g^+(I, \alpha) = \max_{n \in I} D(n, \alpha) \quad \text{and} \quad g^-(I, \alpha) = \min_{n \in I} D(n, \alpha),$$

and let

$$h(I, \alpha) = g^+(I, \alpha) - g^-(I, \alpha).$$

For any $\alpha, \beta \in \mathbb{R}$, write

$$D(n, \alpha, \beta) = D(n, \beta) - D(n, \alpha),$$

and for any interval I, let

$$g^+(I, \alpha, \beta) = \max_{n \in I} D(n, \alpha, \beta) \quad \text{and} \quad g^-(I, \alpha, \beta) = \min_{n \in I} D(n, \alpha, \beta).$$

For any pair of intervals J, J', write

$$h(J, J', \alpha, \beta) = \max\{0, g^-(J, \alpha, \beta) - g^+(J', \alpha, \beta), g^-(J', \alpha, \beta) - g^+(J, \alpha, \beta)\}.$$

The theorems proved in §§4.2–4.3 all depend on the following key inequality.

Lemma 4.1. *Suppose that I is an interval, and J, J' are subintervals of I. Then* *for any* $\alpha, \beta \in \mathbb{R}$,

$$h(I,\alpha)+h(I,\beta)\geqslant h(J,J',\alpha,\beta)+\tfrac{1}{2}(h(J,\alpha)+h(J,\beta)+h(J',\alpha)+h(J',\beta)). \quad (1)$$

Proof. The lemma is trivial if $h(J,J',\alpha,\beta)=0$, so we may assume, without loss of generality, that $h(J,J',\alpha,\beta)=g^-(J,\alpha,\beta)-g^+(J',\alpha,\beta)>0$. Then for all $n\in J$ and $n'\in J'$, we have $D(n,\alpha,\beta)-D(n',\alpha,\beta)\geqslant h(J,J',\alpha,\beta)$, i.e.

$$D(n,\beta)-D(n,\alpha)-D(n',\beta)+D(n',\alpha)\geqslant h(J,J',\alpha,\beta). \quad (2)$$

Choose $m_\alpha, n_\alpha, m_\beta, n_\beta \in J$ such that

$$g^+(J,\alpha)=D(m_\alpha,\alpha) \quad \text{and} \quad g^-(J,\alpha)=D(n_\alpha,\alpha),$$
$$g^+(J,\beta)=D(m_\beta,\beta) \quad \text{and} \quad g^-(J,\beta)=D(n_\beta,\beta).$$

Then

$$D(m_\alpha,\alpha)-D(n_\alpha,\alpha)=h(J,\alpha), \quad (3)$$

$$D(m_\beta,\beta)-D(n_\beta,\beta)=h(J,\beta). \quad (4)$$

Similarly, choose $m'_\alpha, n'_\alpha, m'_\beta, n'_\beta \in J'$ such that

$$D(m'_\alpha,\alpha)-D(n'_\alpha,\alpha)=h(J',\alpha), \quad (5)$$

$$D(m'_\beta,\beta)-D(n'_\beta,\beta)=h(J',\beta). \quad (6)$$

Applying (2) with $n=m_\alpha$ and $n'=m'_\beta$, we have

$$D(m_\alpha,\beta)-D(m_\alpha,\alpha)-D(m'_\beta,\beta)+D(m'_\beta,\alpha)\geqslant h(J,J',\alpha,\beta). \quad (7)$$

Applying (2) with $n=n_\beta$ and $n'=n'_\alpha$, we have

$$D(n_\beta,\beta)-D(n_\beta,\alpha)-D(n'_\alpha,\beta)+D(n'_\alpha,\alpha)\geqslant h(J,J',\alpha,\beta). \quad (8)$$

On the other hand,

$$h(I,\alpha)\geqslant \max\{D(m'_\alpha,\alpha)-D(n_\alpha,\alpha), D(m'_\beta,\alpha)-D(n_\beta,\alpha)\}, \quad (9)$$
$$h(I,\beta)\geqslant \max\{D(m_\beta,\beta)-D(n'_\beta,\beta), D(m_\alpha,\beta)-D(n'_\alpha,\beta)\}. \quad (10)$$

(1) follows on combining (3)–(10). ■

4.2 The problem of van Aardenne-Ehrenfest and a question of Erdös

In this section, we prove Theorems 3B, 7A and 7B. Their proofs depend on the following intermediate result.

Lemma 4.2. *Suppose that the integers s and t satisfy* $s\geqslant 0$ *and* $t\geqslant 1$, *and suppose that* $\beta\in\mathbb{R}$ *is arbitrary. Let I be any interval with* $l(I)\geqslant 6^{s+t}$.

Then

$$6^{-t}\sum_{j=1}^{6^t} h(I,\beta+j6^{-s-t}) \geqslant \frac{t}{120}. \tag{11}$$

Proof. We proceed by induction on t. Consider first the case $t=1$. Writing $S=\frac{1}{2}\cdot 6^{s+1}$, we note that

$$D(n+S,\beta+6^{-s-1})-D(n,\beta+6^{-s-1})-D(n+S,\beta)+D(n,\beta)$$

is an integer minus $\frac{1}{2}$, and so

$$\begin{aligned}&|D(n+S,\beta+6^{-s-1})-D(n,\beta+6^{-s-1})|+|D(n+S,\beta)-D(n,\beta)|\\ &\quad\geqslant|D(n+S,\beta+6^{-s-1})-D(n,\beta+6^{-s-1})-D(n+S,\beta)+D(n,\beta)|\geqslant\tfrac{1}{2}.\end{aligned}$$

Suppose that $l(I)\geqslant 6^{s+1}=2S$. Then there exists an integer n such that $n\in I$ and $n+S\in I$, and so

$$h(I,\beta+6^{-s-1})+h(I,\beta)\geqslant\tfrac{1}{2}. \tag{12}$$

Now (12) is true for every $\beta\in\mathbb{R}$, and so

$$\tfrac{1}{6}\sum_{j=1}^{6} h(I,\beta+j6^{-s-1})\geqslant\tfrac{1}{4}>\tfrac{1}{120},$$

and the proof of (11) for $t=1$ is complete. Suppose now that (11) holds for all intervals I with $l(I)\geqslant 6^{s+t}$ and any $\beta\in\mathbb{R}$. Let I be given such that $l(I)\geqslant 6^{s+t+1}$, and suppose that $I=(a,b]$, where $a,b\in\mathbb{Z}$ and $0\leqslant a<b$. For $r=1,2,\ldots,6$, let J_r be the interval

$$J_r=(a+(r-1)6^{s+t},a+r6^{s+t}]. \tag{13}$$

Then for every $r=1,\ldots,6$, we have that $J_r\subset I$. For $j=1,2,\ldots,$ let

$$\alpha_j=\beta+j6^{-s-t-1}. \tag{14}$$

Writing $T=6^t$, we now consider

$$2\sum_{j=1}^{6T} h(I,\alpha_j)=h(I,\alpha_1)+h(I,\alpha_{6T})+\sum_{j=2}^{6T}(h(I,\alpha_{j-1})+h(I,\alpha_j)). \tag{15}$$

To treat the finite sum on the right-hand side of (15), we extend the definition of $Z(n,\alpha)$ to all $\alpha\in\mathbb{R}$ by writing

$$Z(n,\alpha)=Z(n,\{\alpha\})+n(\alpha-\{\alpha\}).$$

For $j=2,3,\ldots,6T$, let

$$\begin{aligned}z_j=Z(a+4\cdot 6^{s+t},\alpha_j)-Z(a+4\cdot 6^{s+t},\alpha_{j-1})-Z(a+6^{s+t},\alpha_j)\\ +Z(a+6^{s+t},\alpha_{j-1});\end{aligned} \tag{16}$$

and let

$$Z = \sum_{j=2}^{6T} z_j. \tag{17}$$

There are two cases: (I) $Z > 2T/3$; and (II) $Z \leqslant 2T/3$. Consider first of all case (I), so that $Z > 2T/3$. For all $n \in J_5$ and $n' \in J_1$ and all j satisfying $2 \leqslant j \leqslant 6T$, we have

$$Z(n, \alpha_j) - Z(n', \alpha_j) - Z(n, \alpha_{j-1}) + Z(n', \alpha_{j-1}) \geqslant z_j,$$

and so

$$D(n, \alpha_{j-1}, \alpha_j) - D(n', \alpha_{j-1}, \alpha_j) \geqslant z_j - (n - n')(\alpha_j - \alpha_{j-1}) \geqslant z_j - \tfrac{5}{6}.$$

It follows that for $j = 2, 3, \ldots, 6T$,

$$h(J_5, J_1, \alpha_{j-1}, \alpha_j) \geqslant z_j - \tfrac{5}{6}.$$

Since $h(J_5, J_1, \alpha_{j-1}, \alpha_j) \geqslant 0$ always and z_j is an integer, we have, for $j = 2, 3, \ldots, 6T$, that

$$h(J_5, J_1, \alpha_{j-1}, \alpha_j) \geqslant z_j/6. \tag{18}$$

By Lemma 4.1 and (18), we have, for $j = 2, 3, \ldots, 6T$, that

$$\begin{aligned} h(I, \alpha_{j-1}) + h(I, \alpha_j) \geqslant z_j/6 + \tfrac{1}{2}(&h(J_1, \alpha_{j-1}) + h(J_1, \alpha_j) \\ &+ h(J_5, \alpha_{j-1}) + h(J_5, \alpha_j)). \end{aligned} \tag{19}$$

Combining (15), (17), (19) and the trivial estimates

$$h(I, \alpha_1) \geqslant \tfrac{1}{2}(h(J_1, \alpha_1) + h(J_5, \alpha_1))$$

and

$$h(I, \alpha_{6T}) \geqslant \tfrac{1}{2}(h(J_1, \alpha_{6T}) + h(J_5, \alpha_{6T})),$$

we have

$$2 \sum_{j=1}^{6T} h(I, \alpha_j) \geqslant \frac{Z}{6} + \sum_{j=1}^{6T} h(J_1, \alpha_j) + \sum_{j=1}^{6T} h(J_5, \alpha_j). \tag{20}$$

Consider the finite sum on the right-hand side of (20) involving J_1. We have

$$\sum_{j=1}^{6T} h(J_1, \alpha_j) = \sum_{j=1}^{6T} h(J_1, \beta + j6^{-s-t-1}) = \sum_{i=0}^{5} \left(\sum_{\substack{j=1 \\ j \equiv i \pmod 6}}^{6T} h(J_1, \beta + j6^{-s-t-1}) \right).$$

The inner sum is of the form (11) except for the factor 6^{-t} on the left-hand side of (11). By the induction hypothesis, we therefore have, since $l(J_1) \geqslant 6^{s+t}$,

$$\sum_{j=1}^{6T} h(J_1, \alpha_j) \geqslant \sum_{i=0}^{5} \frac{Tt}{120} = 6^{t+1} \frac{t}{120}. \tag{21}$$

Similarly, $l(J_5) \geqslant 6^{s+t}$, so

$$\sum_{j=1}^{6T} h(J_5, \alpha_j) \geqslant 6^{t+1} \frac{t}{120}. \tag{22}$$

Combining (20)–(22), and since $Z > 2T/3$, we have

$$2 \sum_{j=1}^{6T} h(I, \alpha_j) \geqslant \frac{T}{9} + 2 \cdot 6^{t+1} \frac{t}{120} = 2 \cdot 6^{t+1} \left(\frac{t}{120} + \frac{1}{108} \right) > 2 \cdot 6^{t+1} \frac{t+1}{120},$$

and the induction step is complete. We now consider case (II), so that now $Z \leqslant 2T/3$. For all $n \in J_4$ and $n' \in J_2$ and all j satisfying $2 \leqslant j \leqslant 6T$, we have

$$Z(n, \alpha_j) - Z(n', \alpha_j) - Z(n, \alpha_{j-1}) + Z(n', \alpha_{j-1}) \leqslant z_j,$$

and so

$$D(n', \alpha_{j-1}, \alpha_j) - D(n, \alpha_{j-1}, \alpha_j) \geqslant -z_j + (n - n')(\alpha_j - \alpha_{j-1}) \geqslant \tfrac{1}{6} - z_j.$$

It follows that for $j = 2, 3, \ldots, 6T$,

$$h(J_2, J_4, \alpha_{j-1}, \alpha_j) \geqslant \tfrac{1}{6} - z_j. \tag{23}$$

By Lemma 4.1 and (23), we have, for $j = 2, 3, \ldots, 6T$, that

$$\begin{aligned} &h(I, \alpha_{j-1}) + h(I, \alpha_j) \\ &\quad \geqslant \tfrac{1}{6} - z_j + \tfrac{1}{2}(h(J_2, \alpha_{j-1}) + h(J_2 . \alpha_j) + h(J_4, \alpha_{j-1}) + h(J_4, \alpha_j)). \end{aligned} \tag{24}$$

Combining (15), (17), (24) and the trivial estimates

$$h(I, \alpha_1) \geqslant \tfrac{1}{2}(h(J_2, \alpha_1) + h(J_4, \alpha_1))$$

and

$$h(I, \alpha_{6T}) \geqslant \tfrac{1}{2}(h(J_2, \alpha_{6T}) + h(J_4, \alpha_{6T})),$$

we have

$$2 \sum_{j=1}^{6T} h(I, \alpha_j) \geqslant \frac{6T - 1}{6} - Z + \sum_{j=1}^{6T} h(J_2, \alpha_j) + \sum_{j=1}^{6T} h(J_4, \alpha_j). \tag{25}$$

In analogy to (21) and (22), we have that for $r = 2, 4$,

$$\sum_{j=1}^{6T} h(J_r, \alpha_j) \geqslant 6^{t+1} \frac{t}{120}. \tag{26}$$

Combining (25) and (26), and since $Z \leqslant 2T/3$, we have

$$2 \sum_{j=1}^{6T} h(I, \alpha_j) \geqslant \frac{5T}{6} - \frac{2T}{3} + 2 \cdot 6^{t+1} \frac{t}{120} = 2 \cdot 6^{t+1} \left(\frac{t}{120} + \frac{1}{72} \right) > 2 \cdot 6^{t+1} \frac{t+1}{120},$$

and this completes the proof of Lemma 4.2. ∎

We are now in a position to complete Schmidt's proof of Theorem 3B. We state the theorem in the formulation of van Aardenne-Ehrenfest.

Theorem 3B. *Let $s_1, s_2, s_3, \ldots$ be an infinite sequence of real numbers in U_0. For any $n \in \mathbb{N}$ and for any real number $\alpha \in U_1$, let $Z(n, \alpha)$ denote the number of s_ν $(\nu = 1, \ldots, n)$ satisfying $0 \leqslant s_\nu < \alpha$, and write $D(n, \alpha) = Z(n, \alpha) - n\alpha$. Let $D(n) = \sup_{\alpha \in U_1} |D(n, \alpha)|$. There exists a positive absolute constant c_1 such that for every natural number N,*

$$\sup_{1 \leqslant n \leqslant N} D(n) > c_1 \log N.$$

Proof. We use Lemma 4.2 with $s = 0$. Since $|D(1, \frac{1}{2})| = \frac{1}{2}$, we need only prove Theorem 3B for $N \geqslant 6$. Let $t \geqslant 1$ satisfy $6^t \leqslant N < 6^{t+1}$, and let $I = (0, 6^t]$. By Lemma 4.2, there exists $\alpha \in U_1$ such that $h(I, \alpha) \geqslant t/120$. Hence there exists $n \in I$ such that

$$|D(n, \alpha)| \geqslant \frac{t}{240} \geqslant \frac{t+1}{480} \geqslant c_2 \log N,$$

where c_2 is a positive absolute constant. Theorem 3B follows. ■

The proof of Theorem 7A depends on Lemma 4.2 as well as the following simple

Lemma 4.3. *Let $\alpha \in U_1$ and let $v \in \mathbb{R}$ satisfy $0 < v < n$. Then there exists a closed subinterval $\mathscr{I}$ of U_1 containing α such that $|\mathscr{I}| = v/n$ and for every $\beta \in \mathscr{I}$,*

$$|D(n, \beta)| \geqslant |D(n, \alpha)| - v. \tag{27}$$

Proof. If $|D(n, \alpha)| \leqslant v$, the result is trivial; so we can assume that $|D(n, \alpha)| > v$. Then either $D(n, \alpha) > v$ or $D(n, \alpha) < -v$. In the first case, let $\mathscr{I} = [\alpha, \alpha + v/n]$. In the second case, let $\mathscr{I} = [\alpha - v/n, \alpha]$. We leave the details to the reader. ■

We now prove that there exist a nested sequence $\mathscr{I}_1 \supset \cdots \supset \mathscr{I}_m \supset \cdots$ of closed intervals in U_1 and a sequence $n_1 < \cdots < n_m < \cdots$ of positive integers such that for every $\beta \in \mathscr{I}_m$, we have

$$|D(n_m, \beta)| \geqslant c_3 \log n_m, \tag{28}$$

where c_3 is a positive absolute constant. Theorem 7A follows immediately on noting that $\bigcap_{m=1}^{\infty} \mathscr{I}_m \neq \varnothing$.

We prove this assertion by induction on m. The case $m = 1$ is trivial if we take $n_1 = 1$. Assume now that intervals $\mathscr{I}_1 \supset \cdots \supset \mathscr{I}_{m-1}$ and integers

$n_1 < \cdots < n_{m-1}$ have already been chosen. Let $\mathscr{I}^*_{m-1}$ denote the interval of length $\frac{1}{2}|\mathscr{I}_{m-1}|$ and having the same midpoint as $\mathscr{I}_{m-1}$, and choose s so large that

$$|\mathscr{I}^*_{m-1}| \geqslant 6^{-s}. \tag{29}$$

Let

$$t > \max\{s, c_4 n_{m-1}\}, \tag{30}$$

where c_4 is a constant satisfying $c_4 > 240$. We now apply Lemma 4.2 with β being the left-hand end-point of $\mathscr{I}^*_{m-1}$ and $I = (0, 6^{s+t}]$ to obtain (11). Note that in view of (29), $\beta + j6^{-s-t} \in \mathscr{I}^*_{m-1}$ for $j = 1, \ldots, 6^t$. It follows that there exists $\alpha \in \mathscr{I}^*_{m-1}$ such that

$$h(I, \alpha) > \frac{t}{120},$$

and so there exists $n_m \in I$ such that

$$|D(n_m, \alpha)| > \frac{t}{240}.$$

Since $\alpha \in \mathscr{I}^*_{m-1}$, α is in the interior of $\mathscr{I}_{m-1}$. Hence we can apply Lemma 4.3 with a sufficiently small v to show that there exists a closed interval $\mathscr{I}_m \subset \mathscr{I}_{m-1}$ such that for all $\beta \in \mathscr{I}_m$,

$$|D(n_m, \beta)| > \frac{t}{c_4}. \tag{31}$$

In view of (31) and (30), we clearly have $n_m \geqslant t/c_4 > n_{m-1}$. Also, for all $\beta \in \mathscr{I}_m$, since $n_m \leqslant 6^{s+t}$, we have

$$|D(n_m, \beta)| > \frac{t}{c_4} > \frac{s+t}{2c_4} \geqslant c_3 \log n_m$$

if c_3 is a sufficiently small fixed constant, and our induction step is complete.

We complete this section by proving Theorem 7B. We say that an interval $\mathscr{I}$ is of order k if $\mathscr{I}$ is of the form $[(u-1)/k!, u/k!]$, where $u \in \mathbb{Z}$ satisfies $1 \leqslant u \leqslant k!$. We denote such an interval by $\mathscr{I}_k$.

Lemma 4.4. *Let $k > 72$ be fixed and let $N = k!$. There exist positive absolute constants c_5 and c_6 such that for any interval $\mathscr{I}_{k-1}$ of order $(k-1)$, there exist an integer n satisfying*

$$c_5 \log k \leqslant n \leqslant N \tag{32}$$

and a subinterval $\mathscr{I}_k$ (of order k) of $\mathscr{I}_{k-1}$ such that for all $\alpha \in \mathscr{I}_k$,

$$|D(n, \alpha)| \geqslant c_6 \log\log n. \tag{33}$$

Theorem 7B follows easily. For any $k > 72$ and each interval of order $(k-1)$, we can choose a subinterval of order k such that the conclusion of Lemma 4.4 holds. We let E_k denote the union of all such subintervals chosen in this way. Since we choose precisely one subinterval of order k from each interval of order $(k-1)$, we have $\mu(E_k) = 1/k$. If α lies in infinitely many of the sets

$$E_{73}, E_{74}, \ldots, \tag{34}$$

then (33) holds for infinitely many n. It therefore suffices to show that the set of values α that lie in only finitely many of the sets (34) has measure 0. If α lies in only finitely many of the sets (34), then there exists $K \geqslant 73$ such that $\alpha \in T_k$, where T_k is the complement of $\bigcup_{k=K}^{\infty} E_k$. But for every $K \geqslant 73$,

$$\mu(T_K) = \prod_{k=K}^{\infty} (1 - \mu(E_k)) = \prod_{k=K}^{\infty} \left(1 - \frac{1}{k}\right) = 0.$$

This completes the proof of Theorem 7B.

Proof of Lemma 4.4. We choose s to satisfy

$$6^{s-1} \leqslant 2(k-1)! < 6^s \tag{35}$$

and t to satisfy

$$6^{s+t} \leqslant N < 6^{s+t+1}.$$

Then since $k > 72$, we have that $t \geqslant 1$. Let β be the left-hand end-point of $\mathscr{I}_{k-1}^*$, where $\mathscr{I}_{k-1}^*$ is the closed interval of length $\frac{1}{2}|\mathscr{I}_{k-1}|$ and having the same midpoint as $\mathscr{I}_{k-1}$; and let $I = (0, 6^{s+t}]$. By Lemma 4.2, we have (11). In view of the second inequality in (35), we have that $\beta + j6^{-s-t} \in \mathscr{I}_{k-1}^*$ for all $j = 1, \ldots, 6^t$. It follows that there exist $\xi \in \mathscr{I}_{k-1}^*$ and $n \leqslant 6^{s+t} \leqslant N$ such that

$$|D(n, \xi)| \geqslant \frac{t}{240}. \tag{36}$$

We now apply Lemma 4.3 with $v = 2$ to establish the existence of a closed interval $\mathscr{I}$ such that $\xi \in \mathscr{I}$,

$$|\mathscr{I}| = \frac{2}{n} \geqslant \frac{2}{k!} \tag{37}$$

and that for all $\alpha \in \mathscr{I}$,

$$|D(n, \alpha)| \geqslant \frac{t}{240} - 2.$$

Since $6^t \geqslant k/72$ and $\log k! \leqslant k \log k \leqslant k^2$, we conclude that for all $\alpha \in \mathscr{I}$,

$$|D(n,\alpha)| \geqslant c_6 \log\log n,$$

where c_6 is a sufficiently small positive absolute constant. Clearly, in view of (37), $\mathscr{I}$ contains a closed subinterval $\mathscr{I}_k$ of order k. It remains to establish the first inequality in (32). By (36), clearly $n \geqslant t/240$. Combining this with $6^t \geqslant k/72$ gives the desired result. ■

4.3 The scarcity of intervals with bounded error

Let S be a set of real numbers. The derivative of S, denoted by $S^{(1)}$, is the set of all limit points of S. We define the higher derivatives inductively by writing

$$S^{(d)} = (S^{(d-1)})^{(1)}$$

for $d \geqslant 2$. One can then show easily by induction on d that if $S^{(d)}$ is empty for some d, then the set S is at most countable and is nowhere dense. Theorem 8A therefore follows immediately from

Lemma 4.5. *Suppose that $d > 4\kappa$. Then $(S(\kappa))^{(d)}$ is empty.*

We remark that the inequality $d > 4\kappa$ can be somewhat relaxed, but cannot be replaced by $d \geqslant \kappa - 1$. To prove Lemma 4.5, it is necessary to express it in the following more general form suitable for proof by induction.

For convenience, we write $U_* = (0,1)$. Also, by a neighbourhood of a point $\alpha \in U_*$, we mean an open interval in U_* containing α.

Lemma 4.6. *Let $d \geqslant 0$ and $\varepsilon > 0$ be given, and let R be a set such that $R^{(d)} \cap U_*$ is non-empty. Then there exist $w = 2^d$ elements $\lambda_1, \ldots, \lambda_w$ in $R \cap U_*$ with respective neighbourhoods $L_1, \ldots, L_w$ and an integer p such that for all intervals I with $l(I) \geqslant p$ and for all $\mu_1 \in L_1, \ldots, \mu_w \in L_w$, we have*

$$w^{-1} \sum_{j=1}^{w} h(I, \mu_j) > \tfrac{1}{2}(d+1) - \varepsilon.$$

We deduce Lemma 4.5 by contradiction. Suppose that $(S(\kappa))^{(d)}$ is non-empty, where $d > 4\kappa$. Then $(S(\kappa))^{(d-1)} \cap U_*$ is non-empty. Let $\varepsilon = \tfrac{1}{2}d - 2\kappa > 0$. Then by Lemma 4.6, there exists $\lambda \in S(\kappa)$ and an interval I such that

$$h(I, \lambda) > \tfrac{1}{2}d - \varepsilon.$$

It follows that there exists $n \in I$ such that

$$|D(n, \lambda)| > \tfrac{1}{4}d - \tfrac{1}{2}\varepsilon = \kappa,$$

whence $E(\lambda) > \kappa$. This is a contradiction.

We shall prove Lemma 4.6 by induction on d. For $d=0$, we prove

Lemma 4.7. *Let $\alpha\in U_*$ and $\varepsilon>0$ be given. Then there exist a neighbourhood A of α and an integer p such that for all intervals I with $l(I)\geqslant p$ and for all $\beta\in A$, we have*

$$h(I,\beta)>\tfrac{1}{2}-\varepsilon.$$

Proof. We first of all assume that $0<\alpha\leqslant\frac{1}{2}$. Consider the $p=[1/\alpha]+1$ numbers $\alpha,2\alpha,\ldots,p\alpha$. Then every number in the range $[\frac{1}{2}\alpha,(p+\frac{1}{2})\alpha]$ has distance not exceeding $\frac{1}{2}\alpha$ from one of these p numbers. Since $[\frac{1}{2}\alpha,(p+\frac{1}{2})\alpha]$ is of length $p\alpha>1$, we can conclude that for every real number ψ, there exist integers m and n such that $0<n\leqslant p$ and

$$|n\alpha-m-\psi|\leqslant\tfrac{1}{2}\alpha\leqslant\tfrac{1}{4}.$$

It is also not too difficult to see that the restriction $0<\alpha\leqslant\frac{1}{2}$ can be removed. Let $A=\{\beta\in U_*:p|\beta-\alpha|<\frac{1}{2}\varepsilon\}$. Then for every real number ψ, there exist integers m and n such that $0<n\leqslant p$ and for all $\beta\in A$,

$$|n\beta-m-\psi|<\tfrac{1}{4}+\tfrac{1}{2}\varepsilon. \tag{38}$$

Since ψ is arbitrary, for every real number ψ, there exist integers m and n such that $0<n\leqslant p$ and for all $\beta\in A$,

$$0<n\beta-m-\psi<\tfrac{1}{2}+\varepsilon$$

(simply replace ψ in (38) by $\psi+\frac{1}{4}+\frac{1}{2}\varepsilon$). It follows that for $\beta\in A$ and for any interval I with $l(I)\geqslant p$, we can choose integers $n\in I$ and m such that

$$0<n\beta-m+g^-(I,\beta)<\tfrac{1}{2}+\varepsilon. \tag{39}$$

Then

$$Z(n,\beta)=D(n,\beta)+n\beta\geqslant g^-(I,\beta)+n\beta>m,$$

and since $Z(n,\beta)$ is an integer, we must have $Z(n,\beta)\geqslant m+1$. Combining this with (39), we have

$$\begin{aligned}h(I,\beta)&=g^+(I,\beta)-g^-(I,\beta)\geqslant D(n,\beta)-g^-(I,\beta)\\&=Z(n,\beta)-n\beta-g^-(I,\beta)>\tfrac{1}{2}-\varepsilon\end{aligned}$$

as required. ■

We next prove a result which is in the same spirit as Lemma 4.7.

Lemma 4.8. *Let $\varepsilon\in(0,1)$ and $q\in\mathbb{N}$ be given, and suppose that $\alpha,\beta\in U_*$ satisfy $0<|\alpha-\beta|<\varepsilon/8q$. Then there exist neighbourhoods A of α and B of β and an integer p such that for all $\gamma\in A$, $\delta\in B$ and for all intervals I with $l(I)\geqslant p$,* *there exist subintervals J and J' of I with $l(J) = l(J') = q$ and*

$$g^-(J,\gamma,\delta) - g^+(J',\gamma,\delta) > 1 - \varepsilon. \tag{40}$$

Proof. We assume, without loss of generality, that $\alpha < \beta$. Writing

$$p_0 = [1/(\beta-\alpha)] + 1,$$

we can show, as in the proof of Lemma 4.7, that for all real numbers ψ, there exist integers m and n with $0 < n \leqslant p_0$ and

$$|n(\beta-\alpha) - m - \psi| \leqslant \tfrac{1}{2}|\beta-\alpha| < \varepsilon/8.$$

Let

$$A = \{\gamma \in U_*: |\gamma - \alpha| \max\{q, p_0\} < \varepsilon/16\}$$

and

$$B = \{\delta \in U_*: |\delta - \beta| \max\{q, p_0\} < \varepsilon/16\}.$$

It is not too difficult to show that for all $\gamma \in A$ and $\delta \in B$,

$$\delta > \gamma \tag{41}$$

and

$$|\gamma-\delta| \leqslant |\gamma-\alpha| + |\alpha-\beta| + |\beta-\delta| < \varepsilon/4q. \tag{42}$$

Also, for all $\gamma \in A$, $\delta \in B$ and for all real numbers ψ, there exist integers m and n with $0 < n \leqslant p_0$ and

$$|n(\delta-\gamma) - m - \psi| < \tfrac{1}{4}\varepsilon.$$

As in the proof of Lemma 4.7, since ψ is arbitrary, we conclude that for all $\gamma \in A$, $\delta \in B$ and all real numbers ψ, and for all intervals I_0 with $l(I_0) \geqslant p_0$, there exist integers m and n with $n \in I_0$ and

$$0 < n(\delta-\gamma) - m - \psi < \tfrac{1}{2}\varepsilon.$$

Let I_0 be any interval with $l(I_0) \geqslant p_0$. Let $\gamma \in A$ and $\delta \in B$ be fixed. We choose $n_0' \in I_0$ such that

$$g^-(I_0,\gamma,\delta) = D(n_0',\gamma,\delta).$$

Then there exist integers $n_0 \in I_0$ and m such that

$$0 < n_0(\delta-\gamma) - m + g^-(I_0,\gamma,\delta) < \tfrac{1}{2}\varepsilon.$$

It follows that

$$Z(n_0,\delta) - Z(n_0,\gamma) = D(n_0,\gamma,\delta) + n_0(\delta-\gamma) \geqslant g^-(I_0,\gamma,\delta) + n_0(\delta-\gamma) > m,$$

and so

$$Z(n_0,\delta) - Z(n_0,\gamma) \geqslant m+1,$$

and so

$$D(n_0,\gamma,\delta)-D(n_0',\gamma,\delta)$$
$$=Z(n_0,\delta)-Z(n_0,\gamma)-n_0(\delta-\gamma)-g^-(I_0,\gamma,\delta)>1-\tfrac{1}{2}\varepsilon. \qquad (43)$$

Now set $p=p_0+2q$. For any interval $I=(a,b]$ with $l(I)\geqslant p$, let $I_0=(a+q,b-q]$. Then $l(I_0)\geqslant p_0$, and so there exist $n_0,n_0'\in I_0$ such that (43) holds. Let

$$J=(n_0,n_0+q] \quad\text{and}\quad J'=(n_0'-q,n_0'].$$

Then $J,J'\subset I$ and $l(J)=l(J')=q$. For every $n\in J$,

$$D(n,\gamma,\delta)-D(n_0,\gamma,\delta)\geqslant-(n-n_0)(\delta-\gamma)\geqslant-q(\delta-\gamma)>-\tfrac{1}{4}\varepsilon \qquad (44)$$

in view of (41) and (42). Similarly, for every $n'\in J'$,

$$D(n_0',\gamma,\delta)-D(n',\gamma,\delta)>-\tfrac{1}{4}\varepsilon. \qquad (45)$$

Combining (43)–(45), we conclude that for every $n\in J$ and $n'\in J'$,

$$D(n,\gamma,\delta)-D(n',\gamma,\delta)>1-\varepsilon.$$

(40) follows immediately. ■

Lemma 4.9. *Let $R\subset U_*$ be given. Suppose that $\theta_1,\ldots,\theta_t\in R^{(1)}\cap U_*$ and have neighbourhoods $D_1,\ldots,D_t$ respectively. Let $\varepsilon>0$ and $q\in\mathbb{N}$ be given. Then there exist $\alpha_1,\beta_1,\ldots,\alpha_t,\beta_t\in R$ with neighbourhoods $A_1,B_1\subset D_1,\ldots,A_t,B_t\subset D_t$, respectively, and there exists a number r such that for all $\gamma_1\in A_1$, $\delta_1\in B_1,\ldots,\gamma_t\in A_t,\delta_t\in B_t$ and for all intervals I and I' satisfying $l(I)\geqslant r$ and $l(I')\geqslant r$, there exist subintervals $J\subset I$ and $J'\subset I'$ such that*

$$l(J)=l(J')=q$$

and for every $i=1,\ldots,t$,

$$h(J,J',\gamma_i,\delta_i)>1-\varepsilon. \qquad (46)$$

Remark. We shall see that Lemma 4.9 is more general than what we need. However, this generality is necessary for proof by induction.

Proof of Lemma 4.9. We shall use induction on t. Consider first the case $t=1$. Let $\theta_1\in R^{(1)}\cap U_*, D_1$, $\varepsilon>0$ and $q\in\mathbb{N}$ be given. Since $\theta_1\in R^{(1)}$, there exist $\alpha_1,\beta_1\in R\cap D_1$ such that $0<|\alpha_1-\beta_1|<\varepsilon/8q$. We apply Lemma 4.8. There exist neighbourhoods A_1 of α_1 and B_1 of β_1 and an integer p such that for all $\gamma_1\in A_1,\delta_1\in B_1$ and for all intervals I with $l(I)\geqslant p$, there exist subintervals J_1 and J_2 of I with $l(J_1)=l(J_2)=q$ and

$$g^-(J_1,\gamma_1,\delta_1)-g^+(J_2,\gamma_1,\delta_1)>1-\varepsilon. \qquad (47)$$

It is clear that A_1 and B_1 can be chosen so that $A_1 \subset D_1$ and $B_1 \subset D_1$. Suppose now that I' is another interval satisfying $l(I') \geqslant p$. Then there exist subintervals J'_1 and J'_2 of I' with $l(J'_1) = l(J'_2) = q$ and

$$g^-(J'_1, \gamma_1, \delta_1) - g^+(J'_2, \gamma_1, \delta_1) > 1 - \varepsilon. \tag{48}$$

Adding together (47) and (48), we have either

$$g^-(J_1, \gamma_1, \delta_1) - g^+(J'_2, \gamma_1, \delta_1) > 1 - \varepsilon \tag{49}$$

or

$$g^-(J'_1, \gamma_1, \delta_1) - g^+(J_2, \gamma_1, \delta_1) > 1 - \varepsilon. \tag{50}$$

If (49) holds, let $J = J_1$ and $J' = J'_2$. If (50) holds, let $J = J_2$ and $J' = J'_1$. In either case, we have

$$h(J, J', \gamma_1, \delta_1) > 1 - \varepsilon,$$

so that the case $t = 1$ is true with $r = r_1 = p$. We now proceed to the inductive step. Assume the case $(t-1)$. Then there exist $\alpha_1, \beta_1, \ldots, \alpha_{t-1}, \beta_{t-1}, A_1, B_1, \ldots, A_{t-1}, B_{t-1}$ and $r = r_{t-1}$ such that the conclusion holds for $i = 1, \ldots, t-1$ under the conditions stated in the lemma. We now use the case $t = 1$ to find $\alpha_t, \beta_t \in R$ with neighbourhoods $A_t, B_t \subset D_t$ respectively and a number $\bar{r}_1$ such that for all $\gamma_t \in A_t$, $\delta_t \in B_t$ and for all intervals I and I' satisfying $l(I) \geqslant \bar{r}_1$ and $l(I') \geqslant \bar{r}_1$, there exist subintervals $I_0 \subset I$ and $I'_0 \subset I'$ with $l(I_0) = l(I'_0) = r_{t-1}$ and such that

$$h(I_0, I'_0, \gamma_t, \delta_t) > 1 - \varepsilon.$$

It follows that there exist subintervals $J \subset I_0$ and $J' \subset I'_0$ with $l(J) = l(J') = q$ such that (46) holds for $i = 1, \ldots, t-1$. For $i = t$, we simply note that

$$h(J, J', \gamma_t, \delta_t) \geqslant h(I_0, I'_0, \gamma_t, \delta_t).$$

Hence the case t is true for $r = \bar{r}_1$. ∎

We conclude this section by proving Lemma 4.6. We do this by induction on d. The case $d = 0$ is precisely Lemma 4.7, so we may assume that $d \geqslant 1$ and that Lemma 4.6 is true for $(d-1)$. Suppose that $R^{(d)} \cap U_*$ is non-empty. Then since $R^{(d)} = (R^{(1)})^{(d-1)}$, there exist $t = 2^{d-1}$ elements $\theta_1, \ldots, \theta_t$ in $R^{(1)} \cap U_*$ with respective neighbourhoods $D_1, \ldots, D_t$ and an integer p_{d-1} such that for all intervals I with $l(I) \geqslant p_{d-1}$ and for all $\eta_1 \in D_1, \ldots, \eta_t \in D_t$, we have

$$t^{-1} \sum_{i=1}^{t} h(I, \eta_i) > \tfrac{1}{2}d - \tfrac{1}{2}\varepsilon. \tag{51}$$

We now apply Lemma 4.9 with $q = p_{d-1}$ to construct $\alpha_1, \beta_1, \ldots, \alpha_t, \beta_t \in$

$R \cap U_*$ with respective neighbourhoods $A_1, B_1, \ldots, A_t, B_t$ and a number

$$r = r(\theta_1, \ldots, \theta_t, D_1, \ldots, D_t, p_{d-1}) \tag{52}$$

such that if I is an interval with $l(I) \geqslant r$ (note that we only use the special case $I = I'$ of Lemma 4.9), then there exist subintervals J and J' of I such that $l(J) = l(J') = p_{d-1}$ and such that for all $\gamma_1 \in A_1, \delta_1 \in B_1, \ldots, \gamma_t \in A_t, \delta_t \in B_t$, we have, for $i = 1, \ldots, t$, that

$$h(J, J', \gamma_i, \delta_i) > 1 - \varepsilon.$$

Applying Lemma 4.1, we have that for $i = 1, \ldots, t$,

$$h(I, \gamma_i) + h(I, \delta_i) \geqslant (1 - \varepsilon) + \tfrac{1}{2}(h(J, \gamma_i) + h(J, \delta_i) + h(J', \gamma_i) + h(J', \delta_i)). \tag{53}$$

Summing (53) over $i = 1, \ldots, t$ and dividing by $2t$, we have

$$\frac{1}{2t}\left(\sum_{i=1}^{t} h(I, \gamma_i) + \sum_{i=1}^{t} h(I, \delta_i)\right) \geqslant (\tfrac{1}{2} - \tfrac{1}{2}\varepsilon) + \frac{1}{4t}(\Sigma_1 + \Sigma_2 + \Sigma_3 + \Sigma_4), \tag{54}$$

where

$$\Sigma_1 = \sum_{i=1}^{t} h(J, \gamma_i) \quad \text{and} \quad \Sigma_2 = \sum_{i=1}^{t} h(J, \delta_i),$$

$$\Sigma_3 = \sum_{i=1}^{t} h(J', \gamma_i) \quad \text{and} \quad \Sigma_4 = \sum_{i=1}^{t} h(J', \delta_i).$$

Now $\gamma_i \in D_i$ for all $i = 1, \ldots, t$, so that by (51), we have $\Sigma_1 > t(\tfrac{1}{2}d - \tfrac{1}{2}\varepsilon)$. Similarly for Σ_2, Σ_3 and Σ_4. It follows from (54) that

$$\frac{1}{2t}\left(\sum_{i=1}^{t} h(I, \gamma_i) + \sum_{i=1}^{t} h(I, \delta_i)\right) > (\tfrac{1}{2} - \tfrac{1}{2}\varepsilon) + (\tfrac{1}{2}d - \tfrac{1}{2}\varepsilon) = \tfrac{1}{2}(d+1) - \varepsilon.$$

It follows that the $w = 2t = 2^d$ points $\alpha_1, \beta_1, \ldots, \alpha_t, \beta_t$ and their respective neighbourhoods $A_1, B_1, \ldots, A_t, B_t$ and the number $p = r$ given by (52) have the desired properties stated in Lemma 4.6, and this completes the proof.

4.4 A variation of the method and application to a problem of Erdös in diophantine approximation

In this section, we study a variation of the method of Schmidt which enabled Tijdeman and Wagner to prove Theorem 7C. We then discuss Wagner's solution to a problem of Erdös in diophantine approximation.

Tijdeman and Wagner proved the following result from which Theorem 7C follows easily.

Lemma 4.10. *Let L be an integer satisfying $1 \leqslant L \leqslant N$. Then*

$$\max_{1 \leqslant n \leqslant N} |D(n, \alpha)| \geqslant c_7(\log \tfrac{1}{2}L - \tfrac{1}{2}\log N) - 2$$

for all $\alpha \in U_1$ except for a set of measure at most L/N. Here c_7 is a positive absolute constant.

Theorem 7C follows easily by taking, for example, $L \gg \ll N^{\frac{3}{4}}$.

For the sake of brevity, write

$$G(L, N) = c_7(\log \tfrac{1}{2}L - \tfrac{1}{2}\log N). \tag{55}$$

Lemma 4.11. *Let $A \subset \{1, \dots, N\}$ with $\#A = L$, where $\#A$ denotes the number of elements of the set A. Then*

$$\int_0^1 \left(\max_{n \in A} D(n, \alpha) - \min_{n \in A} D(n, \alpha) \right) d\alpha \geqslant 8G(L, N).$$

For the proof of Lemma 4.11, we shall need the following technical lemma. We state without proof

Lemma 4.12. *Let f be a real-valued function defined on U_0. Suppose that f has no more than W points of discontinuity, with the convention that f is discontinuous at 0 if $\lim_{\alpha \to 1^-} f(\alpha) \neq f(0)$. Suppose further that $f'(\alpha) \geqslant M$ for all but finitely many $\alpha \in U_0$. Then*

$$\int_0^1 |f(\alpha)|\, d\alpha \geqslant \frac{M}{4W}.$$

Proof of Lemma 4.11. Let

$$\rho(L, N) = \inf \int_0^1 \left(\max_{n \in A} D(n, \alpha) - \min_{n \in A} D(n, \alpha) \right) d\alpha, \tag{56}$$

where the infimum is taken over all sequences $s_1, s_2, s_3, \dots$ in U_0 and all $A \subset \{1, \dots, N\}$ with $\#A \geqslant L$. We shall prove that

$$\rho(L, 4^t) \geqslant \frac{\log L - t \log 2}{24 \log 2} \tag{57}$$

for $1 \leqslant L \leqslant 4^t$. Then for arbitrary N, we can choose t satisfying $4^t \leqslant N < 4^{t+1}$. Then clearly

$$\rho(L, N) \geqslant \rho(L, 4^{t+1}) \geqslant \frac{\log \frac{1}{2}L - \frac{1}{2}\log N}{24 \log 2},$$

giving the desired result with a suitable c_7. It therefore remains to prove (57). We do this by induction on t. We let $A \subset \{1, \ldots, 4^t\}$. Let A_i $(i = 0, 1, 2, 3)$ be defined by

$$A_i = A \cap (i4^{t-1}, (i+1)4^{t-1}].$$

We consider two cases.

Case 1: There exist p, q such that $q - p \geqslant 2$, $\# A_p \geqslant L/8$ and $\# A_q \geqslant L/8$. We use the following analogue of Lemma 4.1: We have

$$\begin{aligned}
&\int_0^1 \left(\max_{n \in A} D(n, \alpha) - \min_{n \in A} D(n, \alpha) \right) \mathrm{d}\alpha \\
&\geqslant \tfrac{1}{2} \int_0^1 \left(\max_{n \in A_p} D(n, \alpha) - \min_{n \in A_p} D(n, \alpha) \right) \mathrm{d}\alpha \\
&\quad + \tfrac{1}{2} \int_0^1 \left(\max_{n \in A_q} D(n, \alpha) - \min_{n \in A_q} D(n, \alpha) \right) \mathrm{d}\alpha \\
&\quad + \tfrac{1}{2} \int_0^1 \left| \max_{n \in A_p} D(n, \alpha) - \max_{n \in A_q} D(n, \alpha) \right| \mathrm{d}\alpha \\
&\quad + \tfrac{1}{2} \int_0^1 \left| \min_{n \in A_p} D(n, \alpha) - \min_{n \in A_q} D(n, \alpha) \right| \mathrm{d}\alpha. \qquad (58)
\end{aligned}$$

Here the last two terms on the right-hand side of (58) have a role analogous to that of the term $h(J, J', \alpha, \beta)$ in Lemma 4.1. Now the function

$$\max_{n \in A_p} D(n, \alpha) - \max_{n \in A_q} D(n, \alpha)$$

can have discontinuities only at the points s_v, where $p4^{t-1} < v \leqslant (q+1)4^{t-1}$; also the function is differentiable everywhere with derivative at least $(q - p - 1)4^{t-1}$, except at a finite number of points. It follows from Lemma 4.12 that

$$\int_0^1 \left| \max_{n \in A_p} D(n, \alpha) - \max_{n \in A_q} D(n, \alpha) \right| \mathrm{d}\alpha \geqslant \tfrac{1}{12}. \qquad (59)$$

Similarly

$$\int_0^1 \left| \min_{n \in A_p} D(n, \alpha) - \min_{n \in A_q} D(n, \alpha) \right| \mathrm{d}\alpha \geqslant \tfrac{1}{12}. \qquad (60)$$

Combining (56) and (58)–(60), we have that

$$\int_0^1 \left(\max_{n \in A} D(n, \alpha) - \min_{n \in A} D(n, \alpha) \right) \mathrm{d}\alpha \geqslant \rho\left(\frac{L}{8}, 4^{t-1} \right) + \frac{1}{12}. \qquad (61)$$

Case 2: If Case 1 does not hold, then there exists r such that

$\#(A_r \cup A_{r+1}) \geqslant 3L/4$. Let B_i $(i = 0, 1, 2, 3)$ be defined by

$$B_i = A \cap ((r + \tfrac{1}{2}i)4^{t-1}, (r + \tfrac{1}{2}(i+1))4^{t-1}]$$

(note that $B_0 \cup B_1 = A_r$ and $B_2 \cup B_3 = A_{r+1}$). If there exist p, q such that $q - p \geqslant 2$, $\#B_p \geqslant L/8$ and $\#B_q \geqslant L/8$, then, similar to Case 1, we have

$$\int_0^1 \left(\max_{n \in A} D(n, \alpha) - \min_{n \in A} D(n, \alpha) \right) d\alpha \geqslant \rho\left(\frac{L}{8}, 4^{t-1} \right) + \frac{1}{12}. \tag{62}$$

Otherwise, there exists r' such that $\#(B_{r'} \cup B_{r'+1}) \geqslant \frac{1}{2}L$, and

$$\int_0^1 \left(\max_{n \in A} D(n, \alpha) - \min_{n \in A} D(n, \alpha) \right) d\alpha \geqslant \rho\left(\frac{L}{2}, 4^{t-1} \right) \tag{63}$$

(note that for (63) to occur, we must have $\frac{1}{2}L \leqslant 4^{t-1}$).

Combining (61)–(63), we have that

$$\int_0^1 \left(\max_{n \in A} D(n, \alpha) - \min_{n \in A} D(n, \alpha) \right) d\alpha \geqslant \min\left\{ \rho\left(\frac{L}{2}, 4^{t-1} \right), \rho\left(\frac{L}{8}, 4^{t-1} \right) + \frac{1}{12} \right\}. \tag{64}$$

The right-hand side of (64) is independent of sequences $s_1, s_2, s_3, \ldots$ in U_0 and $A \subset \{1, \ldots, 4^t\}$ with $\#A \geqslant L$, so

$$\rho(L, 4^t) \geqslant \min\left\{ \rho\left(\frac{L}{2}, 4^{t-1} \right), \rho\left(\frac{L}{8}, 4^{t-1} \right) + \frac{1}{12} \right\}$$

subject to $\frac{1}{2}L \leqslant 4^{t-1}$. Arguing inductively, we conclude that

$$\rho(L, 4^t) \geqslant \min_{\substack{0 \leqslant \lambda \leqslant t \\ 2^{\lambda-t}8^{-\lambda}L \leqslant 1}} \left\{ \rho(2^{\lambda-t}8^{-\lambda}L, 1) + \frac{\lambda}{12} \right\}.$$

The condition $2^{\lambda-t}8^{-\lambda}L \leqslant 1$ is equivalent to the condition

$$\lambda \geqslant \frac{\log L - t\log 2}{2\log 2}.$$

It follows that

$$\rho(L, 4^t) \geqslant \frac{\log L - t\log 2}{24\log 2}$$

as required. ■

Let $s_1, \ldots, s_N$ be given. Suppose that $s_{i_1} \leqslant \cdots \leqslant s_{i_N}$. Consider the new sequence $\eta_1, \ldots, \eta_N$, where $\eta_{i_\nu} = (\nu - \frac{1}{2})/N$ for $\nu = 1, \ldots, N$. Note that $\eta_1, \ldots, \eta_N$ are obtained from $s_1, \ldots, s_N$ by 'spacing the points $s_1, \ldots, s_N$ more evenly'. Define $\tilde{Z}(n, \alpha)$ and $\tilde{D}(n, \alpha)$ for $\eta_1, \ldots, \eta_n$ in the same way as $Z(n, \alpha)$ and $D(n, \alpha)$ are defined for $s_1, \ldots, s_n$.

Lemma 4.13. *We have that*

$$\max_{1\leqslant n\leqslant N} \tilde{D}(n,k/N) - \min_{1\leqslant n\leqslant N} \tilde{D}(n,k/N) < 4G(L,N) - 2$$

for less than L integers $k\in\{1,\ldots,N\}$.

Proof. First of all, we want to interchange the roles of k and n (see (65) below). Let $\tilde{\omega}_N = \{\eta_1,\ldots,\eta_N\}$. For each $i=1,\ldots,N$, let $j=j(i)$ be defined by

$$\eta_j = \eta_{j(i)} \in \tilde{\omega}_N \cap [(i-1)/N, i/N).$$

For $x\in U_1$, let

$$R_i(x) = \begin{cases} -x & (0 < x \leqslant (j(i)-\tfrac{1}{2})/N); \\ 1-x & \text{(otherwise).} \end{cases}$$

For $k=1,\ldots,N$, let

$$R(k,x) = \sum_{i=1}^{k} R_i(x).$$

Note that $R(k,x) = \#\{1\leqslant i\leqslant k : (j(i)-\tfrac{1}{2})/N < x\} - kx$ is the discrepancy of the sequence

$$\frac{j(1)-\tfrac{1}{2}}{N},\ldots,\frac{j(k)-\tfrac{1}{2}}{N}.$$

It follows that

$$\begin{aligned} R(k,n/N) &= \#\{1\leqslant i\leqslant k : 0 < (j(i)-\tfrac{1}{2})/N < n/N\} - k(n/N) \\ &= \#\{1\leqslant j\leqslant n : \eta_j \in [0,k/N)\} - n(k/N) = \tilde{D}(n,k/N). \end{aligned} \tag{65}$$

Let

$$A = \left\{ k\in\{1,\ldots,N\} : \sup_{x\in U_1} R(k,x) - \inf_{x\in U_1} R(k,x) < 4G(L,N) \right\}.$$

Then since

$$\inf_{x\in U_1} R(k,x) \leqslant 0 \leqslant \sup_{x\in U_1} R(k,x),$$

it follows that $|R(k,x)| < 4G(L,N)$ for all $k\in A$ and all $x\in U_1$, so that

$$\int_0^1 \left(\max_{k\in A} R(k,x) - \min_{k\in A} R(k,x) \right) dx < 8G(L,N).$$

By Lemma 4.11, we have that

$$\int_0^1 \left(\max_{k\in A} R(k,x) - \min_{x\in A} R(k,x) \right) dx \geqslant 8G(\#A,N).$$

Hence $\#A < L$ and so

$$\sup_{x\in U_1} R(k,x) - \inf_{x\in U_1} R(k,x) < 4G(L,N)$$

holds for less than L integers $k\in\{1,\dots,N\}$, whence

$$\max_{1\leqslant n\leqslant N} R(k,n/N) - \min_{1\leqslant n\leqslant N} R(k,n/N) < 4G(L,N) - 2$$

for less than L integers $k\in\{1,\dots,N\}$. The proof of Lemma 4.13 is now complete, on noting (65). ■

We can now complete the proof of Lemma 4.10. By Lemma 4.13,

$$\max_{1\leqslant n\leqslant N} \tilde{D}(n,k/N) - \min_{1\leqslant n\leqslant N} \tilde{D}(n,k/N) \geqslant 4G(L,N) - 2$$

for at least $(N-L)$ integers $k\in\{1,\dots,N\}$. Clearly, the values of the function

$$\max_{1\leqslant n\leqslant N} \tilde{D}(n,\alpha) - \min_{1\leqslant n\leqslant N} \tilde{D}(n,\alpha)$$

differ by at most 2 on each half-open interval of length $\leqslant 1/N$. Hence

$$\begin{aligned}\max_{1\leqslant n\leqslant N} |\tilde{D}(n,\alpha)| &\geqslant \tfrac{1}{2}\left(\max_{1\leqslant n\leqslant N} \tilde{D}(n,\alpha) - \min_{1\leqslant n\leqslant N} \tilde{D}(n,\alpha)\right)\\ &\geqslant \tfrac{1}{2}(4G(L,N) - 2 - 2) = 2G(L,N) - 2\end{aligned}$$

for a set of numbers $\alpha\in U_1$ of measure at least $1 - L/N$. Suppose now that $\alpha\in U_1$ and $|\tilde{D}(n,\alpha)| \geqslant 2G(L,N) - 2$ for some $n\in\{1,\dots,N\}$. There are two cases.

If $|Z(n,\alpha) - \tilde{Z}(n,\alpha)| < G(L,N)$, then

$$|D(n,\alpha)| \geqslant |\tilde{D}(n,\alpha)| - |Z(n,\alpha) - \tilde{Z}(n,\alpha)| > G(L,N) - 2.$$

If $|Z(n,\alpha) - \tilde{Z}(n,\alpha)| \geqslant G(L,N)$, then $|Z(N,\alpha) - \tilde{Z}(N,\alpha)| \geqslant G(L,N)$. However, since $|\tilde{Z}(N,\alpha) - N\alpha| \leqslant \frac{1}{2}$, we must have $|D(N,\alpha)| > G(L,N) - 2$.

This completes the proof of Lemma 4.10.

We conclude this chapter by studying an application of ideas in the theory of irregularities of distribution to a problem in diophantine approximation.

Let $a_1, a_2, a_3, \dots$ be an arbitrary sequence of points, not necessarily distinct, on the unit circle C. For every $n\in\mathbb{N}$ and every $z\in C$, let

$$f_n(z) = \prod_{i=1}^{n} |z - a_i|.$$

Further, let

$$A_n = \max_{z\in C} f_n(z).$$

Erdös asked whether there exists a sequence $a_1, a_2, a_3, \ldots$ on C such that $\sup A_n < \infty$.

If 1980, Wagner showed that no such sequence exists. In fact,

Theorem 12 (Wagner (1980)). *There exists an absolute constant $\delta > 0$ such that $A_n > (\log n)^\delta$ for infinitely many n.*

Suppose, for a given sequence $a_1, a_2, a_3, \ldots$, that $\sup A_n < \infty$ does not exist. This must be a consequence of the impossibility of getting $a_1, \ldots, a_n$ close to uniform distribution for all n simultaneously. There seems, therefore, to be a connection with the problem of discrepancy of a sequence. Indeed, by changing variables, one can easily pass from the unit circle to the unit interval.

More precisely, let $s_1, s_2, s_3, \ldots$ be a sequence of points in U (endowed with the topology of the torus). For every $x \in U$, let

$$P(x) = \log 2 + \log|\sin \pi x|.$$

For every $n \in \mathbb{N}$, write

$$P(n, x) = \sum_{i=1}^{n} P(x - s_i).$$

Then it is easy to see that

$$\log A_n = \max_{x \in U} P(n, x).$$

Theorem 12 follows easily from

Lemma 4.14. *For all N-tuples $s_1, \ldots, s_N$ in U, we have*

$$\int_0^1 \max\{0, P(1, x), \ldots, P(N, x)\}\, dx > 10^{-5} \log\log N.$$

We deduce Lemma 4.14 from

Lemma 4.15. *Suppose that*

$$\int_0^1 \max\{0, P(1, x), \ldots, P(N, x)\}\, dx \geqslant C_N, \tag{66}$$

where C_N is independent of $s_1, s_2, s_3, \ldots$. Then

$$\int_0^1 \max\{0, P(1, x), \ldots, P(N + N^4 + N, x)\}\, dx \geqslant C_N + 10^{-4}.$$

It is trivial that

$$\int_0^1 \max\{0, P(1, x), P(2, x)\}\, dx \geqslant \int_0^1 \max\{0, P(1, x)\}\, dx \geqslant 10^{-4}.$$

Using Lemma 4.15 and induction on k, one can easily show that for $k = 0, 1, 2, \ldots,$

$$\int_0^1 \max\{0, P(1,x), \ldots, P(2^{5^k}, x)\}\,dx \geqslant (k+1)10^{-4}.$$

Given any $N \in \mathbb{N}$, we simply choose k to satisfy $2^{5^k} \leqslant N < 2^{5^{k+1}}$. Then

$$\int_0^1 \max\{0, P(1,x), \ldots, P(N,x)\}\,dx \geqslant (k+1)10^{-4} > 10^{-5}\log\log N$$

as required. It remains to prove Lemma 4.15.

The proof of Lemma 4.15 is based on the following variation of Lemma 4.1: Let

$$\phi_N(x) = \max\{0, P(1,x), \ldots, P(N,x)\}$$

and

$$\psi_N(x) = \max\{P(N+N^4, x), \ldots, P(N+N^4+N, x)\}.$$

Then

$$\int_0^1 \max\{0, P(1,x), \ldots, P(N+N^4+N, x)\}\,dx \geqslant \int_0^1 \max\{\phi_N(x), \psi_N(x)\}\,dx$$

$$= \tfrac{1}{2}\int_0^1 \phi_N(x)\,dx + \tfrac{1}{2}\int_0^1 \psi_N(x)\,dx + \tfrac{1}{2}\int_0^1 |\psi_N(x) - \phi_N(x)|\,dx. \qquad (67)$$

Here, the last term on the right-hand side of (67) has a role analogous to that of the term $h(J, J', \alpha, \beta)$ in Lemma 4.1. By (66),

$$\tfrac{1}{2}\int_0^1 \phi_N(x)\,dx \geqslant \tfrac{1}{2}C_N.$$

On the other hand,

$$\psi_N(x) = P(N+N^4, x) + \max\{0, \bar{P}(1,x), \ldots, \bar{P}(N,x)\},$$

where, for $\alpha = 1, \ldots, N$,

$$\bar{P}(\alpha, x) = \sum_{i=1}^{\alpha} P(x - s_{N+N^4+i}).$$

Note that C_N is independent of the choice of the sequence $s_1, s_2, s_3, \ldots$. Also,

$$\int_0^1 P(N+N^4, x)\,dx = 0.$$

It follows from (66) that

$$\tfrac{1}{2}\int_0^1 \psi_N(x)\,dx \geqslant \tfrac{1}{2}C_N.$$

It therefore remains to prove that

$$\tfrac{1}{2}\int_0^1 |\psi_N(x) - \phi_N(x)|\,dx \geqslant 10^{-4}. \tag{68}$$

Let $P(0, x) = 0$ for all $x \in U$.

Definition. By a component Y of the function $\phi_N(x)$, we mean an interval $Y \subset U$ such that there exists $\alpha = \alpha(Y)$ satisfying $0 \leqslant \alpha \leqslant N$ and $\phi_N(x) = P(\alpha, x)$ for all $x \in Y$. We say that Y is maximal if no Y' satisfying $Y \subset Y'$ and $Y \neq Y'$ is a component of $\phi_N(x)$.

In other words, we are studying 'local representations' of $\phi_N(x)$. Maximal components for $\psi_N(x)$ and for $\psi_N(x) - \phi_N(x)$ are defined in a similar way.

Lemma 4.16. *The function $\psi_N(x) - \phi_N(x)$ has no more than $4N^2$ maximal components, on each of which $\psi_N(x) - \phi_N(x)$ is of the form $P(N + N^4 + \beta, x) - P(\alpha, x)$, where $0 \leqslant \alpha, \beta \leqslant N$.*

Proof. It is sufficient to prove that each of $\phi_N(x)$ and $\psi_N(x)$ has no more than $2N^2$ maximal components. We shall only show this for $\phi_N(x)$. For $N = 1$, $\phi_1(x)$ has exactly two maximal components. Assume now that $\phi_N(x)$ has no more than $2N^2$ maximal components. We now consider $\phi_{N+1}(x)$. Since $\phi_{N+1}(x) = \max\{\phi_N(x), P(N+1, x)\}$, it follows that $\phi_{N+1}(x) = P(N+1, x)$ if and only if $P(N+1, x) - \phi_N(x) \geqslant 0$. It is not too difficult to see that the function $P(N+1, x) - \phi_N(x)$ has singularities at $s_1, \ldots, s_{N+1}$, a strictly negative second derivative everywhere else except at a finite number of points where the first derivative has negative jumps. It follows that the set $\{x \in U : P(N+1, x) - \phi_N(x) > 0\}$ is a union of at most $(N + 1)$ intervals. Hence the number of maximal components of $\phi_{N+1}(x)$ does not exceed $2N^2 + 2(N + 1) < 2(N + 1)^2$. ■

To prove (68), we shall need the following technical lemma.

Lemma 4.17. *Suppose that f is twice differentiable on an open interval Y of length L and $f''(x) \leqslant -K < 0$ for all $x \in Y$. Then*

$$\int_Y |f(x)|\,dx \geqslant \tfrac{1}{32}KL^3.$$

We partition the interval U by the points $s_{N+1}, \ldots, s_{N+N^4}$, and denote the intervals by $\mathscr{X}_\nu$ $(\nu = 1, \ldots, N')$, where $N' \leqslant N^4$. Furthermore, let $\mathscr{Y}_\mu$ $(\mu = 1, \ldots, N'')$, where $N'' \leqslant 4N^2$, denote the maximal components of $\psi_N(x) - \phi_N(x)$. We shall consider all those $\mathscr{X}_\nu$ which are subsets of some maximal component $\mathscr{Y}_\mu$, and let

$$\mathscr{L} = \sum_{\mu=1}^{N''} \sum_{\mathscr{X}_\nu \subset \mathscr{Y}_\mu} |\mathscr{X}_\nu|.$$

There are two cases.

Case 1: $\mathscr{L} \geqslant \frac{1}{2}$. If $\mathscr{X}_\nu \subset \mathscr{Y}_\mu$ for some μ, then in $\mathscr{X}_\nu$, we have

$$\psi_N(x) - \phi_N(x) = P(N + N^4 + \beta, x) - P(\alpha, x) = \sum_{i=\alpha+1}^{N+N^4+\beta} P(x - s_i) \quad (69)$$

for some α, β satisfying $0 \leqslant \alpha$, $\beta \leqslant N$. Suppose that $\mathscr{X}_\nu = (s_{\sigma(\nu)}, s_{\tau(\nu)})$. Then

$$-(\psi_N''(x) - \phi_N''(x)) = -\sum_{i=\alpha+1}^{N+N^4+\beta} P''(x - s_i) = \sum_{i=\alpha+1}^{N+N^4+\beta} \frac{\pi^2}{\sin^2 \pi(x - s_i)}$$

$$\geqslant \frac{\pi^2}{\sin^2 \pi(x - s_{\sigma(\nu)})} + \frac{\pi^2}{\sin^2 \pi(x - s_{\tau(\nu)})} \geqslant \frac{8}{|\mathscr{X}_\nu|^2}.$$

It follows from Lemma 4.17 that

$$\int_{\mathscr{X}_\nu} |\psi_N(x) - \phi_N(x)|\,\mathrm{d}x \geqslant \frac{1}{32} \frac{8}{|\mathscr{X}_\nu|^2} |\mathscr{X}_\nu|^3 = \tfrac{1}{4} |\mathscr{X}_\nu|,$$

and so

$$\int_0^1 |\psi_N(x) - \phi_N(x)|\,\mathrm{d}x \geqslant \tfrac{1}{4} \sum_{\mu=1}^{N''} \sum_{\mathscr{X}_\nu \subset \mathscr{Y}_\mu} |\mathscr{X}_\nu| = \frac{\mathscr{L}}{4} \geqslant \tfrac{1}{8}.$$

Case 2: $\mathscr{L} < \frac{1}{2}$. In this case, the total length of those $\mathscr{X}_\nu$ intersected by more than one maximal component of $\psi_N(x) - \phi_N(x)$ exceeds $\frac{1}{2}$. Consider the intersection of these $\mathscr{X}_\nu$ with the set of maximal components of $\psi_N(x) - \phi_N(x)$. This will give rise to $N^* \leqslant 8N^2$ open intervals $\mathscr{Z}_\kappa$ ($\kappa = 1, \ldots, N^*$), on each of which $\psi_N(x) - \phi_N(x)$ is twice differentiable and of the form (69) for some α, β satisfying $0 \leqslant \alpha$, $\beta \leqslant N$. Then in $\mathscr{Z}_\kappa$, we have

$$-(\psi_N''(x) - \phi_N''(x)) \geqslant \sum_{i=N+1}^{N+N^4} P''(x - s_i) \geqslant N^4 \pi^2.$$

It follows from Lemma 4.17 that

$$\int_0^1 |\psi_N(x) - \phi_N(x)|\,\mathrm{d}x \geqslant \tfrac{1}{32} N^4 \pi^2 \sum_{\kappa=1}^{N^*} |\mathscr{Z}_\kappa|^3 > 2^{-14}\pi^2 > 2 \cdot 10^{-4}.$$

This gives (68), and the proof of Theorem 12 is complete.

Part B: Generalization of the classical problem

5
Schmidt's work

The object of Part A was to study the irregularities of point distribution relative to the family of boxes $B(\mathbf{x})$ with sides parallel to the coordinate axes. In Chapter 3, we showed that for this family the 'maximum error'

$$\sup_{\mathbf{x}\in U_1^K} |D[\mathscr{P}; B(\mathbf{x})]|$$

is less than a power of $(\log N)$ for some distribution $\mathscr{P}$ of N points in U_0^K.

The situation is entirely different if we allow rotation. In 1969, Schmidt proved that for boxes in arbitrary position in the unit torus, the 'maximum error' behaves like a power of N rather than a power of $(\log N)$ (see Chapter 6).

Schmidt also proved the same 'large error' phenomenon for balls (the problem is essentially due to Erdös (1964)) in the unit torus.

Suppose that we have a distribution $\mathscr{P} = \mathscr{P}(K, N)$ of N points, not necessarily distinct, in the K-dimensional unit cube U_0^K. Let $\mathbb{Z}^K$ denote, as usual, the set of integer lattice points in $\mathbb{R}^K$. Denote by $\mathscr{P}^*$ the set of points $\mathbf{x} + \mathbf{l}$ where $\mathbf{x} \in \mathscr{P}$ and $\mathbf{l} \in \mathbb{Z}^K$. Thus $\mathscr{P}^*$ is a periodic set. Let $A \subset \mathbb{R}^K$ be a compact set with usual volume $\mu(A)$. Write $Z[\mathscr{P}^*; A]$ for the number of points of $\mathscr{P}^*$ in A, and

$$D^{\mathrm{tor}}[\mathscr{P}; A] = Z[\mathscr{P}^*; A] - N\mu(A).$$

Let $C(r, \mathbf{c})$ be the ball with radius r and centre $\mathbf{c} = (c_1, \ldots, c_K)$, i.e. the set of all points $\mathbf{x} = (x_1, \ldots, x_K)$ with

$$|\mathbf{x} - \mathbf{c}|^2 = \sum_{i=1}^{K} (x_i - c_i)^2 \leqslant r^2.$$

The following theorem shows that there exists a ball in the unit torus with 'error' very large as compared to that for boxes with sides parallel to the coordinate axes.

Theorem 13A (Schmidt (1969*b*)). *Let $K \geqslant 2$ and $\varepsilon > 0$. Suppose that $\delta > 0$ satisfies $N\delta^K \geqslant 1$. Let $\mathscr{P}$ be a distribution of N points in U_0^K. Then there* *exists a ball $C(r, \mathbf{c})$ with $r \leqslant \delta$ and*

$$|D^{\mathrm{tor}}[\mathscr{P}; C(r, \mathbf{c})]| > c_1(K, \varepsilon)(N\delta^K)^{\frac{1}{2} - 1/2K - \varepsilon}.$$

Note that the ball $C(r, \mathbf{c})$ is not necessarily contained in the unit cube (we interpret it as a subset of the unit torus).

In the opposite direction, we have an infinite sequence with small 'error'.

Theorem 14 (Beck). *There exists an infinite sequence $\mathbf{s}_1, \mathbf{s}_2, \mathbf{s}_3, \ldots$ of points in U_0^K such that for any integer $N \geqslant 2$ and for any ball $C(r, \mathbf{c})$ with $r \leqslant 1$ and $Nr^K \geqslant 1$,*

$$|D^{\mathrm{tor}}[\{\mathbf{s}_1, \ldots, \mathbf{s}_N\}; C(r, \mathbf{c})]| < c_2(K)(Nr^K)^{\frac{1}{2} - 1/2K}(\log N)^{\frac{3}{2}}.$$

In other words, the exponent $(\frac{1}{2} - 1/2K)$ is best possible.

The proof of Theorem 13A is based on the so-called 'integral equation method' due to Schmidt. The method is rather complicated and mysterious at first sight. We therefore illustrate first the key ideas of the method by proving the following simpler result in §5.1.

Theorem 13B (Schmidt (1969*a*)). *Let $\delta > 0$ satisfy $N\delta^3 \geqslant 1$. Let $\mathscr{P}$ be an arbitrary distribution of N points in the 3-dimensional unit cube U_0^3. Then there exists a box B in arbitrary position with diameter δ and such that*

$$|D^{\mathrm{tor}}[\mathscr{P}; B]| > c_3(N\delta^3)^{\frac{1}{6}}.$$

Note that the exponent $\frac{1}{6}$ here can be improved to $\frac{1}{3}$ (see Corollary 17C in Chapter 6).

We shall sketch the proof of Theorem 13B in §5.1 and prove Theorem 13A in §§5.2–5.3. Since Theorem 14 is a particular case of Corollary 18D in Chapter 6, we shall only prove the latter (see Chapter 8).

We conclude the section by discussing a result concerning the family of convex sets. Here the exponent is greater than in the case of balls.

Theorem 15 (Schmidt (1975)). *Let $\mathscr{P}$ be a distribution of N points in U_0^K. Then there exists a convex set A in U_0^K such that*

$$|D[\mathscr{P}; A]| = \left| \sum_{\mathbf{x} \in \mathscr{P} \cap A} 1 - N\mu(A) \right| > c_4(K)N^{1 - 2/(K+1)}.$$

This is much stronger than an earlier estimate due to Zaremba (1970).

In the opposite direction, Stute (1977) investigated the behaviour of the 'maximum error' of convex sets in U_0^K ($K \geqslant 3$) for a distribution of N points selected at random from U_0^K (random N-set, in short) and obtained the following metric results. For $K = 3$, the 'maximum error' of convex sets is $O(N^{\frac{1}{2}}(\log N)^{\frac{3}{2}})$ almost surely, and for $K \geqslant 4$, the 'maximum error' of convex sets is $O(N^{1-2/(K+1)}(\log N)^{2/(K+1)})$ almost surely. After Stute's work only the case $K = 2$ remained open (note that by the Central limit theorem the random N-set in the unit square U_0^2 has 'maximum error' at least $N^{\frac{1}{2}}$ on the set of rectangles $B(\mathbf{x})$ ($\mathbf{x} \in U_1^2$), so Stute's approach certainly fails to achieve the desired upper bound $N^{\frac{1}{3}+\varepsilon}$). Recently, Beck (TA*b*) proved the existence of an N-element set in U_0^2 such that the 'maximum error' of convex sets in U_0^2 is $O(N^{\frac{1}{3}}(\log N)^4)$.

These results indicate that Schmidt's lower bound (see Theorem 15) is remarkably accurate for arbitrary K.

We conclude this introduction by giving the simple but ingenious proof of Theorem 15.

We may clearly suppose that N is large. Let B be the ball of diameter 1 contained in U^K, and let S be the surface of B. Given a spherical cap C on S, the spherical diameter of C is the supremum of the spherical distances between points of C (for example, a half-sphere of S has spherical diameter π). Let $\bar{C}$ denote the convex hull of C, i.e. $\bar{C}$ is a solid spherical cap.

Let C be a spherical cap on S with spherical diameter $\delta < \pi$. Then

$$\mu(\bar{C}) \gg \ll \delta^{K+1},$$

where the implicit constants depend only on K. If N is sufficiently large, there is a real $\delta_0 = \delta_0(N)$ satisfying $0 < \delta_0 < \pi$ and such that a cap $C \subset S$ of spherical diameter δ_0 has $\mu(\bar{C}) = 1/2N$.

We now pick as many pairwise disjoint caps with spherical diameter δ_0 as possible, say $C_1, C_2, \ldots, C_M$. Elementary calculation gives

$$M \gg (\delta_0)^{-K+1} \gg N^{(K-1)/(K+1)}.$$

Given a sequence of numbers $\varepsilon_1, \varepsilon_2, \ldots, \varepsilon_M \in \{\pm 1\}$, let $B(\varepsilon_1, \varepsilon_2, \ldots, \varepsilon_M)$ consist of all $\mathbf{x} \in B$ which do not lie in a solid cap $\overline{C_i}$ with $\varepsilon_i = -1$. In other words, $B(\varepsilon_1, \varepsilon_2, \ldots, \varepsilon_M)$ is obtained from B by removing the solid caps $\overline{C_i}$ for which $\varepsilon_i = -1$. Observe that

$$D[\mathscr{P}; B(\varepsilon_1, \varepsilon_2, \ldots, \varepsilon_M)] - D[\mathscr{P}; B(-\varepsilon_1, -\varepsilon_2, \ldots, -\varepsilon_M)] = \sum_{i=1}^{M} \varepsilon_i D[\mathscr{P}; \overline{C_i}].$$

But

$$D[\mathscr{P}; \overline{C_i}] = \sum_{\mathbf{p} \in \mathscr{P} \cap \bar{C}_i} 1 - N\mu(\overline{C_i}) = \sum_{\mathbf{p} \in \mathscr{P} \cap \bar{C}_i} 1 - \tfrac{1}{2}.$$

Hence for every i, either $D[\mathscr{P}; \overline{C_i}] \geqslant \frac{1}{2}$ or $D[\mathscr{P}; \overline{C_i}] \leqslant -\frac{1}{2}$. We now choose $\varepsilon_i = \pm 1$ so that $\varepsilon_i D[\mathscr{P}; \overline{C_i}] \geqslant \frac{1}{2}$ $(1 \leqslant i \leqslant M)$. Then

$$D[\mathscr{P}; B(\varepsilon_1, \varepsilon_2, \ldots, \varepsilon_M)] - D[\mathscr{P}; B(-\varepsilon_1, -\varepsilon_2, \ldots, -\varepsilon_M)] \geqslant \tfrac{1}{2}M,$$

and so either $A = B(\varepsilon_1, \varepsilon_2, \ldots, \varepsilon_M)$ or $A = B(-\varepsilon_1, -\varepsilon_2, \ldots, -\varepsilon_M)$ satisfies

$$|D[\mathscr{P}; A]| = \left| \sum_{\mathbf{p} \in \mathscr{P} \cap A} 1 - N\mu(A) \right| \geqslant M/4 \gg N^{(K-1)/(K+1)}.$$

Theorem 15 follows.

5.1 An illustration of the method of integral equations

We recall that $B(\mathbf{x})$, where $\mathbf{x} \in U_1^K$, denotes the box consisting of all points $\mathbf{y}$ such that $0 \leqslant y_i < x_i$ $(i = 1, \ldots, K)$.

In this section, we restrict ourselves to the case $K = 3$. If τ is a rotation in $\mathbb{R}^3$ and $\mathbf{v} \in \mathbb{R}^3$, let $B(\mathbf{x}; \mathbf{v}, \tau)$ denote the tilted box consisting of the points $\tau\mathbf{y} + \mathbf{v}$ where $\mathbf{y} \in B(\mathbf{x})$. The diameter of this box is given by $|\mathbf{x}| = (x_1^2 + x_2^2 + x_3^2)^{\frac{1}{2}}$.

Let S^2 be the unit sphere in $\mathbb{R}^3$ consisting of points $|\mathbf{u}| = 1$, and let $\mathrm{d}\sigma(\mathbf{u})$ represent an element of surface area on S^2, normalized such that $\int_{S^2} \mathrm{d}\sigma(\mathbf{u}) = 1$. Any box $B(\mathbf{x}; \mathbf{v}, \tau)$ of diameter δ has $\mathbf{u} \in S^2$ of the type $\mathbf{x} = \delta\mathbf{u}$.

Let T be the group of proper orthogonal transformations in $\mathbb{R}^3$ and $\mathrm{d}\tau$ the volume element of the invariant measure on T, normalized such that $\int_T \mathrm{d}\tau = 1$.

In this section, we outline the proof of the following theorem.

Theorem 13C. *Suppose that $N\delta^3 \geqslant 1$. Let $\mathscr{P}$ be an arbitrary distribution of N points in U_0^3. Then*

$$\int_T \int_{U^3} \int_{S^2} |D^{\mathrm{tor}}[\mathscr{P}; B(\delta\mathbf{u}; \mathbf{v}, \tau)]|^2 \mathrm{d}\sigma(\mathbf{u}) \mathrm{d}\mathbf{v}\, \mathrm{d}\tau > c_5 (N\delta^3)^{\frac{1}{3}}.$$

Theorem 13B follows immediately from Theorem 13C by a standard averaging argument.

We begin the proof of Theorem 13C by setting up an integral equation. Suppose that

$$0 < |\mathbf{x}| < \tfrac{1}{4}. \tag{1}$$

Let

$$f_B(\mathbf{x}; \mathbf{v}, \tau | \mathbf{y}) = \begin{cases} 1 & (\mathbf{y} \in B(\mathbf{x}; \mathbf{v}, \tau)), \\ 0 & (\text{otherwise}); \end{cases} \tag{2}$$

$$g_B(\mathbf{x}; \mathbf{v}, \tau | \mathbf{y}) = \sum_{\mathbf{l} \in \mathbb{Z}^3} f_B(\mathbf{x}; \mathbf{v}, \tau | \mathbf{y} + \mathbf{l}) \tag{3}$$

and

$$h_B(\mathbf{x}; \mathbf{y}, \mathbf{z}) = \int_T \int_{U^3} g_B(\mathbf{x}; \mathbf{v}, \tau | \mathbf{y}) g_B(\mathbf{x}; \mathbf{v}, \tau | \mathbf{z}) \mathrm{d}\mathbf{v}\, \mathrm{d}\tau. \tag{4}$$

Writing ('distance modulo 1')

$$w(\mathbf{y},\mathbf{z})=\min_{\mathbf{l}\in Z^3}|\mathbf{y}+\mathbf{l}-\mathbf{z}|,$$

it easily follows that $h_B(\mathbf{x};\mathbf{y},\mathbf{z})$ as a function of $\mathbf{y}$ and $\mathbf{z}$ depends only on $w(\mathbf{y},\mathbf{z})$, i.e.

$$h_B(\mathbf{x};\mathbf{y},\mathbf{z})=k_B(\mathbf{x};w(\mathbf{y},\mathbf{z})). \tag{5}$$

Note that $0\leqslant w(\mathbf{y},\mathbf{z})\leqslant\frac{1}{2}\sqrt{3}$, and (5) defines $k_B(\mathbf{x};w)$ only if $\mathbf{x}$ satisfies (1) and if $0\leqslant w\leqslant\frac{1}{2}\sqrt{3}$. We now extend this definition by setting $k_B(\mathbf{x};w)=0$ if $w>\frac{1}{2}\sqrt{3}$. We have

$$\int_{U^3}\int_{U^3}k_B(\mathbf{x};w(\mathbf{y},\mathbf{z}))\,d\mathbf{y}\,d\mathbf{z}$$
$$=\int_T\int_{U^3}\int_{U^3}\int_{U^3}g_B(\mathbf{x};\mathbf{v},\tau|\mathbf{y})g_B(\mathbf{x};\mathbf{v},\tau|\mathbf{z})\,d\mathbf{y}\,d\mathbf{z}\,d\mathbf{v}\,d\tau=(\mu(B(\mathbf{x})))^2,$$

where μ denotes the usual volume in $\mathbb{R}^3$.

Suppose now that the elements of $\mathscr{P}$ are $\mathbf{p}_1,\mathbf{p}_2,\ldots,\mathbf{p}_N$. Then

$$\int_T\int_{U^3}|D^{\text{tor}}[\mathscr{P};B(\mathbf{x};\mathbf{v},\tau)]|^2\,d\mathbf{v}\,d\tau$$
$$=N^2(\mu(B(\mathbf{x})))^2-2N\mu(B(\mathbf{x}))\int_T\int_{U^3}Z[\mathscr{P}^*;B(\mathbf{x};\mathbf{v},\tau)]\,d\mathbf{v}\,d\tau$$
$$+\int_T\int_{U^3}(Z[\mathscr{P}^*;B(\mathbf{x};\mathbf{v},\tau)])^2\,d\mathbf{v}\,d\tau$$
$$=-N^2(\mu(B(\mathbf{x})))^2+\sum_{i=1}^{N}\sum_{j=1}^{N}\int_T\int_{U^3}g_B(\mathbf{x};\mathbf{v},\tau|\mathbf{p}_i)g_B(\mathbf{x};\mathbf{v},\tau|\mathbf{p}_j)\,d\mathbf{v}\,d\tau$$
$$=\sum_{i=1}^{N}\sum_{j=1}^{N}(h_B(\mathbf{x};\mathbf{p}_i,\mathbf{p}_j)-(\mu(B(\mathbf{x})))^2)$$
$$=\sum_{i=1}^{N}\sum_{j=1}^{N}\left(k_B(\mathbf{x};w(\mathbf{p}_i,\mathbf{p}_j))-\int_{U^3}\int_{U^3}k_B(\mathbf{x};w(\mathbf{y},\mathbf{z}))\,d\mathbf{y}\,d\mathbf{z}\right). \tag{6}$$

We recall that $C(r,\mathbf{v})$ denotes the ball of radius r and centre $\mathbf{v}$. Assume that

$$0<r<\tfrac{1}{8}.$$

Let

$$f_C(r;\mathbf{v}|\mathbf{y})=\begin{cases}1 & (\mathbf{y}\in C(r,\mathbf{v})),\\ 0 & (\text{otherwise});\end{cases}$$

$$g_C(r;\mathbf{v}|\mathbf{y})=\sum_{\mathbf{l}\in Z^3}f_C(r;\mathbf{v}|\mathbf{y}+\mathbf{l})$$

and

$$h_C(r; \mathbf{y}, \mathbf{z}) = \int_{U^3} g_C(r; \mathbf{v}|\mathbf{y}) g_C(r; \mathbf{v}|\mathbf{z}) \, d\mathbf{v}.$$

Again, $h_C(r; \mathbf{y}, \mathbf{z})$ as a function of $\mathbf{y}$ and $\mathbf{z}$ depends only on $w(\mathbf{y}, \mathbf{z})$, i.e.

$$h_C(r; \mathbf{y}, \mathbf{z}) = k_C(r; w(\mathbf{y}, \mathbf{z})).$$

Also, $k_C(r; w)$ is defined if $0 \leqslant w \leqslant \frac{1}{2}\sqrt{3}$. For $w > \frac{1}{2}\sqrt{3}$, let $k_C(r; w) = 0$. With these definitions, an argument similar to that for (6) gives

$$\int_{U^3} |D^{\mathrm{tor}}[\mathscr{P}; C(r, \mathbf{v})]|^2 \, d\mathbf{v}$$

$$= \sum_{i=1}^{N} \sum_{j=1}^{N} \left(k_C(r; w(\mathbf{p}_i, \mathbf{p}_j)) - \int_{U^3} \int_{U^3} k_C(r; w(\mathbf{y}, \mathbf{z})) \, d\mathbf{y} \, d\mathbf{z} \right). \tag{7}$$

Lemma 5.1. *Suppose that $0 < \delta < \frac{1}{4}$. Suppose further that the function $F(x)$ is non-negative and continuous in the interval $(0, \frac{1}{2}]$ and satisfies the integral equation*

$$\int_0^{\frac{1}{2}} F(r) k_C(\delta r; w) \, dr = \int_{S^2} k_B(\delta \mathbf{u}; w) \, d\sigma(\mathbf{u}) \tag{8}$$

for all $w \geqslant 0$. Then

$$\int_{U^3} \int_0^{\frac{1}{2}} F(r) |D^{\mathrm{tor}}[\mathscr{P}; C(\delta r; \mathbf{v})]|^2 \, dr \, d\mathbf{v}$$

$$= \int_T \int_{U^3} \int_{S^2} |D^{\mathrm{tor}}[\mathscr{P}; B(\delta \mathbf{u}; \mathbf{v}, \tau)]|^2 \, d\sigma(\mathbf{u}) \, d\mathbf{v} \, d\tau. \tag{9}$$

Proof. The relation (9) follows immediately from (7), (8) and (6). ■

Remark. The substitution $\delta \to c\delta$ and $w \to cw$ has the effect of multiplying both sides of (8) by c^3, since both k_B and k_C are homogeneous of degree 3. Hence if ϕ solves the equation for some particular δ, then it solves it for every δ in the range $0 < \delta < \frac{1}{4}$.

The proof of Theorem 13C is based on the fact that we shall find a positive solution of (8) with

$$F(x) > c_6 x^{-2} \quad (c_6 > 0) \tag{10}$$

in some interval $(0, c_7)$. Surprisingly, the trivial estimate

$$|D^{\mathrm{tor}}[\mathscr{P}; C(\rho, \mathbf{v})]| \geqslant \| N\mu(C(\rho, \mathbf{v})) \| = \| 4\pi N \rho^3 / 3 \|, \tag{11}$$

where $\|\cdot\|$ denotes the distance from the nearest integer, will be all we need to finish the proof. Indeed, by (9), (10) and (11), the right-hand side of (9) exceeds

$$c_6 \int_0^{c_7} r^{-2} \| 4\pi N\delta^3 r^3/3 \|^2 \, dr.$$

Putting $y = 4\pi N\delta^3 r^3/3$, we have the lower bound

$$c_8 N^{\frac{1}{3}}\delta \int_0^{c_9 N\delta^3} y^{-\frac{4}{3}} \| y \|^2 \, dy \geqslant c_8 N^{\frac{1}{3}}\delta \int_0^{c_9} y^{-\frac{4}{3}} \| y \|^2 \, dy,$$

which is at least $c_5(N\delta^3)^{\frac{1}{3}}$ provided $N\delta^3 \geqslant 1$.

We now return to the integral equation (8). Suppose that (1) holds. From the definitions (2), (3), (4) and (5), it easily follows that

$$k_B(\mathbf{x}; w) = \int_{S^2} \int_{\mathbb{R}^3} f_B(\mathbf{x}; \mathbf{v}, \tau_* \,|\, w\mathbf{u}) f_B(\mathbf{x}; \mathbf{v}, \tau_* \,|\, \mathbf{0}) \, d\mathbf{v} \, d\sigma(\mathbf{u}), \tag{12}$$

where τ_* is the identity map. Note that both sides of (12) are zero whenever $w > \frac{1}{2}\sqrt{3} > |\mathbf{x}|$. In the range $0 < |\mathbf{x}| < \frac{1}{4}$, the function $k_B(\mathbf{x}; w)$ is homogeneous of degree 3, i.e. $k_B(t\mathbf{x}; tw) = t^3 k_B(\mathbf{x}; w)$. Let $k_B^*(\mathbf{x}; w)$ be the unique function defined for $\mathbf{x} \neq \mathbf{0}$ and $w \geqslant 0$ which is homogeneous of degree 3 and coincides with $k_B(\mathbf{x}; w)$ in $0 < |\mathbf{x}| < \frac{1}{4}$. Then $k_B^*(\mathbf{x}; w)$ is equal to the right-hand side of (12). It is zero when $w > |\mathbf{x}|$.

Similarly, the function

$$k_C^*(r; w) = \int_{S^2} \int_{\mathbb{R}^3} f_C(r; \mathbf{v}, \tau_* \,|\, w\mathbf{u}) f_C(r; \mathbf{v}, \tau_* \,|\, \mathbf{0}) \, d\mathbf{v} \, d\sigma(\mathbf{u})$$

is homogeneous and coincides with $k_C(r; w)$ for $0 < r < \frac{1}{8}$. It vanishes when $w > 2r$.

We may replace k_B and k_C in (8) by k_B^* and k_C^* respectively, and then set $\delta = 1$. Then (8) becomes

$$\int_0^{\frac{1}{2}} F(x) k_C^*(x; y) \, dx = \int_{S^2} k_B^*(\mathbf{u}; y) \, d\sigma(\mathbf{u}) \quad (0 \leqslant y \leqslant 1). \tag{13}$$

To solve this equation, we need information on the function

$$l(y) = \int_{S^2} k_B^*(\mathbf{u}; y) \, d\sigma(\mathbf{u}). \tag{14}$$

In view of (12), we have

$$l(y) = \int_{S^2} \int_{S^2} \int_{\mathbb{R}^3} f_B(\mathbf{u}_1; \mathbf{v}, \tau_* \,|\, y\mathbf{u}_2) f_B(\mathbf{u}_1; \mathbf{v}, \tau_* \,|\, \mathbf{0}) \, d\mathbf{v} \, d\sigma(\mathbf{u}_1) \, d\sigma(\mathbf{u}_2). \tag{15}$$

Note that the product

$$f_B(\mathbf{z};\mathbf{v},\tau_*|\mathbf{x})f_B(\mathbf{z};\mathbf{v},\tau_*|\mathbf{0})$$

is equal to 1 if $|x_i|\leqslant|z_i|$ $(i=1,2,3)$ and if $\mathbf{v}$ lies in a certain box of volume $\prod_{i=1}^{3}(|z_i|-|x_i|)$, where $\mathbf{z}=(z_1,z_2,z_3)$ and $\mathbf{x}=(x_1,x_2,x_3)$; otherwise the product is zero. Therefore the inner integral in (15) can be evaluated as follows:

$$\int_{\mathbf{R}^3} f_B(\mathbf{z};\mathbf{v},\tau_*|\mathbf{x})f_B(\mathbf{z};\mathbf{v},\tau_*|\mathbf{0})\,\mathrm{d}\mathbf{v}=\prod_{i=1}^{3}(|z_i|-|x_i|)^+,$$

where

$$y^+=\begin{cases} y & (y\geqslant 0)\\ 0 & (\text{otherwise}).\end{cases}$$

Hence, using polar coordinates,

$$l(y)=(2/\pi)^2\int_0^{\frac{1}{2}\pi}\int_0^{\frac{1}{2}\pi}\sin\phi\sin\psi(\cos\phi-y\cos\psi)^+I(\phi,\psi)\,\mathrm{d}\phi\,\mathrm{d}\psi,$$

where

$$I(\phi,\psi)$$
$$=\int_0^{\frac{1}{2}\pi}\int_0^{\frac{1}{2}\pi}(\sin\phi\cos\eta-y\sin\psi\cos\theta)^+\times(\sin\phi\sin\eta-y\sin\psi\sin\theta)^+\,\mathrm{d}\eta\,\mathrm{d}\theta.$$

Using this explicit expression of $l(y)$, it is not hard to prove the following lemma.

Lemma 5.2.

(*i*) *$l(y)$ is continuous in* $[0,1]$, *and* $l(1)=0$;
(*ii*) *$l'(y)$ is continuous in* $(0,1)$, *and* $l'(y)\to 0$ *as* $y\to 1$;
(*iii*) *$l''(y)$ is positive and continuous in* $(0,1)$, *and* $l''(y)\to 1$ *as* $y\to 1$, *and* $l''(y)\to c_{10}>0$ *when* $y\to 0$;
(*iv*) *$l'''(y)$ is negative and continuous in* $(0,1)$, *and* $l'''(y)\to 0$ *as* $y\to 1$, *and* $|l'''(y)|\ll y^{-\frac{1}{2}}$ *as* $y\to 0$.

Proof. We refer the reader to the proof of Lemma 4 of Schmidt (1969*a*). ∎

We now return to (13). The 'kernel' $k_C^*(x;y)$ is equal to the volume of the intersection of two balls of radius x and whose centres have distance y. Hence

$$k_C^*(x;y)=2\pi\int_{\frac{1}{2}y}^{x}(x^2-t^2)\,\mathrm{d}t=\frac{4\pi x^3}{3}-\pi x^2 y+\frac{\pi y^3}{12}$$

if $0 \leqslant y \leqslant 2x$, and $k_C^*(x;y)=0$ otherwise. In particular,

$$k_C^*(x;y) \ll x^3. \tag{16}$$

In the interval $0 \leqslant y < 2x$,

$$\frac{\partial}{\partial y} k_C^*(x;y) = -\pi\left(x^2 - \frac{y^2}{4}\right)$$

and

$$\frac{\partial^2}{\partial y^2} k_C^*(x;y) = \tfrac{1}{2}\pi y.$$

Suppose that a function $F(x)$ is continuous in $(0,\frac{1}{2}]$ and satisfies

$$|F(x)| \ll x^{-2} \quad \text{as } x \to 0. \tag{17}$$

Then by (16), the function $F(x)k_C^*(x;y)$ tends to zero as $x \to 0$ and it can therefore be extended to a continuous function defined for $x \in [0,\frac{1}{2}]$ and $y \in [0,1]$. Therefore the left-hand side, and hence both sides, of (13) are continuous functions of y, and it will suffice to verify (13) for $y \in (0,1)$. The integral equation may therefore be rewritten (see (13) and (14)) as

$$\int_{\frac{1}{2}y}^{\frac{1}{2}} F(x)k_C^*(x;y)\,\mathrm{d}x = l(y) \quad (0<y<1). \tag{18}$$

Remark. The kernel $k_C^*(x;y)$ vanishes on the line $x=\frac{1}{2}y$, and hence the equation is a Volterra equation of the first kind.

Suppose that (18) holds. Taking the derivative on both sides, we obtain

$$-\pi\int_{\frac{1}{2}y}^{\frac{1}{2}} F(x)\left(x^2-\frac{y^2}{4}\right)\mathrm{d}x = l'(y) \quad (0<y<1). \tag{19}$$

Conversely, this equation implies (18), since both sides of (18) tend to zero as $y \to 1$. It will therefore suffice to find a solution of (19). Differentiating again it is not hard to see that it will suffice to find a solution $F(x)$ of

$$\tfrac{1}{2}\pi\int_{\frac{1}{2}y}^{\frac{1}{2}} F(x)y\,\mathrm{d}x = l''(y) \quad (0<y<1),$$

or, of

$$\int_{\frac{1}{2}y}^{\frac{1}{2}} F(x)\,\mathrm{d}x = \frac{2}{\pi y}l''(y) \quad (0<y<1). \tag{20}$$

Differentiating both sides of (20) we obtain

$$-\tfrac{1}{2}F(\tfrac{1}{2}y) = \frac{2}{\pi y}l'''(y) - \frac{2}{\pi y^2}l''(y) \quad (0<y<1),$$

whence

$$F(x) = \frac{1}{\pi x^2} l''(2x) - \frac{2}{\pi x} l'''(2x) \quad (0 < x < \tfrac{1}{2}). \tag{21}$$

In view of Lemma 5.2, both sides of (20) approach zero as $y \to 1$, and hence (21) does in fact give a solution of the integral equation. The function $F(x)$ is positive and continuous in $(0, \frac{1}{2})$, and tends to zero as $x \to \frac{1}{2}$. This also follows from Lemma 5.2. Hence $F(x)$ may be extended to a function which is continuous in $(0, \frac{1}{2}]$. Finally, again by Lemma 5.2, we have

$$x^{-2} \ll F(x) \ll x^{-2} \quad \text{as } x \to 0,$$

and hence (10) and (17) hold. The proof of Theorem 13C is now complete.

Remark. We emphasize again that the machinery of the 'integral equation method' was used to 'blow up' the following 'trivial error': if $0 < \varepsilon < 1$ and $N^{-1}\varepsilon < \mu(C(\rho, \mathbf{v})) < N^{-1}(1-\varepsilon)$, then $|D^{\text{tor}}[\mathscr{P}; C(\rho, \mathbf{v})]| > \varepsilon$.

5.2 Balls in the unit torus

Let $\mathscr{P}$ be an arbitrary but fixed distribution of N points in U_0^K, and set

$$E(r) = \int_{U^K} |D^{\text{tor}}[\mathscr{P}; C(r, \mathbf{c})]|^2 \, d\mathbf{c}.$$

We shall deduce Theorem 13A from the following estimate.

Theorem 13D. *Suppose that $\delta > 0$ satisfies $N\delta^K > N^{\varepsilon}$. Then*

$$\int_0^{\delta} r^{-1} E(r) \, dr \gg (N\delta^K)^{1-1/K-\varepsilon}.$$

Remark. Here and later, the implicit constants in $\gg$ depend on K and ε, but are independent of δ and N.

To prove that Theorem 13D implies Theorem 13A, we require the simple inequality

$$E(r) \ll N^2 \max\{r^K, r^{2K}\}. \tag{22}$$

To see this, note that

$$E(r) \leqslant \int_{U^K} (Z[\mathscr{P}^*; C(r, \mathbf{c})] + N\mu(C(r, \mathbf{c})))(Z[\mathscr{P}^*; C(r, \mathbf{c})] + N\mu(C(r, \mathbf{c})))d\mathbf{c},$$

and observe that

$$Z[\mathscr{P}^*; C(r, \mathbf{c})] + N\mu(C(r, \mathbf{c})) \ll N \max\{1, r^K\},$$

and

$$\int_{U^K} Z[\mathscr{P}^*; C(r,\mathbf{c})]\,\mathrm{d}\mathbf{c} = N\mu(C(r,\mathbf{0})) \ll Nr^K.$$

Using these relations, we obtain

$$\begin{aligned} E(r) &\ll N \max\{1, r^K\} \int_{U^K} (Z[\mathscr{P}^*; C(r,\mathbf{c})] + N\mu(C(r,\mathbf{c})))\,\mathrm{d}\mathbf{c} \\ &\ll N \max\{1, r^K\} Nr^K = N^2 \max\{r^K, r^{2K}\}, \end{aligned}$$

which proves (22).

We deduce Theorem 13A as follows: If $N\delta^K \ll 1$, take a ball with centre $\mathbf{p}\in\mathscr{P}$ and a very small radius ρ. Then

$$\begin{aligned} |D^{\mathrm{tor}}[\mathscr{P}; C(\rho,\mathbf{p})]| &\geqslant Z[\mathscr{P}^*; C(\rho,\mathbf{p})] - N\mu(C(\rho,\mathbf{p})) \\ &\geqslant 1 - \tfrac{1}{2} = \tfrac{1}{2} \gg (N\delta^K)^{\frac{1}{2} - 1/2K - \varepsilon}. \end{aligned}$$

We may therefore assume that $N\delta^K > c_{11}(K,\varepsilon)$, where $c_{11}(K,\varepsilon)$ is a large constant depending only on K and ε. Put

$$\eta = N^{-2/K}.$$

Then $\eta < \delta$, since $N\eta^K = N^{-1} \leqslant c_{11}(K,\varepsilon) < N\delta^K$. By (22),

$$\int_0^\eta r^{-1} E(r)\,\mathrm{d}r \ll N^2 \int_0^\eta r^{K-1}\,\mathrm{d}r \ll N^2\eta^K = 1. \tag{23}$$

From Theorem 13D and (23), we obtain

$$\int_\eta^\delta r^{-1} E(r)\,\mathrm{d}r \gg (N\delta^K)^{1-1/K-\varepsilon} - \mathrm{O}(1) \gg (N\delta^K)^{1-1/K-\varepsilon} \tag{24}$$

if $c_{11}(K,\varepsilon)$ is sufficiently large. Note that

$$\int_\eta^\delta r^{-1} E(r)\,\mathrm{d}r \leqslant \left(\max_{\eta\leqslant r\leqslant\delta} E(r)\right) \int_\eta^\delta r^{-1}\,\mathrm{d}r. \tag{25}$$

But

$$\begin{aligned} \int_\eta^\delta r^{-1}\,\mathrm{d}r &= \log(\delta/\eta) = \log(\delta N^{2/K}) = K^{-1}\log(N^2\delta^K) \\ &\ll \log(N\delta^K) \ll (N\delta^K)^\varepsilon, \end{aligned} \tag{26}$$

since $N\delta^K > N^\varepsilon$ and $N\delta^K > c_{11}(K,\varepsilon)$. Combining (24)–(26), we conclude that there is an r_0 with $\eta \leqslant r_0 \leqslant \delta$ and

$$E(r_0) \gg (N\delta^K)^{1-1/K-2\varepsilon}.$$

Therefore there exists $\mathbf{c}_0 \in U^K$ such that

$$|D^{\text{tor}}[\mathscr{P}; C(r_0, \mathbf{c}_0)]| \gg (N\delta^K)^{\frac{1}{2}-1/2K-\varepsilon}.$$

This completes the deduction of Theorem 13A from Theorem 13D. It remains to prove Theorem 13D.

Put

$$\nu = \nu(K) = \begin{cases} 1 & (K \text{ is odd}), \\ 0 & (K \text{ is even}). \end{cases}$$

Let $\alpha \in (0, 1)$, and let β be such that $\alpha + \beta = 1 + \nu$ (note in advance that α will be specified as $\alpha = 1 - \varepsilon$; also refer to §5.3 for the explanation for the definition of ν). Write

$$E(r, s) = \int_{U^K} D^{\text{tor}}[\mathscr{P}; C(r, \mathbf{c})] D^{\text{tor}}[\mathscr{P}; C(s, \mathbf{c})] \, d\mathbf{c},$$

and let

$$A = A(\alpha, \beta, \delta) = \underset{r+s\leqslant 1}{\int_0^1 \int_0^1} E(\delta r, \delta s) |r - s|^{-\alpha} |r + s|^{-\beta} \, dr \, ds.$$

We shall first show that

$$|A| \ll \int_0^\delta r^{-1} E(r) \, dr, \tag{27}$$

where the constant implied by $\ll$ depends only on α and K. Note that

$$\begin{aligned} |2E(r,s)| &\leqslant 2 \int_{U^K} |D^{\text{tor}}[\mathscr{P}; C(r, \mathbf{c})] D^{\text{tor}}[\mathscr{P}; C(s, \mathbf{c})]| \, d\mathbf{c} \\ &\leqslant \int_{U^K} \{|D^{\text{tor}}[\mathscr{P}; C(r, \mathbf{c})]|^2 + |D^{\text{tor}}[\mathscr{P}; C(s, \mathbf{c})]|^2\} \, d\mathbf{c} \\ &= E(r) + E(s). \end{aligned}$$

Therefore,

$$|A| \leqslant \tfrac{1}{2} \underset{r+s\leqslant 1}{\int_0^1 \int_0^1} (E(\delta r) + E(\delta s)) |r - s|^{-\alpha} |r + s|^{-\beta} \, dr \, ds.$$

Since $|r - s|^{-\alpha} |r + s|^{-\beta}$ is symmetric in r and s, by introducing a new variable $t = s/r$ we have

$$\begin{aligned} |A| &\leqslant \underset{r+s\leqslant 1}{\int_0^1 \int_0^1} E(\delta r) |r - s|^{-\alpha} |r + s|^{-\beta} \, dr \, ds \\ &= \int_0^1 E(\delta r) r^{-\nu} \left(\int_0^{1/r-1} |1 - t|^{-\alpha} |1 + t|^{-\beta} \, dt \right) dr. \end{aligned} \tag{28}$$

For $0<r\leqslant 1$, observe that the inner integral

$$\int_0^{1/r-1} |1-t|^{-\alpha}|1+t|^{-\beta}\mathrm{d}t \ll \begin{cases} 1 & (\alpha+\beta=1+\nu=2), \\ 1 & (\alpha+\beta=1 \text{ and } 1/r\leqslant 10), \\ \log(1/r) & (\alpha+\beta=1 \text{ and } 1/r>10). \end{cases}$$

So in general this inner integral is

$$\ll (1+\log(1/r))^{1-\nu} \ll r^{\nu-1} \quad (0<r\leqslant 1).$$

Hence by (28),

$$|A| \ll \int_0^1 r^{-1}E(\delta r)\,\mathrm{d}r = \int_0^{\delta} r^{-1}E(r)\,\mathrm{d}r,$$

and (27) is proved.

We then set up an integral equation. Put

$$f(r,\mathbf{c}\,|\,\mathbf{x}) = \begin{cases} 1 & (\mathbf{x}\in C(r,\mathbf{c})), \\ 0 & (\text{otherwise}), \end{cases}$$

so that $f(r,\mathbf{c}\,|\,\mathbf{x})$ for fixed r and $\mathbf{c}$ is the characteristic function of the ball $C(r,\mathbf{c})$. Write

$$g(r,\mathbf{c}\,|\,\mathbf{x}) = \sum_{\mathbf{l}\in\mathbb{Z}^K} f(r,\mathbf{c}\,|\,\mathbf{x}+\mathbf{l})$$

(note that $f(r,\mathbf{c}\,|\,\mathbf{x}+\mathbf{l})=0$ except for finitely many lattice points $\mathbf{l}\in\mathbb{Z}^K$). Also, write

$$h(r,s,\mathbf{x},\mathbf{y}) = \int_{U^K} g(r,\mathbf{c}\,|\,\mathbf{x})g(s,\mathbf{c}\,|\,\mathbf{y})\,\mathrm{d}\mathbf{c}.$$

Similarly as in the previous section, write $w(\mathbf{x},\mathbf{y})$ for the minimum of the distances $|\mathbf{x}+\mathbf{l}-\mathbf{y}|$, taken over all $\mathbf{l}\in\mathbb{Z}^K$ ('distance modulo 1 of $\mathbf{x}$ any $\mathbf{y}$'). From the definitions above it easily follows that $h(r,s,\mathbf{x},\mathbf{y})$ as a function of $\mathbf{x}$ and $\mathbf{y}$ depends only on $w(\mathbf{x},\mathbf{y})$, i.e.

$$h(r,s,\mathbf{x},\mathbf{y}) = k(r,s,w(\mathbf{x},\mathbf{y})).$$

Moreover, it is easy to see that $k(r,s,w)$ is equal to the volume of the intersection of two K-dimensional balls with radii r and s, whose centres have distance w.

Clearly

$$E(r,s) = \int_{U^K} D^{\mathrm{tor}}[\mathscr{P};C(r,\mathbf{c})]D^{\mathrm{tor}}[\mathscr{P};C(s,\mathbf{c})]\,\mathrm{d}\mathbf{c}$$

$$= \int_{U^K}\Bigg(\sum_{\mathbf{p}\in\mathscr{P}^*\cap C(r,\mathbf{c})} 1\Bigg)\Bigg(\sum_{\mathbf{p}\in\mathscr{P}^*\cap C(s,\mathbf{c})} 1\Bigg)\mathrm{d}\mathbf{c} - N^2\mu(C(r,\mathbf{0}))\mu(C(s,\mathbf{0})).$$

Let $\mathscr{P} = \{\mathbf{p}_1, \mathbf{p}_2, \ldots, \mathbf{p}_N\}$. Since

$$\sum_{\mathbf{p} \in \mathscr{P}^* \cap C(r,\mathbf{c})} 1 = \sum_{i=1}^{N} g(r, \mathbf{c} | \mathbf{p}_i),$$

we have

$$\begin{aligned} E(r,s) &= \int_{U^K} \left(\sum_{i=1}^{N} g(r,\mathbf{c}|\mathbf{p}_i) \right) \left(\sum_{i=1}^{N} g(s,\mathbf{c}|\mathbf{p}_i) \right) d\mathbf{c} - N^2 \mu(C(r,\mathbf{0}))\mu(C(s,\mathbf{0})) \\ &= \sum_{i=1}^{N} \sum_{j=1}^{N} k(r,s,w(\mathbf{p}_i,\mathbf{p}_j)) - N^2 \mu(C(r,\mathbf{0}))\mu(C(s,\mathbf{0})). \end{aligned} \tag{29}$$

Furthermore,

$$\begin{aligned} \int_{U^K} \int_{U^K} k(r,s,w(\mathbf{x},\mathbf{y}))\, d\mathbf{x}\, d\mathbf{y} &= \int_{U^K} \int_{U^K} \int_{U^K} g(r,\mathbf{c}|\mathbf{x}) g(s,\mathbf{c}|\mathbf{y})\, d\mathbf{c}\, d\mathbf{x}\, d\mathbf{y} \\ = \int_{U^K} \left(\int_{U^K} g(r,\mathbf{c}|\mathbf{x})\, d\mathbf{x} \right) &\left(\int_{U^K} g(s,\mathbf{c}|\mathbf{y})\, d\mathbf{y} \right) d\mathbf{c} = \mu(C(r,\mathbf{0}))\mu(C(s,\mathbf{0})), \end{aligned} \tag{30}$$

since

$$\int_{U^K} g(r,\mathbf{c}|\mathbf{x})\, d\mathbf{x} = \sum_{\mathbf{l} \in \mathbf{Z}^K} \int_{U^K} f(r,\mathbf{c}|\mathbf{x}+\mathbf{l})\, d\mathbf{x} = \int_{\mathbf{R}^K} f(r,\mathbf{c}|\mathbf{x})\, d\mathbf{x} = \mu(C(r,\mathbf{c})).$$

Combining (29) and (30), we obtain

$$E(r,s) = \sum_{i=1}^{N} \sum_{j=1}^{N} \left(k(r,s,w(\mathbf{p}_i,\mathbf{p}_j)) - \int_{U^K} \int_{U^K} k(r,s,w(\mathbf{x},\mathbf{y}))\, d\mathbf{x}\, d\mathbf{y} \right). \tag{31}$$

We now state the analogue of Lemma 5.1 (see §5.1).

Lemma 5.3. *Suppose that $0 < \delta < \frac{1}{2}$. Suppose further that the function $F(r)$ is continuous in $(0, \frac{1}{2}]$ and satisfies the integral equation*

$$\int_0^{\frac{1}{2}} F(r) k(\delta r, \delta r, w)\, dr = \underset{r+s \leqslant 1}{\int_0^1 \int_0^1} k(\delta r, \delta s, w) |r-s|^{-\alpha} |r+s|^{-\beta}\, dr\, ds \tag{32}$$

for all $w \geqslant 0$. Then

$$\int_0^{\frac{1}{2}} F(r) E(\delta r)\, dr = \underset{r+s \leqslant 1}{\int_0^1 \int_0^1} E(\delta r, \delta s) |r-s|^{-\alpha} |r+s|^{-\beta}\, dr\, ds, \tag{33}$$

provided that all the functions occurring in the integrals are summable.

Proof. The lemma follows immediately from (31). ■

Remark. If (32) is true for some $\delta > 0$, then (32) and (33) are true for every $\delta > 0$. This is seen as follows: For a positive integer m, denoted by $\mathscr{P}_m^*$ the set

$$\mathscr{P}_m^* = \{m^{-1}\mathbf{p}:\mathbf{p} = \mathbf{p}_i + \mathbf{l}, \mathbf{p}_i \in \mathscr{P}, \mathbf{l} \in \mathbb{Z}^K\}.$$

Define $E_m(r, s)$ with reference to $\mathscr{P}_m^*$. Note that for every r, $s > 0$,

$$E_m(r/m, s/m) = E(r, s), \tag{34}$$

since the set $\mathscr{P}^*$ is periodic and

$$\sum_{\mathbf{x} \in \mathscr{P}_m^* \cap C(r/m, m^{-1}\mathbf{c})} 1 - Nm^K \mu(\mathbf{C}(r/m, m^{-1}\mathbf{c})) = \sum_{\mathbf{p} \in \mathscr{P}^* \cap C(r, \mathbf{c})} 1 - N\mu(C(r, \mathbf{c})).$$

Suppose that (32) is true for some $\delta > 0$. Since for any $c > 0$,

$$k(cr, cs, cw) = c^K k(r, s, w),$$

it follows that (32) is true for every $\delta > 0$. Now, given $\delta > 0$, choose an integer m such that $0 < \delta/m < \frac{1}{2}$. Then by Lemma 5.3,

$$\int_0^{\frac{1}{2}} F(r) E_m(\delta r/m)\, \mathrm{d}r = \underset{r+s \leqslant 1}{\int_0^1 \int_0^1} E_m(\delta r/m, \delta s/m) |r - s|^{-\alpha} |r + s|^{-\beta}\, \mathrm{d}r\, \mathrm{d}s. \tag{35}$$

Now (34) and (35) yield (33).

In view of the remark above, we may restrict ourselves to the equation (32) with $\delta = 1$, i.e.

$$\int_0^{\frac{1}{2}} F(r) k(r, r, w)\, \mathrm{d}r = \underset{r+s \leqslant 1}{\int_0^1 \int_0^1} k(r, s, w) |r - s|^{-\alpha} |r + s|^{-\beta}\, \mathrm{d}r\, \mathrm{d}s. \tag{36}$$

We shall determine a solution $F(r)$ of this integral equation.

We recall that

$$v = \begin{cases} 1 & (K \text{ is odd}), \\ 0 & (K \text{ is even}). \end{cases}$$

Lemma 5.4. *There is a continuous solution $F(r)$, where $r \in (0, \frac{1}{2}]$, of (36) such that $F(r) = F_0(r) - F_*(r)$, where*

$$\begin{cases} F_0(r) \gg \ll r^{1-v-\alpha} & \text{as } r \to 0, \\ F_*(r) \gg \ll r^{1-K-\alpha} & \text{as } r \to 0. \end{cases} \tag{37}$$

We postpone the proof of Lemma 5.4 to §5.3. Instead, we now derive Theorem 13D from Lemmas 5.3 and 5.4.

Since

$$|F(r)| \ll r^{1-K-\alpha} \quad \text{as } r \to 0,$$

and since $\alpha < 1$, all the functions occurring in the integrals (32) and (33) are summable. Hence by Lemma 5.3 and the remark below it,

$$\int_0^{\frac{1}{2}} F(r)E(\delta r)\,dr = \underset{r+s\leqslant 1}{\int_0^1 \int_0^1} E(\delta r, \delta s)|r-s|^{-\alpha}|r+s|^{-\beta}\,dr\,ds = A.$$

We have

$$\int_0^{\frac{1}{2}} F_0(r)E(\delta r)\,dr - A = \int_0^{\frac{1}{2}} F_*(r)E(\delta r)\,dr. \tag{38}$$

Recall (27). We have

$$|A| \ll \int_0^{\delta} r^{-1}E(r)\,dr.$$

Furthermore,

$$\int_0^{\frac{1}{2}} F_0(r)E(\delta r)\,dr \ll \int_0^1 r^{-1}E(\delta r)\,dr = \int_0^{\delta} r^{-1}E(r)\,dr,$$

since $F_0(r) \ll r^{1-\nu-\alpha}$ as $r \to 0$ and since $\alpha < 1$. Therefore, the left-hand side of (38) has absolute value

$$\left|\int_0^{\frac{1}{2}} F_0(r)E(\delta r)\,dr - A\right| \ll \int_0^{\delta} r^{-1}E(r)\,dr. \tag{39}$$

We shall now show that the right-hand side of (38) is large. Note that (trivial error)

$$|D^{\text{tor}}[\mathscr{P}; C(r, \mathbf{c})]| \geqslant \| N\mu(r) \|,$$

where $\mu(r) = \mu(C(r, \mathbf{0}))$ and where $\|\cdot\|$ denotes the distance from the nearest integer. It follows that

$$E(r) \geqslant \| N\mu(r) \|^2.$$

In view of this and (37), the right-hand side of (38) is

$$\int_0^{\frac{1}{2}} F_*(r)E(\delta r)\,dr \gg \int_0^{c_{12}} \| N\mu(\delta r) \|^2 r^{1-K-\alpha}\,dr, \tag{40}$$

where $c_{12} = c_{12}(K, \alpha) > 0$. Let the interval J be defined by

$$J = \{r > 0 : \tfrac{1}{3} \leqslant N\mu(\delta r) \leqslant \tfrac{2}{3}\}.$$

Then every $r \in J$ satisfies the following relations:

$$\begin{cases} \| N\mu(\delta r)\| \geqslant \frac{1}{3}, \\ c_{13}(K)(N\delta^K)^{-1/K} \leqslant r \leqslant 2^{1/K} c_{13}(K)(N\delta^K)^{-1/K}, \\ r^{1-K-\alpha} \gg (N\delta^K)^{-(1-K-\alpha)/K}. \end{cases}$$

Observe that $J \subset [0, c_{12}]$ if N is sufficiently large, and that

$$\text{length}\,(J) \gg (N\delta^K)^{-1/K}.$$

Therefore

$$\int_0^{c_{12}} \| N\mu(\delta r)\|^2 r^{1-K-\alpha}\,\mathrm{d}r \geqslant \int_J \| N\mu(\delta r)\|^2 r^{1-K-\alpha}\,\mathrm{d}r$$

$$\gg \int_J r^{1-K-\alpha}\,\mathrm{d}r \gg (N\delta^K)^{1+\alpha/K-2/K}. \tag{41}$$

Combining (38)–(41), we get

$$\int_0^{\delta} r^{-1} E(r)\,\mathrm{d}r \gg (N\delta^K)^{1+\alpha/K-2/K}.$$

This is true for every α with $0 < \alpha < 1$. Putting $\alpha = 1 - \varepsilon$, we obtain

$$\int_0^{\delta} r^{-1} E(r)\,\mathrm{d}r \gg (N\delta^K)^{1-1/K-\varepsilon},$$

and so Theorem 13D follows.

It remains to prove Lemma 5.4.

5.3 Solving the integral equations

In this section, we complete the proof of Theorem 13D by proving Lemma 5.4.

It is sufficient to consider (36) for $0 \leqslant w < 1$, since both sides of (36) vanish when $w \geqslant 1$. We may further assume that $0 < w < 1$, since both sides of (36) are continuous functions of w. In fact, they satisfy a Lipschitz condition; we have

$$|k(r, s, w) - k(r, s, w')| \ll |w - w'| \min\{r^{K-1}, s^{K-1}\}, \tag{42}$$

since

$$\int_0^{\frac{1}{2}} |F(r)| r^{K-1}\,\mathrm{d}r \ll 1 \tag{43}$$

provided $|F(r)| \ll r^{1-K-\alpha}$, and since

$$\int_0^1 \int_0^1_{r+s\leqslant 1} \min\{r^{K-1}, s^{K-1}\}|r-s|^{-\alpha}|r+s|^{-\beta}\,dr\,ds \ll 1. \tag{44}$$

Suppose now that $|r-s|<w<|r+s|$. Let $C(r,\mathbf{c})$ and $C(s,\mathbf{d})$ be K-dimensional balls with $|\mathbf{c}-\mathbf{d}|=w$. The boundaries of $C(r,\mathbf{c})$ and $C(s,\mathbf{d})$ intersect in a $(K-2)$-dimensional sphere S^{K-2}. Denote the radius of S^{K-2} by ρ, the distance from the centre of S^{K-2} to $\mathbf{c}$ and $\mathbf{d}$ by a and b respectively (at this point, the reader may want to draw a sketch). We have

$$a+b=w,\ a^2+\rho^2=r^2 \quad \text{and} \quad b^2+\rho^2=s^2.$$

Eliminating a and b, we obtain

$$\rho^2=\frac{1}{4}\left(1-\frac{(r-s)^2}{w^2}\right)((r+s)^2-w^2). \tag{45}$$

We also have

$$k(r,s,w)=c_{14}\int_0^\rho z^{K-2}((r^2-z^2)^{\frac{1}{2}}+(s^2-z^2)^{\frac{1}{2}}-w)\,dz. \tag{46}$$

The constant c_{14} and the subsequent constants $c_{15}, c_{16},\ldots$ in this section are positive and depend only on K and $\alpha=1-\varepsilon$. Since the integrand in (46) vanishes for $z=\rho$, we obtain

$$\frac{\partial}{\partial w}k(r,s,w)=-c_{14}\int_0^\rho z^{K-2}\,dz=-c_{15}\rho^{K-1}.$$

All these hold for $|r-s|<w<|r+s|$. For $w\leqslant|r-s|$, $k(r,s,w)$ is independent of w, and for $w\geqslant r+s$, we have $k(r,s,w)=0$. It therefore follows that

$$\frac{\partial}{\partial w}k(r,s,w)=\begin{cases}-c_{15}\rho^{K-1} & (|r-s|<w<|r+s|),\\ 0 & \text{(otherwise)}.\end{cases} \tag{47}$$

We now differentiate both sides of (36) with respect to w. By (42), (43) and Lebesgue's theorem on dominated convergence, the left-hand side of (36) can be differentiated inside the integral. By (42), (44) and the Dominated convergence theorem, the same can be done for the right-hand side of (36). In view of (45) and (47), the derivative of the left-hand side of (36) is

$$-c_{15}\int_{\frac{1}{2}w}^{\frac{1}{2}} F(r)\left(r^2-\frac{w^2}{4}\right)^{\frac{1}{2}(K-1)}dr,$$

and the derivative of the right-hand side of (36) is

$$-c_{15}\int_0^1\int_0^1_{|r-s|\leqslant w\leqslant|r+s|\leqslant 1}\left(\frac{1}{4}\left(1-\frac{(r-s)^2}{w^2}\right)((r+s)^2-w^2)\right)^{\frac{1}{2}(K-1)}$$

$$\times|r-s|^{-\alpha}|r+s|^{-\beta}\,\mathrm{d}r\,\mathrm{d}s. \tag{48}$$

Putting $x=r+s$ and $y=|r-s|$, (48) becomes

$$-c_{16}\left(\int_w^1(x^2-w^2)^{\frac{1}{2}(K-1)}x^{-\beta}\,\mathrm{d}x\right)\left(\int_0^w\left(1-\frac{y^2}{w^2}\right)^{\frac{1}{2}(K-1)}y^{-\alpha}\,\mathrm{d}y\right)$$

$$=-c_{16}c_{17}w^{1-\alpha}\int_w^1 x^{-\beta}(x^2-w^2)^{\frac{1}{2}(K-1)}\,\mathrm{d}x,$$

since, substituting $y=tw$, we get

$$\int_0^w\left(1-\frac{y^2}{w^2}\right)^{\frac{1}{2}(K-1)}y^{-\alpha}\,\mathrm{d}y=w^{1-\alpha}\int_0^1(1-t^2)^{\frac{1}{2}(K-1)}t^{-\alpha}\,\mathrm{d}t=c_{17}w^{1-\alpha}.$$

Hence after differentiation, the integral equation (36) becomes

$$\int_{\frac{1}{2}w}^{\frac{1}{2}}F(r)\left(r^2-\frac{w^2}{4}\right)^{\frac{1}{2}(K-1)}\mathrm{d}r=c_{18}w^{1-\alpha}\int_w^1 x^{-\beta}(x^2-w^2)^{\frac{1}{2}(K-1)}\,\mathrm{d}x$$

$$(0<w<1). \tag{49}$$

Any solution of (49) is also a solution of (36), since both sides of (36) are continuous and are zero for $w=1$. Putting $w^2=t$ in (49), we get

$$\int_{\frac{1}{2}\sqrt{t}}^{\frac{1}{2}}F(r)\left(r^2-\frac{t}{4}\right)^{\frac{1}{2}(K-1)}\mathrm{d}r=c_{18}t^{\frac{1}{2}(1-\alpha)}\int_{\sqrt{t}}^1 x^{-\beta}(x^2-t)^{\frac{1}{2}(K-1)}\,\mathrm{d}x\quad(0<t<1). \tag{50}$$

We now differentiate (50)

$$m=[\tfrac{1}{2}K]=\tfrac{1}{2}(K-v)$$

times with respect to t. The left-hand side of (50) becomes

$$c_{19}(-1)^m\int_{\frac{1}{2}\sqrt{t}}^{\frac{1}{2}}F(r)\left(r^2-\frac{t}{4}\right)^{\frac{1}{2}(v-1)}\mathrm{d}r.$$

On the other hand,

$$\left(\frac{\mathrm{d}^j}{\mathrm{d}t^j}t^{\frac{1}{2}(1-\alpha)}\right)\left(\frac{\mathrm{d}^{m-j}}{\mathrm{d}t^{m-j}}\int_{\sqrt{t}}^1 x^{-\beta}(x^2-t)^{\frac{1}{2}(K-1)}\,\mathrm{d}x\right)$$

is equal to

$$(-1)^m c_{20}^{(0)} t^{\frac{1}{2}(1-\alpha)} \int_{\sqrt{t}}^{1} x^{-\beta}(x^2-t)^{\frac{1}{2}(\nu-1)}\,\mathrm{d}x = (-1)^m c_{20}^{(0)} \int_{1}^{1/\sqrt{t}} y^{-\beta}(y^2-1)^{\frac{1}{2}(\nu-1)}\,\mathrm{d}y$$

if $j=0$, and equal to

$$(-1)^{m-1} c_{20}^{(j)} t^{\frac{1}{2}(1-\alpha-2j)} \int_{\sqrt{t}}^{1} x^{-\beta}(x^2-t)^{\frac{1}{2}(K-1-2m+2j)}\,\mathrm{d}x$$

$$= (-1)^{m-1} c_{20}^{(j)} \int_{1}^{1/\sqrt{t}} y^{-\beta}(y^2-1)^{\frac{1}{2}(\nu-1+2j)}\,\mathrm{d}y$$

if $1 \leqslant j \leqslant m$. Here all the constants $c_{20}^{(j)}$ $(0 \leqslant j \leqslant m)$ are positive. Hence (50) becomes, on differentiation,

$$\int_{\frac{1}{2}\sqrt{t}}^{\frac{1}{2}} F(r)\left(r^2-\frac{t}{4}\right)^{\frac{1}{2}(\nu-1)} dr = c_{21}^{(0)} \int_{1}^{1/\sqrt{t}} y^{-\beta}(y^2-1)^{\frac{1}{2}(\nu-1)}\,\mathrm{d}y$$

$$- \sum_{j=1}^{m} c_{21}^{(j)} \int_{1}^{1/\sqrt{t}} y^{-\beta}(y^2-1)^{\frac{1}{2}(\nu-1+2j)}\,\mathrm{d}y \quad (0<t<1). \tag{51}$$

Now both sides of (50) and their first $(m-1)$ derivatives are continuous in t and vanish for $t=1$. It follows therefore that every solution $F(r)$ of (51) is also a solution of (50). We now write $\sqrt{t}=w$ and rewrite (51) as

$$\int_{\frac{1}{2}w}^{\frac{1}{2}} F(r)\left(r^2-\frac{w^2}{4}\right)^{\frac{1}{2}(\nu-1)} dr$$

$$= c_{21}^{(0)} \int_{1}^{1/w} y^{-\beta}(y^2-1)^{\frac{1}{2}(\nu-1)}\,\mathrm{d}y - \sum_{j=1}^{m} c_{21}^{(j)} \int_{0}^{1/w} y^{-\beta}(y^2-1)^{\frac{1}{2}(\nu-1+2j)}\,\mathrm{d}y$$

$$= l_0(w) - \sum_{j=1}^{m} l_j(w), \tag{52}$$

say. This equation holds for $0<w<1$.

For every j with $0 \leqslant j \leqslant m$, we shall find a solution $F_j(r)$ of the integral equation

$$\int_{\frac{1}{2}w}^{\frac{1}{2}} F_j(r)\left(r^2-\frac{w^2}{4}\right)^{\frac{1}{2}(\nu-1)} dr = l_j(w) \quad (0<w<1). \tag{53}$$

Then

$$F(r) = F_0(r) - \sum_{j=1}^{m} F_j(r) = F_0(r) - F_*(r)$$

will be a solution of (52). We distinguish two cases.

Case 1: K is odd, i.e. $\nu=1$. We differentiate (53) with respect to w. In view of (52), we obtain

$$l'_j(w) = -c_{21}^{(j)} w^{\beta - \nu - 2j - 1}(1 - w^2)^{\frac{1}{2}(\nu - 1 + 2j)}, \tag{54}$$

and so

$$F_j(\tfrac{1}{2}w) = 2c_{21}^{(j)} w^{-2+\beta-2j}(1 - w^2)^j,$$

whence

$$F_j(r) = c_{22}^{(j)} r^{-\alpha-2j}(1 - 4r^2)^j \quad (0 < r \leqslant \tfrac{1}{2}).$$

It is clear that $F_j(r)$ is in fact a solution of the integral equation (53). Observe that for $0 \leqslant j \leqslant m$, the function $F_j(r)$ is continuous in $(0, \frac{1}{2}]$. As $r \to 0$,

$$F_0(r) \gg \ll r^{-\alpha}$$

and

$$F_*(r) = \sum_{j=1}^{m} F_j(r) \gg \ll \sum_{j=1}^{\frac{1}{2}(K-1)} r^{-\alpha-2j} \gg \ll r^{1-K-\alpha}.$$

Case 2: K is even, i.e. $\nu = 0$. Putting $\nu = 0$ in (53), we have

$$\int_{\frac{1}{2}w}^{\frac{1}{2}} F_j(r)\left(r^2 - \frac{w^2}{4}\right)^{-\frac{1}{2}} \mathrm{d}r = l_j(w) \quad (0 < w < 1). \tag{55}$$

This is an Abel integral equation. We check that

$$F_j(r) = -(4/\pi)\int_{2r}^{1} r(t^2 - 4r^2)^{-\frac{1}{2}} l'_j(t)\,\mathrm{d}t \quad (0 < r < \tfrac{1}{2}) \tag{56}$$

is a solution. Indeed, substituting (56) into the left-hand side of (55), we get

$$-(4/\pi)\int_{\frac{1}{2}w}^{\frac{1}{2}}\left(r^2 - \frac{w^2}{4}\right)^{-\frac{1}{2}}\left(\int_{2r}^{1} r(t^2 - 4r^2)^{-\frac{1}{2}} l'_j(t)\,\mathrm{d}t\right)\mathrm{d}r$$

$$= -(4/\pi)\int_{w}^{1} l'_j(t)\left(\int_{\frac{1}{2}w}^{\frac{1}{2}t} r\left(r^2 - \frac{w^2}{4}\right)^{-\frac{1}{2}}(t^2 - 4r^2)^{-\frac{1}{2}}\mathrm{d}r\right)\mathrm{d}t.$$

Substituting $u = (r^2 - w^2/4)^{\frac{1}{2}}$, the inner integral becomes

$$\int_0^{\frac{1}{2}(t^2 - w^2)^{\frac{1}{2}}} (t^2 - 4u^2 - w^2)^{-\frac{1}{2}}\,\mathrm{d}u = \tfrac{1}{2}\int_0^1 (1 - z^2)^{-\frac{1}{2}}\,\mathrm{d}z = \pi/4.$$

So the left-hand side of (55) is equal to

$$-\int_w^1 l'_j(t)\,\mathrm{d}t = l_j(w) - l_j(1) = l_j(w)$$

as required. Note that for $0 \leqslant j \leqslant m$, $F_j(r)$ is continuous in $(0, \frac{1}{2})$, and can be extended continuously to $(0, \frac{1}{2}]$. Furthermore, $F_j(r) \geqslant 0$, since $l'_j(t) < 0$

in (0, 1] (see (54)). Using (56) and (54), we have, for $0 \leqslant j \leqslant m$, that

$$F_j(r) = c_{23}^{(j)} \int_{2r}^{1} r(t^2 - 4r^2)^{-\frac{1}{2}} t^{-\alpha - 2j}(1 - t^2)^{\frac{1}{2}(2j-1)}\,\mathrm{d}t = c_{23}^{(j)} \int_{2r}^{1} rH_j(r, t)\,\mathrm{d}t,$$

say. Suppose that r is small, say $r \in (0, \frac{1}{4})$. Write

$$F_j(r) = c_{23}^{(j)} \int_{2r}^{\frac{1}{2}} rH_j(r, t)\,\mathrm{d}t + c_{23}^{(j)} \int_{\frac{1}{2}}^{1} rH_j(r, t)\,\mathrm{d}t.$$

The second integral is $\ll r$. The first integral is

$$\gg\ll r \int_{2r}^{\frac{1}{2}} (t^2 - 4r^2)^{-\frac{1}{2}} t^{-\alpha - 2j}\,\mathrm{d}t$$

$$\gg\ll r^{1-\alpha-2j} \int_{1}^{1/(4r)} \frac{\mathrm{d}u}{u^{\alpha+2j}(u^2 - 1)^{\frac{1}{2}}} \gg\ll r^{1-\alpha-2j}.$$

It follows that as $r \to 0$,

$$F_j(r) \gg\ll r^{1-\alpha-2j}.$$

This holds for $0 \leqslant j \leqslant m$, and so the proof of Lemma, 5.4 is complete.

6
A Fourier transform approach

Consider, instead of the unit torus, the geometrically more natural underlying set: the unit cube U^K. The problem is to find a ball contained in U^K with large discrepancy. The pioneering work in this direction is again due to Schmidt (1969*b*). He showed that there exists a ball $C(r, \mathbf{c})$ contained in U^K with discrepancy

$$|D[\mathscr{P}; C(r,\mathbf{c})]| = \left| \sum_{\mathbf{x} \in \mathscr{P} \cap C(r,\mathbf{c})} 1 - N\mu(C(r,\mathbf{c})) \right| > c_1(K,\varepsilon) N^{(K-1)/2K(K+2)-\varepsilon}. \quad (1)$$

Here $\varepsilon > 0$ is arbitrarily small but fixed, and $\mathscr{P}$ is a distribution of N points in U^K.

We begin with an essential improvement of Schmidt's bound (1) (observe that in (1) the exponent of N tends to 0 as K tends to infinity).

Theorem 16A (Beck (TA*a*)). *Let $\mathscr{P}$ be a distribution of N points in U^K. Then there exists a ball $C(r, \mathbf{c})$ contained in U^K such that*

$$|D[\mathscr{P}; C(r,\mathbf{c})]| > c_2(K,\varepsilon) N^{\frac{1}{2} - 1/2K - \varepsilon}. \quad (2)$$

Comparing Theorem 14 and Theorem 16A, we see that the estimate (2) is essentially best possible (in (2) the exponent of N tends to $\frac{1}{2}$ as K tends to infinity).

Here and in Chapter 7, we attempt to explain the fascinating fact that balls have much greater discrepancy than boxes with sides parallel to the coordinate axes (the problem was raised in §16 of Schmidt (1977*b*)).

To avoid the technical difficulties caused by the requirement 'contained in U^K', we shall study a non-compact and renormalized model.

Let $\mathscr{Q} = \{\mathbf{q}_1, \mathbf{q}_2, \mathbf{q}_3, \ldots\}$ be a completely arbitrary infinite discrete set of points in $\mathbb{R}^K$, where $K \geqslant 2$. Given a compact set $A \subset \mathbb{R}^K$, write

$$\mathscr{D}[\mathscr{Q}; A] = \sum_{\mathbf{q}_i \in A} 1 - \mu(A),$$

where μ denotes the K-dimensional Lebesgue measure (i.e. the usual volume). Observe that here the normalization is different from that in the previous results (compare the definitions of discrepancy here and in (1)).

For any real number $\lambda \in (0,1]$, any vector $\mathbf{v} \in \mathbb{R}^K$ and any proper orthogonal transformation τ in $\mathbb{R}^K$, let

$$A(\lambda, \mathbf{v}, \tau) = \{\tau(\lambda \mathbf{x} + \mathbf{v}) : \mathbf{x} \in A\}.$$

Clearly $A(\lambda, \mathbf{v}, \tau)$ and A are similar to each other. Let

$$\Omega[\mathscr{Q}; A] = \sup_{\lambda, \mathbf{v}, \tau} |\mathscr{D}[\mathscr{Q}; A(\lambda, \mathbf{v}, \tau)]|,$$

where the supremum is taken over all contractions λ, translations $\mathbf{v}$ and rotations τ; and let the rotation-discrepancy $\Omega[A]$ of the set A be defined by

$$\Omega[A] = \inf_{\mathscr{Q}} \Omega[\mathscr{Q}; A],$$

where the infimum is extended over all infinite discrete sets $\mathscr{Q} \subset \mathbb{R}^K$.

Assume now that A is convex. Let $\sigma(\partial A)$ denote the surface area of the boundary ∂A of A. Moreover, let $d(A)$ denote the diameter of A and let $r(A)$ denote the radius of the largest inscribed ball of A.

We shall show that for convex bodies the rotation-discrepancy is always large and behaves like the square-root of the surface area of the boundary.

Theorem 17A (Beck (TA*a*)). *Let $\mathscr{Q} \subset \mathbb{R}^K$ be an arbitrary infinite discrete set and $A \subset \mathbb{R}^K$ be a compact and convex body with $r(A) \geqslant 1$. Then*

$$\Omega[\mathscr{Q}; A] > c_3(K)(\sigma(\partial A))^{\frac{1}{2}},$$

i.e. there exist $\lambda_0 \in (0,1]$, $\mathbf{v}_0 \in \mathbb{R}^K$ and τ_0 such that

$$|\mathscr{D}[\mathscr{Q}; A(\lambda_0, \mathbf{v}_0, \tau_0)]| > c_3(K)(\sigma(\partial A))^{\frac{1}{2}}.$$

Note that in the proof of Theorem 17A, we actually estimate from below the L^2-norm of the discrepancy $\mathscr{D}[\mathscr{Q}; A(\lambda, \mathbf{v}, \tau)]$. More precisely, we prove that

$$\liminf_{M \to \infty} (2M)^{-K} \int_{[-M,M]^2} \int_0^1 \int_T |\mathscr{D}[\mathscr{Q}; A(\lambda, \mathbf{v}, \tau)]|^2 \, d\tau \, d\lambda \, d\mathbf{v} \gg \sigma(\partial A),$$

where T is the group of proper orthogonal transformations in $\mathbb{R}^K$, $d\tau$ is the volume element of the invariant measure on T, normalized such that $\int_T d\tau = 1$, and the implicit constant in Vinogradov's notation $\gg$ depends on the dimension K only.

On the other hand, the same random construction as in the proof of Theorem 18A (see below) will establish that this stronger L^2-norm version of Theorem 17A is sharp apart from constant factors (see the argument in the proof of Theorem 24A in Chapter 8). In other words, there exists an infinite discrete set $\mathcal{Q}_0 = \mathcal{Q}_0(A) \subset \mathbb{R}^K$ such that

$$\limsup_{M\to\infty} (2M)^{-K} \int_{[-M,M]^2} \int_0^1 \int_T |\mathcal{D}[\mathcal{Q}_0; A(\lambda, \mathbf{v}, \tau)]|^2 \, d\tau \, d\lambda \, d\mathbf{v} \ll \sigma(\partial A).$$

Our next result shows that Theorem 17A is nearly best possible.

Theorem 18A (Beck). *Let $A \subset \mathbb{R}^K$ be a compact and convex body with $r(A) \geqslant 1$. Then there exists an infinite discrete set $\mathcal{Q}_0 = \mathcal{Q}_0(A) \subset \mathbb{R}^K$ such that*

$$\Omega[\mathcal{Q}_0; A] < c_4(K)(\sigma(\partial A))^{\frac{1}{2}}(\log \sigma(\partial A))^{\frac{1}{2}}.$$

From Theorem 17A, one can immediately obtain results concerning the unit torus. Let $\mathcal{P}$ be a distribution of N points in the unit cube U_0^K. Extend $\mathcal{P}$ periodically over the whole of $\mathbb{R}^K$ (modulo U_0^K). In other words, let

$$\mathcal{P}^* = \{\mathbf{x} + \mathbf{l} : \mathbf{x} \in \mathcal{P}, \mathbf{l} \in \mathbb{Z}^K\}.$$

Given a compact set A, write $Z[\mathcal{P}^*; A]$ for the number of points of $\mathcal{P}^*$ in A, and put

$$D^{\text{tor}}[\mathcal{P}; A] = Z[\mathcal{P}^*; A] - N\mu(A).$$

Finally, let

$$\Omega^{\text{tor}}[\mathcal{P}; A] = \sup_{0<\lambda\leqslant 1, \mathbf{v}, \tau} |D^{\text{tor}}[\mathcal{P}; A(\lambda, \mathbf{v}, \tau)]|$$

and

$$\Omega_N^{\text{tor}}[A] = \inf \Omega^{\text{tor}}[\mathcal{P}; A],$$

where the infimum is taken over all N-element sets $\mathcal{P} \subset U_0^K$.

If we rescale the periodic set $\mathcal{P}^*$ in the ratio $N^{1/K}:1$ and apply Theorem 17A, then we conclude that

Corollary 17B (Beck). *Let $\mathcal{P}$ be an arbitrary distribution of N points in U_0^K, and let A be a compact and convex body in $\mathbb{R}^K$. Suppose that $r(A) \geqslant N^{-1/K}$. Then*

$$\Omega^{\text{tor}}[\mathcal{P}; A] > c_5(K) N^{\frac{1}{2} - 1/2K} (\sigma(\partial A))^{\frac{1}{2}}.$$

In the case $K = 2$, a similar result was independently proved by Montgomery (1985).

In the opposite direction we have

Theorem 18B (Beck). *Let A be a compact and convex body in $\mathbb{R}^K$. Then there exists an infinite sequence $\mathbf{s}_1, \mathbf{s}_2, \mathbf{s}_3, \ldots$ of points in U_0^K such that for* *every integer $N \geqslant 2$ and for every real number $\lambda \in (0, 1]$ satisfying $N\lambda^K \geqslant 1$,*

$$\sup_{\mathbf{v},\tau} |D^{\mathrm{tor}}[\{\mathbf{s}_1, \mathbf{s}_2, \ldots, \mathbf{s}_N\}; A(\lambda, \mathbf{v}, \tau)]|$$
$$< c_6(A)(N\lambda^K)^{\frac{1}{2}-1/2K}(\log N)^{\frac{1}{2}} \log(1 + 1/\lambda),$$

where the positive constant $c_6(A)$ depends only on A.

The following two results can easily be deduced from Theorem 18B.

Corollary 18C (Beck). *Let A be a compact convex body in $\mathbb{R}^K$. There exists an infinite sequence $\mathbf{s}_1, \mathbf{s}_2, \mathbf{s}_3, \ldots$ of points in U_0^K such that for every integer $N \geqslant 2$, the set $\mathscr{P}_N = \{\mathbf{s}_1, \mathbf{s}_2, \ldots, \mathbf{s}_N\}$ satisfies*

$$\Omega^{\mathrm{tor}}[\mathscr{P}_N; A] < c_6(A)N^{\frac{1}{2}-1/2K}(\log N)^{\frac{1}{2}}. \tag{3}$$

Corollary 18D (Beck). *Let A be a compact convex body in $\mathbb{R}^K$. There exists an infinite sequence $\mathbf{s}_1, \mathbf{s}_2, \mathbf{s}_3, \ldots$ of points in U_0^K such that for every integer $N \geqslant 2$ and for every real number $\lambda \in (0, 1]$ satisfying $N\lambda^K \geqslant 1$, the set $\mathscr{P}_N = \{\mathbf{s}_1, \mathbf{s}_2, \ldots, \mathbf{s}_N\}$ satisfies*

$$\sup_{\mathbf{v},\tau} |D^{\mathrm{tor}}[\mathscr{P}_N; A(\lambda, \mathbf{v}, \tau)]| < c_6(A)(N\lambda^K)^{\frac{1}{2}-1/2K}(\log N)^{\frac{3}{2}}.$$

Remark. Comparing Theorem 15 and Corollary 18C, we see that there is no universal sequence $\mathbf{s}_1, \mathbf{s}_2, \mathbf{s}_3, \ldots$ which will give upper bound (3) simultaneously for all convex bodies $A \subset \mathbb{R}^K$.

We now return to Corollary 17B. In the particular cases where A is a cube or a ball, we get respectively Corollary 17C and Corollary 17D.

Corollary 17C (Beck). *Let $\mathscr{P}$ be an arbitrary distribution of N points in U_0^K. Then there exists a cube A in arbitrary position with diameter $\leqslant \delta$ and with discrepancy*

$$|D^{\mathrm{tor}}[\mathscr{P}; A]| > c_7(K)(N\delta^K)^{\frac{1}{2}-1/2K}.$$

We should mention here the pioneering result of Schmidt. For boxes in arbitrary position and $K = 2, 3$ he proved the slightly weaker lower bound $(N\delta^K)^{\frac{1}{2}-1/2K-\varepsilon}$ (see Schmidt (1969*b*)). For arbitrary K it was hopeless to handle the very difficult integral equations that arise.

Corollary 17D (Beck). *Let $\mathscr{P}$ be an arbitrary distribution of N points in U_0^K. Then there exists a ball $C(r, \mathbf{c})$ with $r \leqslant \delta$ and*

$$|D^{\mathrm{tor}}[\mathscr{P}; C(r, \mathbf{c})]| > c_8(K)(N\delta^K)^{\frac{1}{2}-1/2K}.$$

We recall the slightly weaker Theorem 13A in Chapter 5: There exists a ball $C(r, \mathbf{c})$ with $r \leqslant \delta$ and

$$|D^{\mathrm{tor}}[\mathscr{P}; C(r, \mathbf{c})]| > c_9(K, \varepsilon)(N\delta^K)^{\frac{1}{2} - 1/2K - \varepsilon}.$$

We comment here that Corollaries 17C and 17D were proved independently by Montgomery (for a direct proof of the case $K = 2$ of Corollary 17D, see Montgomery (1985)).

We note without proof that using the 'truncation' technique in the proof of Theorem 16A, it is not hard to show the following 'contained in U_0^K' version of Corollary 17B: Let $\mathscr{P}$ be an arbitrary distribution of N points in U^K. Let $A \subset \mathbb{R}^K$ be a convex body of diameter less than 1. Suppose further that $r(A) \geqslant N^{-1/K}$. Then there exist an orthogonal transformation τ_0, a real number $\lambda_0 \in (0, 1]$ and a vector $\mathbf{v}_0$ such that $A(\lambda_0, \mathbf{v}_0, \tau_0)$ is contained in U^K and has discrepancy

$$|D[\mathscr{P}; A(\lambda_0, \mathbf{v}_0, \tau_0)]| = \left| \sum_{\mathbf{x} \in \mathscr{P} \cap A(\lambda_0, \mathbf{v}_0, \tau_0)} 1 - N\mu(A(\lambda_0, \mathbf{v}_0, \tau_0)) \right|$$
$$> c_{10}(K, \varepsilon) N^{\frac{1}{2} - 1/2K - \varepsilon} (\sigma(\partial A))^{\frac{1}{2}}.$$

We shall discuss Theorem 16A in §6.2 and Theorem 17A in §6.3. We postpone the proofs of Theorems 18A and 18B to Chapter 8.

Since the proofs of Theorems 16A and 17A are rather complicated, we shall illustrate in §6.1 the basic ideas of the Fourier transform method on two simpler results (see Theorems 17E and 19A below).

As far as we know, the first application of this method is in Roth (1964) (see also Theorem 25A in Chapter 9). Later the same basic idea was independently employed by Baker, Beck and Montgomery.

The use of Fourier analysis in the opposite direction (i.e. to show the uniformity of sequences) is a classical idea and goes back to Weyl. Let $\mathbf{x}_1, \mathbf{x}_2, \mathbf{x}_3, \ldots$ be an infinite sequence of points in the K-dimensional unit torus. By Weyl's criterion we know that the points $\mathbf{x}_n$ ($n = 1, 2, 3, \ldots$) are approximately uniformly distributed if the exponential sums

$$S(\mathbf{l}; N) = \sum_{n=1}^{N} \mathrm{e}^{2\pi i \mathbf{l} \cdot \mathbf{x}_n}$$

are 'small' when $\mathbf{l} \in \mathbb{Z}^K \setminus \{\mathbf{0}\}$ is a non-zero lattice point. This can be made quantitative, so that we can establish an upper bound for the discrepancy of aligned boxes in terms of the $S(\mathbf{l}; N)$. The well-known 'Erdös–Turán inequality' is an example of this type (see, for example, Kuipers and Niederreiter (1974)).

We shall first demonstrate the Fourier transform method on the following two theorems.

Theorem 17E (Beck (TA*a*)). *Let $\mathscr{P}$ be a distribution of N points in U^K. Then there exists a cube A in arbitrary position and with $d(A) \leqslant 1$ such that*

$$|D[\mathscr{P}; A \cap U^K]| = \left| \sum_{\mathbf{x} \in \mathscr{P} \cap A} 1 - N\mu(A \cap U^K) \right| > c_{11}(K) N^{\frac{1}{2} - 1/2K}.$$

Observe that Theorem 17E is an easy consequence of Corollary 17C. However, we give a simple direct proof in §6.1.

Theorem 19A (Beck). *Let $\mathscr{P}$ be a distribution of N points in U^K. Then there exists a cube A with sides parallel to the coordinate axes such that A is contained in U^K and*

$$|D[\mathscr{P}; A]| = \left| \sum_{\mathbf{x} \in \mathscr{P} \cap A} 1 - N\mu(A) \right| > c_{12}(K)(\log N)^{\frac{1}{2}(K-1)}.$$

In other words, we have the same lower bound as in Roth's much earlier result for boxes $B(\mathbf{x})$ (see Theorem 3A in Chapter 2). Following Roth, we also estimate the L^2-norm of the discrepancy.

A very similar result concerning squares in the unit torus was proved independently by Montgomery (1985).

The family of cubes with sides parallel to the coordinate axes is invariant under contraction and translation, but certainly not invariant under rotation. For further results of this type, see Chapter 7.

6.1 Demonstration of the method on two examples

The proofs of both Theorem 17E and Theorem 19A are based on a machinery to 'blow up the trivial error'.

We first of all prove Theorem 17E. Let $\mathscr{P} = \{\mathbf{p}_1, \ldots, \mathbf{p}_N\}$. We introduce two measures. For any $H \subset \mathbb{R}^K$, let

$$Z_0(H) = \sum_{\mathbf{p}_j \in H} 1;$$

in other words, Z_0 denotes the counting measure generated by the distribution $\mathscr{P}$. Also, for any Lebesgue measurable $H \subset \mathbb{R}^K$, let

$$\mu_0(H) = \mu(H \cap U^K);$$

i.e. μ_0 denotes the restriction of the usual volume to the unit cube.

Given any proper orthogonal transformation (rotation in short) τ and any real number $r > 0$, let $\chi_{r,\tau}$ denote the characteristic function of the rotated cube

$$\tau[-r, r]^K = \{\tau\mathbf{x} : \mathbf{x} \in [-r, r]^K\}.$$

Consider now the function

$$F_{r,\tau} = \chi_{r,\tau} * (\mathrm{d}Z_0 - N\mathrm{d}\mu_0), \tag{4}$$

where $*$ denotes the convolution operation. More explicitly,

$$\begin{aligned} F_{r,\tau}(\mathbf{x}) &= \int_{\mathbb{R}^K} \chi_{r,\tau}(\mathbf{x}-\mathbf{y})(\mathrm{d}Z_0(\mathbf{y}) - N\mathrm{d}\mu_0(\mathbf{y})) \\ &= \sum_{\mathbf{p}_j \in \tau[-r,r]^K + \mathbf{x}} 1 - N\mu((\tau[-r,r]^K + \mathbf{x}) \cap U^K). \end{aligned} \tag{5}$$

In other words, $F_{r,\tau}(\mathbf{x})$ represents the discrepancy of the set $(\tau[-r,r]^K + \mathbf{x}) \cap U^K$. Since the function $F_{r,\tau}$ has the form of a convolution, it is natural to employ the theory of Fourier transformation.

We recall some well-known facts. If $f \in L^2(\mathbb{R}^K)$, then

$$\hat{f}(\mathbf{t}) = \frac{1}{(2\pi)^{\frac{1}{2}K}} \int_{\mathbb{R}^K} \mathrm{e}^{-\mathrm{i}\mathbf{x}\cdot\mathbf{t}} f(\mathbf{x})\,\mathrm{d}\mathbf{x}$$

denotes the Fourier transform of f. Here $\mathrm{i} = \sqrt{-1}$ and $\mathbf{x}\cdot\mathbf{t}$ is the standard inner product. It is well-known that

$$\widehat{f * g} = \hat{f}\hat{g} \tag{6}$$

and (Parseval–Plancherel identity)

$$\int_{\mathbb{R}^K} |f(\mathbf{x})|^2\,\mathrm{d}\mathbf{x} = \int_{\mathbb{R}^K} |\hat{f}(\mathbf{t})|^2\,\mathrm{d}\mathbf{t}. \tag{7}$$

Let T be the group of proper orthogonal transformations in $\mathbb{R}^K$ and let $\mathrm{d}\tau$ be the volume element of the invariant measure on T, normalized such that $\int_T \mathrm{d}\tau = 1$.

Let $q > 0$ be a real parameter. We write

$$\Omega_0(q) = \frac{1}{q}\int_q^{2q}\int_T\int_{\mathbb{R}^K} |F_{r,\tau}(\mathbf{x})|^2\,\mathrm{d}\mathbf{x}\,\mathrm{d}\tau\,\mathrm{d}r. \tag{8}$$

By (4), (6) and (7), we have

$$\Omega_0(q) = \int_{\mathbb{R}^K}\left(\frac{1}{q}\int_q^{2q}\int_T |\hat{\chi}_{r,\tau}(\mathbf{t})|^2\,\mathrm{d}\tau\,\mathrm{d}r\right)|(\widehat{\mathrm{d}Z_0 - N\mathrm{d}\mu_0})(\mathbf{t})|^2\,\mathrm{d}\mathbf{t}. \tag{9}$$

For the sake of brevity, let

$$\omega_q(\mathbf{t}) = \frac{1}{q}\int_q^{2q}\int_T |\hat{\chi}_{r,\tau}(\mathbf{t})|^2\,\mathrm{d}\tau\,\mathrm{d}r \tag{10}$$

and

$$\phi(\mathbf{t}) = (\widehat{\mathrm{d}Z_0 - N\mathrm{d}\mu_0})(\mathbf{t}) = \frac{1}{(2\pi)^{\frac{1}{2}K}} \int_{\mathbf{R}^K} \mathrm{e}^{-\mathrm{i}\mathbf{x}\cdot\mathbf{t}} (\mathrm{d}Z_0 - N\mathrm{d}\mu_0)(\mathbf{x}).$$

We can therefore rewrite (9) in the form

$$\Omega_0(q) = \int_{\mathbf{R}^K} \omega_q(\mathbf{t}) |\phi(\mathbf{t})|^2 \,\mathrm{d}\mathbf{t}. \tag{11}$$

First of all, we note the following 'trivial error'.

Lemma 6.1. *Suppose that* $B \subset U^K$ *satisfies* $0 < \delta/N < \mu(B) < (1-\delta)/N$. *Then*

$$\int_{\mathbf{R}^K} \left| \sum_{\mathbf{p}_j \in B+\mathbf{x}} 1 - N\mu((B+\mathbf{x}) \cap U^K) \right|^2 \mathrm{d}\mathbf{x} > \delta^3.$$

Proof. We clearly have

$$\left| \sum_{\mathbf{p}_j \in B+\mathbf{x}} 1 - N\mu((B+\mathbf{x}) \cap U^K) \right| > \delta \sum_{\mathbf{p}_j \in B+\mathbf{x}} 1,$$

and so

$$\int_{\mathbf{R}^K} \left| \sum_{\mathbf{p}_j \in B+\mathbf{x}} 1 - N\mu((B+\mathbf{x}) \cap U^K) \right|^2 \mathrm{d}\mathbf{x} > \delta^2 \int_{\mathbf{R}^K} \left(\sum_{\mathbf{p}_j \in B+\mathbf{x}} 1 \right)^2 \mathrm{d}\mathbf{x}$$

$$\geqslant \delta^2 \int_{\mathbf{R}^K} \left(\sum_{\mathbf{p}_j \in B+\mathbf{x}} 1 \right) \mathrm{d}\mathbf{x} = \delta^2 \sum_{j=1}^{N} \mu(\mathbf{p}_j - B) = \delta^2 N\mu(B) > \delta^3. \qquad \blacksquare$$

The next lemma contains the machinery to 'blow up' the estimate obtained in Lemma 6.1.

Lemma 6.2. *If* $0 < q < p$, *then*

$$\frac{\omega_p(\mathbf{t})}{\omega_q(\mathbf{t})} \gg \left(\frac{p}{q}\right)^{K-1} \tag{12}$$

uniformly for all $\mathbf{t} \in \mathbb{R}^K$.

Here and in what follows, the implicit constants in Vinogradov's notation $\gg$ are positive and depend only on the dimension K.

Before we prove Lemma 6.2, we shall first deduce Theorem 17E from Lemmas 6.1 and 6.2.

Let $q = (5N^{1/K})^{-1}$. Combining Lemma 6.1, (5) and (8), we have the trivial estimate

$$\Omega_0(q) \gg 1. \tag{13}$$

Next, let $p=(4K^{\frac{1}{2}})^{-1}$. From Lemma 6.2, (11) and (13), we have

$$\Omega_0(p)=\int_{\mathbf{R}^K}\omega_p(\mathbf{t})|\phi(\mathbf{t})|^2\,d\mathbf{t}\gg(N^{1/K})^{K-1}\int_{\mathbf{R}^K}\omega_q(\mathbf{t})|\phi(\mathbf{t})|^2\,d\mathbf{t}$$

$$=N^{(K-1)/K}\Omega_0(q)\gg N^{(K-1)/K}. \tag{14}$$

In view of (5),

$$F_{r,\tau}(\mathbf{x})=0 \quad \text{if} \quad x\notin[-1-K^{\frac{1}{2}}r,1+K^{\frac{1}{2}}r]^K. \tag{15}$$

It follows from (8), (14) and (15) that there exists a cube A in arbitrary position such that the diameter of A is less than 1 and such that

$$\left|\sum_{\mathbf{p}_j\in A}1-N\mu(A\cap U^K)\right|\gg N^{(K-1)/2K}=N^{\frac{1}{2}-1/2K}.$$

Theorem 17E follows.

Proof of Lemma 6.2. By definition,

$$\hat{\chi}_{r,\tau}(\mathbf{t})=\hat{\chi}_r(\tau^{-1}\mathbf{t}), \tag{16}$$

where χ_r denotes the characteristic function of the cube $[-r,r]^K$. It therefore suffices to study the function $\hat{\chi}_r(\mathbf{u})$, where $\mathbf{u}\in\mathbb{R}^K$. From the definition, it follows from elementary calculation and writing $\mathbf{u}=(u_1,\ldots,u_K)$ that

$$\hat{\chi}_r(\mathbf{u})=\prod_{j=1}^{K}\frac{2\sin(ru_j)}{(2\pi)^{\frac{1}{2}}u_j}.$$

Since

$$\left|\frac{\sin(ru)}{u}\right|>\frac{r}{2} \quad \text{if} \quad |u|<\frac{1}{r},$$

it follows that if $|u_1|<1/q,\ldots,|u_{K-1}|<1/q$, then

$$\frac{1}{q}\int_q^{2q}|\hat{\chi}_r(\mathbf{u})|^2\,dr\gg\left(q^{K-1}\min\left\{q,\frac{1}{|\mathbf{u}|}\right\}\right)^2, \tag{17}$$

where $|\mathbf{u}|=(u_1^2+\cdots+u_K^2)^{\frac{1}{2}}$. Let

$$V\left(\frac{1}{q},K-1\right)=\left\{\mathbf{u}=(u_1,\ldots,u_K)\in\mathbb{R}^K:|u_j|<\frac{1}{q},j=1,\ldots,K-1\right\}, \tag{18}$$

and for any $\mathbf{t}\in\mathbb{R}^K$, let

$$W\left(\mathbf{t},\frac{1}{q},K-1\right)=\left\{\tau\in T:\tau\mathbf{t}\in V\left(\frac{1}{q},K-1\right)\right\}. \tag{19}$$

Elementary calculation gives

$$\int_{W(\mathbf{t},1/q,K-1)} d\tau \gg \min\left\{1, \left(\frac{1}{q|\mathbf{t}|}\right)^{K-1}\right\}. \tag{20}$$

On combining (16)–(20), we conclude that

$$\begin{aligned}\omega_q(\mathbf{t}) &= \frac{1}{q}\int_q^{2q}\int_T |\hat{\chi}_{r,\tau}(\mathbf{t})|^2\, d\tau\, dr \\ &\gg \left(q^{K-1}\min\left\{q, \frac{1}{|\mathbf{t}|}\right\}\right)^2 \min\left\{1, \left(\frac{1}{q|\mathbf{t}|}\right)^{K-1}\right\} \\ &= \min\left\{q^{2K}, \frac{q^{K-1}}{|\mathbf{t}|^{K+1}}\right\}. \end{aligned} \tag{21}$$

We next show that similar calculation establishes the opposite inequality

$$\omega_q(\mathbf{t}) \ll \min\left\{q^{2K}, \frac{q^{K-1}}{|\mathbf{t}|^{K+1}}\right\}. \tag{22}$$

For $j = 1, \ldots, K-1$, let $l_j = 2^{s_j}$, where $s_j \geqslant 0$. It is easily seen that

$$\frac{1}{q}\int_q^{2q} |\hat{\chi}_r(\mathbf{u})|^2\, dr \gg\ll \left(\left(\prod_{j=1}^{K-1}\frac{q}{l_j}\right)\min\left\{q, \frac{1}{|\mathbf{u}|}\right\}\right)^2$$

provided that

$$\frac{[\frac{1}{2}l_j]}{q} \leqslant |u_j| < \frac{l_j}{q} \quad \text{for all} \quad j = 1, \ldots, K-1.$$

Let

$$V\left(\frac{1}{q}, l_1, \ldots, l_{K-1}\right) = \{\mathbf{u} = (u_1, \ldots, u_K) \in \mathbb{R}^K : |u_j| < l_j/q, j = 1, \ldots, K-1\}$$

and

$$W\left(\mathbf{t}, \frac{1}{q}, l_1, \ldots, l_{K-1}\right) = \left\{\tau \in T : \tau\mathbf{t} \in V\left(\frac{1}{q}, l_1, \ldots, l_{K-1}\right)\right\}.$$

It is not difficult to see that

$$\int_{W(\mathbf{t},1/q,l_1,\ldots,l_{K-1})} d\tau \gg\ll \prod_{j=1}^{K-1}\min\left\{1, \frac{l_j}{q|\mathbf{t}|}\right\}.$$

Hence

$$\omega_q(\mathbf{t}) = \frac{1}{q}\int_q^{2q}\int_T |\hat{\chi}_{r,\tau}(\mathbf{t})|^2\,\mathrm{d}\tau\,\mathrm{d}r$$

$$\ll \sum_{s_1\geqslant 0}\cdots\sum_{s_{K-1}\geqslant 0} 2^{-(s_1+\cdots+s_{K-1})}\min\left\{q^{2K},\frac{q^{K-1}}{|\mathbf{t}|^{K+1}}\right\}$$

$$=\left(\sum_{s\geqslant 0}2^{-s}\right)^{K-1}\min\left\{q^{2K},\frac{q^{K-1}}{|\mathbf{t}|^{K+1}}\right\}\ll\min\left\{q^{2K},\frac{q^{K-1}}{|\mathbf{t}|^{K+1}}\right\},$$

which gives (22). Lemma 6.2 follows easily from (21) and (22). ■

We now prove Theorem 19A. Let $\mathscr{P}=\{\mathbf{p}_1,\ldots,\mathbf{p}_N\}$. Let Z_0 and μ_0 be defined as in the proof of Theorem 17E.

For any real number $\lambda\in(0,1]$ and $g\in L^2(\mathbb{R}^K)$, write, for any $\mathbf{x}\in\mathbb{R}^K$,

$$g_\lambda(\mathbf{x}) = g(\lambda^{-1}\mathbf{x}).$$

Consider the function

$$F_g = g*(\mathrm{d}Z_0 - N\mathrm{d}\mu_0). \tag{23}$$

More explicitly,

$$F_g(\mathbf{x}) = \int_{\mathbb{R}^K} g(\mathbf{x}-\mathbf{y})(\mathrm{d}Z_0(\mathbf{y}) - N\mathrm{d}\mu_0(\mathbf{y}))$$

$$= \sum_{j=1}^{N} g(\mathbf{x}-\mathbf{p}_j) - N\int_{\mathbb{R}^K} g(\mathbf{x}-\mathbf{y})\,\mathrm{d}\mu_0(\mathbf{y}). \tag{24}$$

Let

$$\Delta_1(g) = \int_{\mathbb{R}^K}|F_g(\mathbf{x})|^2\,\mathrm{d}\mathbf{x} \quad\text{and}\quad \Delta_2(g) = \int_0^1\int_{\mathbb{R}^K}|F_{g_\lambda}(\mathbf{x})|^2\,\mathrm{d}\mathbf{x}\,\mathrm{d}\lambda. \tag{25}$$

It follows from (23), (6) and (7) that

$$\Delta_1(g) = \int_{\mathbb{R}^K}|\hat{g}(\mathbf{t})|^2|(\widehat{\mathrm{d}Z_0 - N\mathrm{d}\mu_0})(\mathbf{t})|^2\,\mathrm{d}\mathbf{t} \tag{26a}$$

and

$$\Delta_2(g) = \int_{\mathbb{R}^K}\left(\int_0^1|\hat{g}_\lambda(\mathbf{t})|^2\,\mathrm{d}\lambda\right)|(\widehat{\mathrm{d}Z_0 - N\mathrm{d}\mu_0})(\mathbf{t})|^2\,\mathrm{d}\mathbf{t}. \tag{26b}$$

We shall use the following well-known elementary results:

(I) if $f(x)=\mathrm{e}^{-\frac{1}{2}a^2x^2}$, then $\hat{f}(t)=a^{-1}\mathrm{e}^{-t^2/2a^2}$; and

(II) $\int_{-\infty}^{\infty}\mathrm{e}^{-\frac{1}{2}a^2x^2}\,\mathrm{d}x=(2\pi)^{\frac{1}{2}}/a$ for all real numbers $a>0$.

Given $\mathbf{l}\in\mathbb{Z}^K$, let

$$h_{\mathbf{l}}(\mathbf{x})=\prod_{j=1}^{K}\mathrm{e}^{-\frac{1}{2}l_j^2x_j^2}.$$

By (I), we have

$$\hat{h}_{\mathbf{l}}(\mathbf{t})=\prod_{j=1}^{K}\frac{1}{l_j}\mathrm{e}^{-t_j^2/2l_j^2}. \tag{27}$$

Let L be the integer power of 2 satisfying $4(2\pi)^{\frac{1}{2}K}N\leqslant L<8(2\pi)^{\frac{1}{2}K}N$, and let $m\geqslant 1$ be a real parameter to be fixed later. Let

$$\mathbb{Z}^K(L,m)=\left\{\mathbf{l}=(l_1,\ldots,l_K)\in\mathbb{Z}^K: l_j=2^{s_j}\geqslant m,\right.$$
$$\left. s_j\in\mathbb{Z}, j=1,\ldots,K \text{ and } \prod_{j=1}^{K}l_j=L\right\}.$$

The analogue of Lemma 6.2 in this situation is

Lemma 6.3. *We have*

$$\sum_{\mathbf{l}\in\mathbb{Z}^K(L,m)}\Delta_1(h_{\mathbf{l}})\ll\Delta_2(\chi_{1/m}), \tag{28}$$

where χ_β denotes the characteristic function of the cube $[-\beta,\beta]^K$.

Before proving Lemma 6.3, we first explain how we can deduce Theorem 19A. We need the following estimate for the 'trivial error'.

Lemma 6.4. *For every $\mathbf{l}\in\mathbb{Z}^K(L,m)$,*

$$\Delta_1(h_{\mathbf{l}})\gg 1. \tag{29}$$

Proof. Let $Y=Y(\mathbf{l})=\{\mathbf{x}\in\mathbb{R}^K: h_{\mathbf{l}}(\mathbf{x})\geqslant\frac{1}{2}\}$. By (24) and (II), we have

$$F_{h_{\mathbf{l}}}(\mathbf{x})=\sum_{j=1}^{N}h_{\mathbf{l}}(\mathbf{x}-\mathbf{p}_j)-N\int_{U^K}h_{\mathbf{l}}(\mathbf{x}-\mathbf{y})\,\mathrm{d}\mathbf{y}$$
$$\geqslant\frac{1}{2}\sum_{\mathbf{p}_j\in Y+\mathbf{x}}1-N\int_{\mathbb{R}^K}h_{\mathbf{l}}(\mathbf{x}-\mathbf{y})\,\mathrm{d}\mathbf{y}$$
$$=\frac{1}{2}\sum_{\mathbf{p}_j\in Y+\mathbf{x}}1-N\prod_{j=1}^{K}\frac{(2\pi)^{\frac{1}{2}}}{l_j}\geqslant\frac{1}{2}\sum_{\mathbf{p}_j\in Y+\mathbf{x}}1-\frac{1}{4}.$$

Hence

$$|F_{h_{\mathbf{l}}}(\mathbf{x})|\geqslant\frac{1}{4}\sum_{\mathbf{p}_j\in Y+\mathbf{x}}1.$$

It follows that

$$\Delta_1(h_1) = \int_{\mathbb{R}^K} |F_{h_1}(\mathbf{x})|^2 \, d\mathbf{x} \geqslant \tfrac{1}{16} \int_{\mathbb{R}^K} \left(\sum_{\mathbf{p}_j \in Y + \mathbf{x}} 1 \right)^2 d\mathbf{x}$$

$$\geqslant \tfrac{1}{16} \int_{\mathbb{R}^K} \left(\sum_{\mathbf{p}_j \in Y + \mathbf{x}} 1 \right) d\mathbf{x} = \frac{N\mu(Y)}{16}.$$

The result follows, since $\mu(Y) \gg 1/N$. ■

We now observe that the cardinality of $\mathbb{Z}^K(L, m)$ is $\gg (\log(N^{1/K} m^{-1}))^{K-1}$. Thus by (28) and (29),

$$\Delta_2(\chi_{1/m}) \gg (\log(N^{1/K} m^{-1}))^{K-1}. \tag{30}$$

Now let $m = N^{2/(2K+1)}$. We distinguish two cases. First assume that for some $\lambda_0 \in (0, 1]$ and $\mathbf{x}_0 \in \mathbb{R}^K$,

$$|F_{\chi_{\lambda_0/m}}(\mathbf{x}_0)| = \left| \sum_{j=1}^{N} \chi_{\lambda_0/m}(\mathbf{x}_0 - \mathbf{p}_j) - N \int_{\mathbb{R}^K} \chi_{\lambda_0/m}(\mathbf{x}_0 - \mathbf{y}) \, d\mu_0(\mathbf{y}) \right|$$

$$> 2N(2/m)^K. \tag{31}$$

Then the result is obvious. Indeed, there is a translated image A of the cube $[-1/m, 1/m]^K$ such that A is contained in U^K and such that A contains $([-\lambda_0/m, \lambda_0/m]^K + \mathbf{x}_0) \cap U^K$. Clearly A has discrepancy at least $N(2/m)^K > N^{1/(2K+1)}$. In the second case, we have (see (24) and (25) and note that (31) does not hold)

$$\Delta_2(\chi_{1/m}) \leqslant \max_{A \subset U^K} (\text{discrepancy of } A)^2 + O(m^{-1}(2N(2/m)^K)^2), \tag{32}$$

where the maximum is taken over all cubes A in U^K with sides parallel to the coordinate axes. By (30) and (32), we conclude that for some cube in U^K and with sides parallel to the coordinate axes, the square of the discrepancy is

$$\gg (\log(N^{1/K} m^{-1}))^{K-1} - O(N^2 m^{-2K-1})$$
$$= (\log(N^{1/(2K+1)K}))^{K-1} - O(1) \gg (\log N)^{K-1}.$$

Theorem 19A follows. It remains to prove Lemma 6.3.

Proof of Lemma 6.3. By (26), it suffices to show that

$$\sum_{\mathbf{l} \in \mathbb{Z}^K(L,m)} |\hat{h}_{\mathbf{l}}(\mathbf{t})|^2 \ll \int_0^1 |\hat{\chi}_{\lambda/m}(\mathbf{t})|^2 \, d\lambda \tag{33}$$

uniformly for all $\mathbf{t}\in\mathbb{R}^K$. Since

$$\hat{\chi}_\beta(\mathbf{t})=\prod_{j=1}^{K}\frac{2\sin(\beta t_j)}{(2\pi)^{\frac{1}{2}}t_j},$$

using elementary estimates, we get

$$\int_0^1|\hat{\chi}_{\lambda/m}(\mathbf{t})|^2\,\mathrm{d}\lambda\gg\prod_{j=1}^{K}\left(\frac{1}{m+|t_j|}\right)^2. \tag{34}$$

On the other hand, by (27), we have

$$\begin{aligned}\sum_{\mathbf{l}\in Z^K(L,m)}|\hat{h}_{\mathbf{l}}(\mathbf{t})|^2&=\sum_{\mathbf{l}\in Z^K(L,m)}\prod_{j=1}^{K}\frac{\mathrm{e}^{-t_j^2/l_j^2}}{l_j^2}\\&\leqslant\sum_{2^{s_1}\geqslant m}\cdots\sum_{2^{s_K}\geqslant m}\prod_{j=1}^{K}\frac{\mathrm{e}^{-t_j^2/4^{s_j}}}{4^{s_j}}=\prod_{j=1}^{K}\left(\sum_{2^s\geqslant m}\frac{\mathrm{e}^{-t_j^2/4^s}}{4^s}\right).\end{aligned} \tag{35}$$

Using the inequality

$$\sum_{2^s\geqslant m}\frac{\mathrm{e}^{-t^2/4^s}}{4^s}=O((m+|t|)^{-2})$$

($m>0$ and $t\in\mathbb{R}$ are arbitrary), it follows from (34) and (35) that

$$\sum_{\mathbf{l}\in Z^K(L,m)}|\hat{h}_{\mathbf{l}}(\mathbf{t})|^2\ll\prod_{j=1}^{K}\frac{1}{(m+|t_j|)^2}\ll\int_0^1|\hat{\chi}_{\lambda/m}(\mathbf{t})|^2\,\mathrm{d}\lambda$$

as required. ■

6.2 Balls contained in the unit cube – a 'truncation' technique

We first renormalize Theorem 16A as follows.

Theorem 16B. *Let ε be a positive real number. Let $\mathbf{z}_1,\ldots,\mathbf{z}_N$ be points in the cube $[-M,M]^K$, where $M=\frac{1}{2}N^{1/K}$. Then there exists a ball $C(r,\mathbf{c})$ contained in $[-M,M]^K$ such that*

$$\left|\sum_{\mathbf{z}_j\in C(r,\mathbf{c})}1-\mu(C(r,\mathbf{c}))\right|>N^{\frac{1}{2}-1/2K-\varepsilon},$$

provided that $N>c_{13}(K,\varepsilon)$.

We assume throughout that N is sufficiently large, depending on K and ε only. For convenience of notation, let $S=\{\mathbf{z}_1,\ldots,\mathbf{z}_N\}$ and for $a>0$, let

$$Q(a)=[-a,a]^K.$$

We introduce two measures. For any $H \subset \mathbb{R}^K$, let

$$Z_0(H) = \#(S \cap H \cap Q(m_0)),$$

where m_0, satisfying $M \exp(-(\log N)^{\frac{2}{3}}) < m_0 < \frac{1}{2}M$, will be fixed later, and where $\#X$ denotes the number of elements of the set X. Also, for any Lebesgue measurable $H \subset \mathbb{R}^K$, let

$$\mu_0(H) = \mu(H \cap Q(m_0)).$$

Let χ_r denote the characteristic function of the ball

$$C(r, \mathbf{0}) = \left\{ \mathbf{x} = (x_1, \ldots, x_K) \in \mathbb{R}^K : \sum_{j=1}^{K} x_j^2 \leqslant r^2 \right\}$$

centred at the origin and having radius r. The parameter r will be specified later.

Consider the function

$$F_r = \chi_r * (\mathrm{d}Z_0 - \mathrm{d}\mu_0), \tag{36}$$

where $*$ denotes the convolution operation. More explicitly,

$$\begin{aligned} F_r(\mathbf{x}) &= \int_{\mathbb{R}^K} \chi_r(\mathbf{x} - \mathbf{y})(\mathrm{d}Z_0 - \mathrm{d}\mu_0)(\mathbf{y}) \\ &= \#(S \cap C(r, \mathbf{x}) \cap Q(m_0)) - \mu(C(r, \mathbf{x}) \cap Q(m_0)), \end{aligned} \tag{37}$$

where $C(r, \mathbf{x}) = C(r, \mathbf{0}) + \mathbf{x}$.

Let

$$E(\mathbf{x}) = \exp(-|\mathbf{x}|^2 m_1^{-2}), \tag{38}$$

where $|\mathbf{x}| = (x_1^2 + \cdots + x_K^2)^{\frac{1}{2}}$ denotes the usual euclidean distance and where $m_1 = m_0(\log N)^{-1}$. Consider now the following 'truncated' version of F_r: Let

$$G_r = EF_r. \tag{39}$$

Clearly $G_r(\mathbf{x})$ is a good approximation of the discrepancy function

$$\begin{cases} \sum_{\mathbf{z}_j \in C(r, \mathbf{x})} 1 - \mu(C(r, \mathbf{x})) & (C(r, \mathbf{x}) \subset Q(M)), \\ 0 & (\text{otherwise}), \end{cases}$$

since the 'weight' $E(\mathbf{x})$ is extremely small whenever the ball $C(r, \mathbf{x})$ is not contained in $Q(m_0)$ (we mention in advance that $r < \frac{1}{2}m_0$).

In order to estimate the quadratic average of $G_r(\mathbf{x})$, we shall employ the theory of Fourier transformation. Besides identities (6) and (7), we need (see any textbook on harmonic analysis): For $f, g \in L^2(\mathbb{R}^K)$,

$$\widehat{fg} = \hat{f} * \hat{g}. \tag{40}$$

By the Parseval–Plancherel identity (see (7))

$$\int_{\mathbf{R}^K} |G_r(\mathbf{x})|^2 \, \mathrm{d}\mathbf{x} = \int_{\mathbf{R}^K} |\hat{G}_r(\mathbf{t})|^2 \, \mathrm{d}\mathbf{t}.$$

On combining (6), (36), (39) and (40), we have

$$\hat{G}_r = \hat{E} * \hat{F}_r = \hat{E} * (\hat{\chi}_r (\widehat{\mathrm{d}Z_0 - \mathrm{d}\mu_0})). \tag{41}$$

Unfortunately, $\hat{G}_r$ has a very complicated form, so we introduce the auxiliary function

$$H_r = \chi_r * (E(\mathrm{d}Z_0 - \mathrm{d}\mu_0)); \tag{42}$$

in other words,

$$H_r(\mathbf{x}) = \int_{\mathbf{R}^K} \chi_r(\mathbf{x} - \mathbf{y})(E(\mathbf{y}) \, \mathrm{d}Z_0(\mathbf{y}) - E(\mathbf{y}) \, \mathrm{d}\mu_0(\mathbf{y}))$$

$$= \sum_{\mathbf{z}_j \in C(r,\mathbf{x}) \cap Q(m_0)} E(\mathbf{z}_j) - \int_{C(r,\mathbf{x}) \cap Q(m_0)} E(\mathbf{y}) \, \mathrm{d}\mathbf{y}. \tag{43}$$

From (42), (6) and (40), we obtain

$$\hat{H}_r = \hat{\chi}_r (\hat{E} * (\widehat{\mathrm{d}Z_0 - \mathrm{d}\mu_0})). \tag{44}$$

For the sake of brevity, let

$$\phi = (\widehat{\mathrm{d}Z_0 - \mathrm{d}\mu_0}) \quad \text{and} \quad \psi = \hat{E} * (\widehat{\mathrm{d}Z_0 - \mathrm{d}\mu_0}) = \hat{E} * \phi. \tag{45}$$

Then, by (41), (42), (44) and (45),

$$\hat{G}_r = \hat{E} * (\hat{\chi}_r \phi) \quad \text{and} \quad \hat{H}_r = \hat{\chi}_r \psi, \tag{46}$$

and we have

$$\hat{H}_r(\mathbf{t}) - \hat{G}_r(\mathbf{t}) = \int_{\mathbf{R}^K} (\hat{\chi}_r(\mathbf{t}) - \hat{\chi}_r(\mathbf{t} - \mathbf{u}))\phi(\mathbf{t} - \mathbf{u})\hat{E}(\mathbf{u}) \, \mathrm{d}\mathbf{u}. \tag{47}$$

We first give a brief description of the strategy of the proof. Since $\hat{H}_r = \hat{\chi}_r \psi$ has the form of a simple product, it will be easy to handle the L^2-norm of $\hat{H}_r(\mathbf{t})$. We shall also show that the difference $\hat{H}_r(\mathbf{t}) - \hat{G}_r(\mathbf{t})$ is 'small'. Combining these, we shall obtain a good lower bound to the L^2-norm of $\hat{G}_r(\mathbf{t})$ and, via the Parseval–Plancherel identity, to the L^2-norm of the truncated discrepancy function $G_r(\mathbf{x})$.

We begin with the investigation of the difference $\hat{H}_r(\mathbf{t}) - \hat{G}_r(\mathbf{t})$. Using the well-known result

$$f(x) = \exp(-\tfrac{1}{2}a^2x^2) \Leftrightarrow \hat{f}(t) = a^{-1}\exp(-t^2/2a^2), \tag{48}$$

we have, by (38),

$$\hat{E}(\mathbf{u}) = m_1^K 2^{-\frac{1}{2}K}\exp(-|\mathbf{u}|^2 m_1^2/4). \tag{49}$$

Clearly

$$|\phi(\mathbf{t})| = |(\widehat{\mathrm{d}Z_0 - \mathrm{d}\mu_0})(\mathbf{t})| \leqslant \frac{1}{(2\pi)^{\frac{1}{2}K}}\left(\left|\sum_{j=1}^{N} \mathrm{e}^{-\mathrm{i}\mathbf{z}_j\cdot\mathbf{t}}\right| + \int_{Q(M)} \mathrm{d}\mathbf{x}\right) \leqslant N,$$

and since the parameter r will be less than M,

$$|\hat{\chi}_r(\mathbf{t})| = (2\pi)^{-\frac{1}{2}K}\left|\int_{C(r,\mathbf{0})} \mathrm{e}^{-\mathrm{i}\mathbf{x}\cdot\mathbf{t}}\,\mathrm{d}\mathbf{x}\right| \leqslant (2\pi)^{-\frac{1}{2}K}\mu(C(r,\mathbf{0})) \leqslant N.$$

Let $\delta_0 = (\log N)/m_1$. Then by (49),

$$\int_{\mathbf{R}^K\setminus Q(\delta_0)} \hat{E}(\mathbf{u})\,\mathrm{d}\mathbf{u} \ll N^{-2}.$$

Using these upper estimates to (47), we see that

$$\begin{aligned}
&|\hat{H}_r(\mathbf{t}) - \hat{G}_r(\mathbf{t})| \\
&\leqslant \left|\int_{Q(\delta_0)} (\hat{\chi}_r(\mathbf{t}) - \hat{\chi}_r(\mathbf{t}-\mathbf{u}))\phi(\mathbf{t}-\mathbf{u})\hat{E}(\mathbf{u})\,\mathrm{d}\mathbf{u}\right| + c_{14}(K) \\
&\leqslant \max_{\mathbf{u}\in Q(\delta_0)} |\hat{\chi}_r(\mathbf{t}) - \hat{\chi}_r(\mathbf{t}-\mathbf{u})| \int_{Q(\delta_0)} |\phi(\mathbf{t}-\mathbf{u})\hat{E}(\mathbf{u})|\,\mathrm{d}\mathbf{u} + c_{14}(K).
\end{aligned} \tag{50}$$

We next study the Fourier transform $\hat{\chi}_r$ of the characteristic function of the ball $C(r,\mathbf{0})$. For the sake of brevity, let $s = |\mathbf{s}|$. By definition,

$$\begin{aligned}
\hat{\chi}_r(\mathbf{s}) &= (2\pi)^{-\frac{1}{2}K}\int_{\mathbf{R}^K} \mathrm{e}^{-\mathrm{i}\mathbf{x}\cdot\mathbf{s}}\chi_r(\mathbf{x})\,\mathrm{d}\mathbf{x} = (2\pi)^{-\frac{1}{2}K}\int_{|\mathbf{x}|\leqslant r} \mathrm{e}^{-\mathrm{i}\mathbf{x}\cdot\mathbf{s}}\,\mathrm{d}\mathbf{x} \\
&= c_{15}(K)\int_{-r}^{r} \mathrm{e}^{-\mathrm{i}ys}(r^2 - y^2)^{\frac{1}{2}(K-1)}\,\mathrm{d}y \\
&= c_{15}(K)r^K\int_{-1}^{1} (1-h^2)^{\frac{1}{2}(K-1)}\cos(srh)\,\mathrm{d}h.
\end{aligned} \tag{51}$$

The classical Bessel function $J_\nu(x)$ (see, for example, (3) on p. 48 of Watson

(1958)) has, for $\nu > -\frac{1}{2}$, the integral representation (Poisson integral)

$$J_\nu(x) = \frac{1}{\pi^{\frac{1}{2}}\Gamma(\nu+\frac{1}{2})}\left(\frac{x}{2}\right)^\nu \int_{-1}^{1} (1-h^2)^{\nu-\frac{1}{2}} \cos(xh)\,dh. \tag{52}$$

Hence, by (51) and (52),

$$\hat{\chi}_r(\mathbf{s}) = c_{16}(K)(r/s)^{\frac{1}{2}K} J_{\frac{1}{2}K}(rs). \tag{53}$$

By Hankel's asymptotic expansion (see, for example, (1) on p. 199 of Watson (1958)),

$$J_\nu(x) = (2/\pi x)^{\frac{1}{2}} \cos(x - (2\nu+1)\pi/4) + O(x^{-\frac{3}{2}}), \tag{54}$$

where the implicit constant in the O-notation depends only on ν. Therefore, by (53) and (54),

$$\hat{\chi}_r(\mathbf{s}) = c_{17}(K) r^{\frac{1}{2}(K-1)} s^{-\frac{1}{2}(K+1)} \cos(rs - (K+1)\pi/4) + O(r^{\frac{1}{2}(K-3)} s^{-\frac{1}{2}(K+3)}) \tag{55}$$

(here and in what follows, the implicit constants in $\ll$ and the O-notation depend only on the dimension K). Combining (50) and (55), we obtain, via elementary estimates and writing $t = |\mathbf{t}|$,

$$|\hat{H}_r(\mathbf{t}) - \hat{G}_r(\mathbf{t})| \ll \left(\delta_0\left(\frac{r}{t}\right)^{\frac{1}{2}(K+1)} + \frac{r^{\frac{1}{2}(K-3)}}{t^{\frac{1}{2}(K+3)}}\right) \int_{Q(\delta_0)} |\phi(\mathbf{t}-\mathbf{u})\hat{E}(\mathbf{u})|\,d\mathbf{u} + O(1) \tag{56}$$

whenever $rt \geqslant 1$. Since $\delta_0 = (\log N)/m_1$, we have, by (49),

$$\hat{E}(\mathbf{u}) \ll \delta_0^{-K}(\log N)^K \quad \text{for all} \quad \mathbf{u} \in \mathbb{R}^K.$$

Consequently,

$$\int_{Q(\delta_0)} |\phi(\mathbf{t}-\mathbf{u})\hat{E}(\mathbf{u})|\,d\mathbf{u} \ll (\log N)^K (2\delta_0)^{-K} \int_{Q(\delta_0)} |\phi(\mathbf{t}-\mathbf{u})|\,d\mathbf{u}. \tag{57}$$

By the Cauchy–Schwarz inequality,

$$(2\delta_0)^{-K} \int_{Q(\delta_0)} |\phi(\mathbf{t}-\mathbf{u})|\,d\mathbf{u} \leqslant \left((2\delta_0)^{-K} \int_{Q(\delta_0)} |\phi(\mathbf{t}-\mathbf{u})|^2\,d\mathbf{u}\right)^{\frac{1}{2}}. \tag{58}$$

Using the elementary inequality $(a+b)^2 \leqslant 2a^2 + 2b^2$ and (56)–(58), we have

$$|\hat{H}_r(\mathbf{t}) - \hat{G}_r(\mathbf{t})|^2$$
$$\ll \left(\delta_0^{\frac{2}{3}} \left(\frac{r}{t} \right)^{K+1} + \frac{r^{K-3}}{t^{K+3}} \right) (\log N)^{2K} (2\delta_0)^{-K} \int_{Q(\delta_0)} |\phi(\mathbf{t}-\mathbf{u})|^2 \, d\mathbf{u} + O(1) \quad (59)$$

whenever $rt \geqslant 1$.

Let q, where $0 < q < \frac{1}{2} m_0 (\log N)^{-3}$, be a real parameter to be fixed later. Using the general inequality $a^2 \geqslant \frac{1}{2} b^2 - (a-b)^2$, (46) and (59), we have

Lemma 6.5. *Let $\mathscr{T} \subset \mathbb{R}^K$ be a Lebesgue measurable set such that the euclidean distance of the origin $\mathbf{0} \in \mathbb{R}^K$ and $\mathscr{T}$ is greater than $1/q$. Then*

$$\int_q^{2q} \int_{\mathscr{T}} |\hat{G}_r(\mathbf{t})|^2 \, d\mathbf{t} \, dr \geqslant \tfrac{1}{2} I_1 - c_{18}(K) I_2 - c_{19}(K) \int_q^{2q} \int_{\mathscr{T}} d\mathbf{t} \, dr,$$

where

$$I_1 = \int_q^{2q} \int_{\mathscr{T}} |\hat{H}_r(\mathbf{t})|^2 \, d\mathbf{t} \, dr = \int_{\mathscr{T}} \left(\int_q^{2q} |\hat{\chi}_r(\mathbf{t})|^2 \, dr \right) |\phi(\mathbf{t})|^2 \, d\mathbf{t}$$

and

$$I_2 = \int_q^{2q} \int_{\mathscr{T}} \left(\left(\delta_0^{\frac{2}{3}} \left(\frac{r}{t} \right)^{K+1} + \frac{r^{K-3}}{t^{K+3}} \right) (\log N)^{2K} (2\delta_0)^{-K} \right.$$
$$\left. \times \int_{Q(\delta_0)} |\phi(\mathbf{t}-\mathbf{u})|^2 \, d\mathbf{u} \right) d\mathbf{t} \, dr.$$

Let m, satisfying $0 < m \leqslant M = \frac{1}{2} N^{1/K}$, be a real parameter. Let

$$W_m(\mathbf{x}) = \#((U^K + \mathbf{x}) \cap Q(m) \cap S),$$

where $Q(m) = [-m, m]^K$ and $S = \{\mathbf{z}_1, \ldots, \mathbf{z}_N\}$. We need

Lemma 6.6. *For arbitrary, sufficiently large N, there exists a real number m_0, satisfying $M \exp(-(\log N)^{\frac{2}{3}}) < m_0 < \frac{1}{2} M$, such that either*

(i) $\#(Q(m_0) \cap S) < \frac{1}{10}(2m_0)^K$; or
(ii) $\#(Q(m_0) \cap S) \geqslant \frac{1}{10}(2m_0)^K$ and, with $m_1 = m_0 (\log N)^{-1}$,

$$\int_{\mathbb{R}^K} (W_{m_1}(\mathbf{x}))^2 \, d\mathbf{x} > \exp(-(\log N)^{\frac{2}{3}}) \int_{\mathbb{R}^K} (W_{m_0}(\mathbf{x}))^2 \, d\mathbf{x}. \quad (60)$$

We postpone the proof to the end of this section.

If alternative (i) of Lemma 6.6 is true, then Theorem 16B follows immediately. Indeed, the cube $Q(m_0)$ contains less than 10% of the expected number of points $\mathbf{z}_j$, and by a standard averaging argument we establish the existence of a ball B contained in $Q(m_0)$ with radius m_0/K such that $\sum_{\mathbf{z}_j \in B} 1 < \frac{1}{2}\mu(B)$. Thus B is certainly contained in $Q(M) = [-M, M]^K$ and has a huge discrepancy

$$\left|\sum_{\mathbf{z}_j \in B} 1 - \mu(B)\right| \geqslant \tfrac{1}{2}\mu(B) \gg m_0^K \geqslant (M \exp(-(\log N)^{\frac{2}{3}}))^K$$
$$= N(\tfrac{1}{2}\exp(-(\log N)^{\frac{2}{3}}))^K > N^{1-\varepsilon}.$$

We may therefore assume from now on the validity of alternative (ii) of Lemma 6.6, and fix the parameter m_0 in the definition of the measures Z_0 and μ_0 in terms of (ii) in Lemma 6.6.

Furthermore, we need two lemmas concerning $\phi(\mathbf{t})$ and $\psi(\mathbf{t})$. We recall that $m_1 = m_0(\log N)^{-1}$.

Lemma 6.7. *We have*

$$\int_{Q(100)} |\psi(\mathbf{t})|^2 \, d\mathbf{t} \gg \int_{\mathbf{R}^K} (W_{m_1}(\mathbf{x}))^2 \, d\mathbf{x}.$$

Before stating the next lemma, we introduce some notation. For any real number $b \in (0, 200]$, let

$$D_b(\mathbf{x}) = \#((Q(1/b) + \mathbf{x}) \cap Q(m_0) \cap S) - \mu((Q(1/b) + \mathbf{x}) \cap Q(m_0)) \quad (61)$$

and

$$\Delta_b(\mathbf{x}) = \sum_{\mathbf{z}_j \in (Q(1/b) + \mathbf{x}) \cap Q(m_0)} E(\mathbf{x} - \mathbf{z}_j) - \int_{Q(1/b) + \mathbf{x}} E(\mathbf{x} - \mathbf{y}) \, d\mu_0(\mathbf{y}). \quad (62)$$

Lemma 6.8. *Let* $b \in (0, 200]$.

(*i*) *We have*

$$\int_{Q(b)} |\phi(\mathbf{t})|^2 \, d\mathbf{t} \ll \int_{\mathbf{R}^K} |D_b(\mathbf{x})|^2 (\tfrac{1}{2}b)^{2K} \, d\mathbf{x}.$$

(*ii*) *We have*

$$\int_{Q(b)} |\psi(\mathbf{t})|^2 \, d\mathbf{t} \ll \int_{\mathbf{R}^K} |\Delta_b(\mathbf{x})|^2 (\tfrac{1}{2}b)^{2K} \, d\mathbf{x}.$$

We postpone the proof of these lemmas to the end of the section.

Combining Lemma 6.7 and (60), we have

$$\int_{Q(100)} |\psi(\mathbf{t})|^2 \, d\mathbf{t} \gg \int_{\mathbf{R}^K} (W_{m_1}(\mathbf{x}))^2 \, d\mathbf{x} > \exp(-(\log N)^{\frac{2}{3}}) \int_{\mathbf{R}^K} (W_{m_0}(\mathbf{x}))^2 \, d\mathbf{x}. \quad (63)$$

By Lemma 6.6(ii), we have

$$\int_{\mathbb{R}^K} (W_{m_0}(\mathbf{x}))^2 \, d\mathbf{x} \geqslant \int_{\mathbb{R}^K} W_{m_0}(\mathbf{x}) \, d\mathbf{x} = \#(Q(m_0) \cap S) \gg m_0^K. \tag{64}$$

Clearly

$$W_{m_0}(\tilde{\mathbf{x}}) + \mu_0(Q(\tfrac{1}{200}) + \mathbf{x}) \geqslant |D_{200}(\mathbf{x})|,$$

where $\tilde{\mathbf{x}}$ is obtained from $\mathbf{x} = (x_1, \ldots, x_K)$ by writing $\tilde{\mathbf{x}} = (x_1 - \frac{1}{2}, \ldots, x_K - \frac{1}{2})$. Using the inequality $(a+b)^2 \leqslant 2a^2 + 2b^2$, we therefore have

$$2\int_{\mathbb{R}^K} (W_{m_0}(\mathbf{x}))^2 \, d\mathbf{x} + 2\int_{\mathbb{R}^K} (\mu_0(Q(\tfrac{1}{200}) + \mathbf{x}))^2 \, d\mathbf{x} \geqslant \int_{\mathbb{R}^K} |D_{200}(\mathbf{x})|^2 \, d\mathbf{x}. \tag{65}$$

Since

$$\int_{\mathbb{R}^K} (\mu_0(Q(\tfrac{1}{200}) + \mathbf{x}))^2 \, d\mathbf{x} < \int_{\mathbb{R}^K} \mu_0(U^K + \mathbf{x}) \, d\mathbf{x} = (2m_0)^K, \tag{66}$$

it follows from (64)–(66) that

$$\int_{\mathbb{R}^K} (W_{m_0}(\mathbf{x}))^2 \, d\mathbf{x} \gg \int_{\mathbb{R}^K} |D_{200}(\mathbf{x})|^2 \, d\mathbf{x}. \tag{67}$$

Combining (63), (67) and Lemma 6.8(i), we have

$$\int_{Q(100)} |\psi(\mathbf{t})|^2 \, d\mathbf{t} \gg \exp(-(\log N)^{\frac{2}{3}}) \int_{Q(200)} |\phi(\mathbf{t})|^2 \, d\mathbf{t}. \tag{68}$$

Moreover, from (63) and (64), we have

$$\int_{Q(100)} |\psi(\mathbf{t})|^2 \, d\mathbf{t} \gg \exp(-(\log N)^{\frac{2}{3}}) m_0^K. \tag{69}$$

Now let $\eta > 0$ be arbitrarily small but fixed. We distinguish two cases.

Case 1: We have

$$\int_{Q(a_0)} |\psi(\mathbf{t})|^2 \, d\mathbf{t} < \tfrac{1}{2} \int_{Q(100)} |\psi(\mathbf{t})|^2 \, d\mathbf{t},$$

where $a_0 = N^\eta m_0^{-1}$. Then by (68),

$$\int_{Q(100)\setminus Q(a_0)} |\psi(\mathbf{t})|^2 \, d\mathbf{t} \gg \exp(-(\log N)^{\frac{2}{3}}) \int_{Q(200)} |\phi(\mathbf{t})|^2 \, d\mathbf{t}. \tag{70}$$

Also, there exists a real number b_0 with $a_0 \leqslant b_0 \leqslant 50$ and such that

$$\int_{\mathscr{T}_0} |\psi(\mathbf{t})|^2 \, d\mathbf{t} \gg (\log N)^{-1} \int_{Q(100)\setminus Q(a_0)} |\psi(\mathbf{t})|^2 \, d\mathbf{t}, \tag{71}$$

where $\mathscr{T}_0 = Q(2b_0)\backslash Q(b_0)$. Let $q = m_0 N^{-\frac{1}{2n}} = N^{\frac{1}{2n}} a_0^{-1}$. Then clearly

$$\inf_{\mathbf{t}\in\mathscr{T}_0} |\mathbf{t}| = b_0 \geqslant a_0 > q^{-1};$$

in other words, the euclidean distance of the origin $\mathbf{0}\in\mathbb{R}^K$ and $\mathscr{T}_0$ is greater than $1/q$. Consider Lemma 6.5 with $\mathscr{T} = \mathscr{T}_0$. Then, in the terminology of Lemma 6.5, we have

$$I_2 \ll q\left(\delta_0^2\left(\frac{q}{b_0}\right)^{K+1} + \frac{q^{K-3}}{b_0^{K+3}}\right)(\log N)^{2K} \times \int_{\mathscr{T}_0}\left((2\delta_0)^{-K}\int_{Q(\delta_0)} |\phi(\mathbf{t}-\mathbf{u})|^2\,d\mathbf{u}\right)d\mathbf{t}. \tag{72}$$

Since

$$\delta_0 = (\log N)/m_1 = (\log N)^2/m_0 \leqslant (\log N)^2(\tfrac{1}{2}N^{1/K}\exp(-(\log N)^{\frac{2}{3}}))^{-1} \leqslant 100,$$

we have

$$\int_{\mathscr{T}_0}\left((2\delta_0)^{-K}\int_{Q(\delta_0)} |\phi(\mathbf{t}-\mathbf{u})|^2\,d\mathbf{u}\right)d\mathbf{t} \leqslant \int_{Q(200)} |\phi(\mathbf{t})|^2\,d\mathbf{t}. \tag{73}$$

Furthermore, by (55),

$$\inf_{\mathbf{t}\in\mathscr{T}_0}\int_q^{2q} |\hat{\chi}_r(\mathbf{t})|^2\,dr \gg \frac{q^K}{b_0^{K+1}}. \tag{74}$$

We recall that $q = m_0 N^{-\frac{1}{2n}}$, $50 \geqslant b_0 \geqslant a_0 = N^n m_0^{-1}$ and $\delta_0 = (\log N)^2/m_0$. Using elementary calculation, Lemma 6.5 and (72)–(74), we conclude that for all sufficiently large N, we have

$$\int_q^{2q}\int_{\mathscr{T}_0} |\hat{G}_r(\mathbf{t})|^2\,d\mathbf{t}\,dr \geqslant c_{20}(K)\frac{q^K}{b_0^{K+1}}\int_{\mathscr{T}_0} |\psi(\mathbf{t})|^2\,d\mathbf{t} - N^{-\frac{1}{2n}}\frac{q^K}{b_0^{K+1}}\int_{Q(200)} |\phi(\mathbf{t})|^2\,d\mathbf{t} - O(q). \tag{75}$$

It follows from (75), (70) and (71) that

$$\int_q^{2q}\int_{\mathscr{T}_0} |\hat{G}_r(\mathbf{t})|^2\,d\mathbf{t}\,dr \gg (1 - O(N^{-n/4}))\frac{q^K}{b_0^{K+1}}\int_{\mathscr{T}_0} |\psi(\mathbf{t})|^2\,d\mathbf{t} - O(q). \tag{76}$$

By (71) and (69), we get

$$\int_{\mathscr{T}_0} |\psi(\mathbf{t})|^2\,d\mathbf{t} \gg (\log N)^{-1}\exp(-(\log N)^{\frac{2}{3}})m_0^K. \tag{77}$$

Combining (76) and (77), we have

$$\int_q^{2q}\int_{\mathscr{T}_0}|\hat{G}_r(\mathbf{t})|^2\,\mathrm{d}\mathbf{t}\,\mathrm{d}r \gg \frac{q^K}{b_0^{K+1}}m_0^{K-\frac{1}{2}\varepsilon} \gg q^K m_0^{K-\frac{1}{2}\varepsilon} \gg q m_0^{2K-1-\varepsilon}. \tag{78}$$

By the Parseval–Plancherel identity (7) and (78),

$$\int_q^{2q}\int_{\mathbb{R}^K}|G_r(\mathbf{x})|^2\,\mathrm{d}\mathbf{x}\,\mathrm{d}r = \int_q^{2q}\int_{\mathbb{R}^K}|\hat{G}_r(\mathbf{t})|^2\,\mathrm{d}\mathbf{t}\,\mathrm{d}r$$

$$\geqslant \int_q^{2q}\int_{\mathscr{T}_0}|\hat{G}_r(\mathbf{t})|^2\,\mathrm{d}\mathbf{t}\,\mathrm{d}r \gg q m_0^{2K-1-\varepsilon}. \tag{79}$$

We recall that $G_r = EF_r$, where, writing $m_1 = m_0(\log N)^{-1}$,

$$E(\mathbf{x}) = \exp(-|\mathbf{x}|^2 m_1^{-2})$$

and

$$F_r(\mathbf{x}) = \#(S\cap C(r,\mathbf{x})\cap Q(m_0)) - \mu(C(r,\mathbf{x})\cap Q(m_0)).$$

Clearly

$$E(\mathbf{x}) \leqslant N^{-c_{21}(K)\log N} \quad \text{if} \quad q\leqslant r\leqslant 2q \text{ and } C(r,\mathbf{x})\not\subset Q(m_0). \tag{80}$$

It now follows from (79) and (80) that there exists a ball $C(r_0,\mathbf{x}_0)$ such that $C(r_0,\mathbf{x}_0)$ is contained in $Q(m_0)\subset Q(M)$, $q\leqslant r_0\leqslant 2q$ and

$$\left|\sum_{\mathbf{z}_j\in C(r_0,\mathbf{x}_0)} 1 - \mu(C(r_0,\mathbf{x}_0))\right| \gg m_0^{K-1-\varepsilon}.$$

Since $m_0 \geqslant \frac{1}{2}N^{1/K}\exp(-(\log N)^{\frac{2}{3}})$, we conclude that

$$\left|\sum_{\mathbf{z}_j\in C(r_0,\mathbf{x}_0)} 1 - \mu(C(r_0,\mathbf{x}_0))\right| > N^{\frac{1}{2}-1/2K-\varepsilon}$$

if N is sufficiently large depending only on K and ε. This completes Case 1.

Case 2: We have

$$\int_{Q(a_0)}|\psi(\mathbf{t})|^2\,\mathrm{d}\mathbf{t} \geqslant \tfrac{1}{2}\int_{Q(100)}|\psi(\mathbf{t})|^2\,\mathrm{d}\mathbf{t},$$

where $a_0 = N^{\eta}m_0^{-1}$. This is the simpler case. From (69) and Lemma 6.8(ii), we obtain

$$\left(\frac{2}{a_0}\right)^{-2K}\int_{\mathbb{R}^K}|\Delta_{a_0}(\mathbf{x})|^2\,\mathrm{d}\mathbf{x} \gg \int_{Q(a_0)}|\psi(\mathbf{t})|^2\,\mathrm{d}\mathbf{t}$$

$$\geqslant \tfrac{1}{2}\int_{Q(100)}|\psi(\mathbf{t})|^2\,\mathrm{d}\mathbf{t} \gg \exp(-(\log N)^{\frac{2}{3}})(2m_0)^K. \tag{81}$$

Standard averaging arguments then yield that (see (62)) either (2A) there is a vector $\mathbf{x}_1 \in \mathbb{R}^K$ such that the translated image $Q(1/a_0)+\mathbf{x}_1$ of the cube $Q(1/a_0)$ is contained in $Q(m_0)$ and

$$|\Delta_{a_0}(\mathbf{x}_1)| \gg \exp(-\tfrac{1}{2}(\log N)^{\frac{2}{3}})(2/a_0)^K;$$

or (2B) there is another vector $\mathbf{x}_2 \in \mathbb{R}^K$ such that $Q(1/a_0)+\mathbf{x}_2$ is not contained in $Q(m_0)$ but

$$|\Delta_{a_0}(\mathbf{x}_2)| > 4(2/a_0)^K.$$

Since by (38),

$$1 \geqslant E(\mathbf{y}) \geqslant 1 - c_{22}(K)(\log N)^2 N^{-2\eta} \quad \text{whenever} \quad \mathbf{y} \in Q(1/a_0),$$

elementary calculation gives (see (61) and (62))

$$|D_{a_0}(\mathbf{x}_j)| \geqslant \tfrac{1}{2}|\Delta_{a_0}(\mathbf{x}_j)| \begin{cases} \gg \exp(-\tfrac{1}{2}(\log N)^{\frac{2}{3}})(2/a_0)^K & ((2A) \text{ with } j=1), \\ \geqslant 2(2/a_0)^K & ((2B) \text{ with } j=2). \end{cases} \tag{82}$$

If (2A) holds, then let $A_0 = Q(1/a_0)+\mathbf{x}_1$; if (2B) holds, then let A_0 be a translated image of $Q(1/a_0)$ such that $(Q(1/a_0)+\mathbf{x}_2)\cap Q(m_0) \subset A_0 \subset Q(m_0)$. In the latter case, by (82) and (61),

$$\#((Q(1/a_0)+\mathbf{x}_2)\cap Q(m_0)\cap S) \geqslant |D_{a_0}(\mathbf{x}_2)| \geqslant 2(2/a_0)^K = 2\mu(A_0),$$

and so

$$\sum_{\mathbf{z}_j \in A_0} 1 \geqslant 2\mu(A_0).$$

Consequently, we have

$$\left|\sum_{\mathbf{z}_j \in A_0} 1 - \mu(A_0)\right| \begin{cases} \gg \exp(-\tfrac{1}{2}(\log N)^{\frac{2}{3}})\mu(A_0) & ((2A)), \\ \geqslant \mu(A_0) & ((2B)). \end{cases} \tag{83}$$

We can now complete Case 2 as follows: Let $r = m_0 N^{-2\eta} = a_0^{-1} N^{-\eta}$. Again we distinguish two cases: (α) $\sum_{\mathbf{z}_j \in A_0} 1 - \mu(A_0) > 0$; and ($\beta$) $\sum_{\mathbf{z}_j \in A_0} 1 - \mu(A_0) < 0$. In case ($\alpha$), using standard averaging arguments, we conclude that either (α_1) there is a ball $C(r, \mathbf{x}_3)$ contained in $A_0 \subset Q(m_0)$ with

$$\sum_{\mathbf{z}_j \in C(r,\mathbf{x}_3)} 1 - \mu(C(r,\mathbf{x}_3)) > c_{23}(K) N^{-K\eta}\left(\sum_{\mathbf{z}_j \in A_0} 1 - \mu(A_0)\right);$$

or (α_2) there is another ball $C(r, \mathbf{x}_4)$ such that

$$C(r,\mathbf{x}_4)\cap Q(m_0) \neq \varnothing \quad \text{and} \quad \sum_{\mathbf{z}_j \in C(r,\mathbf{x}_4)\cap Q(m_0)} 1 > 2\mu(C(r,\mathbf{x}_4)).$$

In the case (α_2), since $2r < m_0 \leqslant \frac{1}{2}M$, we have that $C(r, \mathbf{x}_4) \subset Q(M)$ and

$$\sum_{\mathbf{z}_j \in C(r,\mathbf{x}_4)} 1 - \mu(C(r,\mathbf{x}_4)) > \mu(C(r,\mathbf{x}_4)).$$

Summarizing, there exists a ball $C \subset Q(M) = [-M, M]^K$ with radius $r = m_0 N^{-2\eta} \geqslant \frac{1}{2} N^{1/K} \exp(-(\log N)^{\frac{2}{3}}) N^{-2\eta}$ such that the discrepancy $\sum_{\mathbf{z}_j \in C} 1 - \mu(C)$ is greater than $N^{1-\varepsilon}$ if $\eta > 0$ is sufficiently small depending only on K and ε. In case (β), there is only one alternative. One can find a ball $C(r, \mathbf{x}_5)$ contained in A_0 with

$$\mu(C(r,\mathbf{x}_5)) - \sum_{\mathbf{z}_j \in C(r,\mathbf{x}_5)} 1 > c_{24}(K) N^{-K\eta} \left(\mu(A_0) - \sum_{\mathbf{z}_j \in A_0} 1 \right).$$

This completes Case 2.

It remains to prove Lemmas 6.6, 6.7 and 6.8.

Remark. We emphasize again the role played by the truncating function $E(\mathbf{x})$. (80) ensures that the contribution to the left-hand side of (79) from $\mathbf{x} \in \mathbb{R}^K$ that are 'far' from the origin $\mathbf{0} \in \mathbb{R}^K$ is 'small'. Also, in case (2B), although $Q(1/a_0) + \mathbf{x}_2$ is not contained in $Q(m_0)$, the small weight carried by the truncating function in (62) means that the discrepancy is in fact so large that we can translate $Q(1/a_0) + \mathbf{x}_2$ back into $Q(m_0)$ and still have a large discrepancy.

Proof of Lemma 6.6. Let $p_1 = \frac{1}{2}M$ and, for $j \geqslant 1$, let $p_{j+1} = p_j (\log N)^{-1}$, and write

$$\mathscr{W}_j = \int_{\mathbb{R}^K} (W_{p_j}(\mathbf{x}))^2 \, d\mathbf{x}.$$

We may assume that for every $p_j \geqslant M \exp(-(\log N)^{\frac{2}{3}})$,

$$\#(Q(p_j) \cap S) \geqslant \tfrac{1}{10}(2p_j)^K; \tag{84}$$

otherwise, (i) holds and the proof is complete. From (84), we obtain by elementary estimates that for every $p_j \geqslant M \exp(-(\log N)^{\frac{2}{3}})$,

$$N(\exp(-(\log N)^{\frac{2}{3}}))^K \leqslant (2p_j)^K \ll \mathscr{W}_j \leqslant N^2. \tag{85}$$

Now suppose, contrary to (ii), that

$$\mathscr{W}_{j+1} \leqslant \exp(-(\log N)^{\frac{2}{3}}) \mathscr{W}_j$$

for all j with $1 \leqslant j \leqslant l = (\log N)^{\frac{1}{2}}$. Then clearly

$$\mathscr{W}_{l+1} \leqslant (\exp(-(\log N)^{\frac{2}{3}}))^l \mathscr{W}_1,$$

and so by (85),

$$\mathscr{W}_{l+1} \leqslant \exp(-(\log N)^{\frac{7}{6}})\mathscr{W}_1 \ll N^{-2}\mathscr{W}_1 \leqslant 1. \tag{86}$$

Let N be sufficiently large. Then (86) contradicts (85), since

$$p_{l+1} = p_1(\log N)^{-l} = \tfrac{1}{2}M(\log N)^{-l} \geqslant M\exp(-(\log N)^{\frac{2}{3}}).$$

Lemma 6.6 follows. ∎

For later use, we prove the following slight generalization of Lemma 6.7.

Lemma 6.7′. *Let $\psi_0 = \phi$ and $\psi_1 = \psi$. Then for $i = 0, 1$,*

$$\int_{Q(100)} |\psi_i(\mathbf{t})|^2\, d\mathbf{t} \gg \int_{\mathbf{R}^K} (W_{m_i}(\mathbf{x}))^2\, d\mathbf{x}.$$

Proof. Let

$$f(\mathbf{x}) = \prod_{j=1}^{K} \left(\frac{2\sin(bx_j)}{(2\pi)^{\frac{1}{2}}x_j}\right)^2,$$

where the real parameter $b > 0$ will be fixed later. From (40), we see that the Fourier transform $\hat{f}$ of f is equal to the convolution of the characteristic function of the cube $Q(b) = [-b, b]^K$ with itself, i.e.

$$\hat{f}(\mathbf{t}) = (\chi_{Q(b)} * \chi_{Q(b)})(\mathbf{t}) = \prod_{j=1}^{K} (2b - |t_j|)^+,$$

where $y^+ = y$ if $y > 0$ and 0 otherwise. Let $E_0(\mathbf{x}) \equiv 1$ and $E_1(\mathbf{x}) = E(\mathbf{x})$. Then from (6), we obtain that the Fourier transform of the convolution

$$g_i(\mathbf{x}) = \int_{\mathbf{R}^K} f(\mathbf{x}-\mathbf{y})(E_i(\mathbf{y})\, dZ_0(\mathbf{y}) - E_i(\mathbf{y})\, d\mu_0(\mathbf{y}))$$

is equal to $\hat{f}\psi_i$ ($i = 0, 1$) (see also (40) and (45)). By the Parseval–Plancherel identity (7), we have, for $i = 0, 1$,

$$\begin{aligned}\int_{\mathbf{R}^K} g_i^2(\mathbf{x})\, d\mathbf{x} &= \int_{\mathbf{R}^K} |\hat{f}(\mathbf{t})|^2 |\psi_i(\mathbf{t})|^2\, d\mathbf{t} \\ &= \int_{Q(2b)} \left(\prod_{j=1}^{K} (2b - |t_j|)^2\right) |\psi_i(\mathbf{t})|^2\, d\mathbf{t} \\ &\leqslant (2b)^{2K} \int_{Q(2b)} |\psi_i(\mathbf{t})|^2\, d\mathbf{t}. \end{aligned} \tag{87}$$

Let $b = 50$. Elementary calculation shows that ($i = 0, 1$)

$$f(\mathbf{x}-\mathbf{z}_j)E_i(\mathbf{z}_j) - \int_{\mathbf{R}^K} f(\mathbf{x}-\mathbf{y})E_i(\mathbf{y})\, d\mu_0(\mathbf{y}) > c_{25}(K) > 0 \tag{88}$$

if $\mathbf{z}_j \in (Q(\frac{1}{50}) + \mathbf{x}) \cap Q(m_i)$. Therefore, using the fact that f is a positive function (a Fejér kernel), we deduce from (88) that ($i = 0, 1$)

$$|g_i(\mathbf{x})| = \left| \int_{\mathbb{R}^K} f(\mathbf{x}-\mathbf{y})(E_i(\mathbf{y})\,\mathrm{d}Z_0(\mathbf{y}) - E_i(\mathbf{y})\,\mathrm{d}\mu_0(\mathbf{y})) \right|$$

$$= \left| \sum_{j=1}^{N} f(\mathbf{x}-\mathbf{z}_j)E_i(\mathbf{z}_j) - \int_{\mathbb{R}^K} f(\mathbf{x}-\mathbf{y})E_i(\mathbf{y})\,\mathrm{d}\mu_0(\mathbf{y}) \right| \gg S_i(\mathbf{x}), \quad (89)$$

where $S_i(\mathbf{x})$ denotes the number of points $\mathbf{z}_j$ which lie in $(Q(\frac{1}{50}) + \mathbf{x}) \cap Q(m_i)$. It is easy to see that ($i = 0, 1$)

$$\int_{\mathbb{R}^K} (S_i(\mathbf{x}))^2\,\mathrm{d}\mathbf{x} \gg \int_{\mathbb{R}^K} (W_{m_i}(\mathbf{x}))^2\,\mathrm{d}\mathbf{x}. \quad (90)$$

The result follows on combining (87), (89) and (90). ■

Proof of Lemma 6.8. We prove only (i). The proof of (ii) is similar, and we leave it to the reader. Let $h(\mathbf{x})$ denote the characteristic function of the cube $Q(1/b)$. Then

$$\hat{h}(\mathbf{t}) = \prod_{j=1}^{K} \frac{2\sin(x_j/b)}{(2\pi)^{\frac{1}{2}} x_j}.$$

It is easily seen that

$$\int_{Q(b)} |\phi(\mathbf{t})|^2\,\mathrm{d}\mathbf{t} \ll b^{2K} \int_{\mathbb{R}^K} |\hat{h}(\mathbf{t})|^2 |\phi(\mathbf{t})|^2\,\mathrm{d}\mathbf{t}. \quad (91)$$

By (6) and (7),

$$\int_{\mathbb{R}^K} |\hat{h}(\mathbf{t})|^2 |\phi(\mathbf{t})|^2\,\mathrm{d}\mathbf{t} = \int_{\mathbb{R}^K} \left(\int_{\mathbb{R}^K} h(\mathbf{x}-\mathbf{y})(\mathrm{d}Z_0 - \mathrm{d}\mu_0)(\mathbf{y}) \right)^2 \mathrm{d}\mathbf{x}. \quad (92)$$

Observe that (see (61))

$$D_b(\mathbf{x}) = \int_{\mathbb{R}^K} h(\mathbf{x}-\mathbf{y})(\mathrm{d}Z_0 - \mathrm{d}\mu_0)(\mathbf{y}). \quad (93)$$

Combining (91)–(93), we conclude that

$$\int_{Q(b)} |\phi(\mathbf{t})|^2\,\mathrm{d}\mathbf{t} \ll b^{2K} \int_{\mathbb{R}^K} |D_b(\mathbf{x})|^2\,\mathrm{d}\mathbf{x}$$

as required. ■

6.3 Similar convex sets

In this section, we prove Theorem 17A.

For convenience of notation, let $Q(a)$ denote the cube $[-a, a]^K$, where $a > 0$ is real.

Let $M > 0$ be a parameter to be fixed later, and let $\mathscr{Q} = \{\mathbf{q}_1, \mathbf{q}_2, \mathbf{q}_3, \ldots\}$ be an infinite discrete set in $\mathbb{R}^K$. We introduce two measures. For any $E \subset \mathbb{R}^K$, let

$$Z_0(E) = \sum_{\mathbf{q}_j \in E \cap Q(M)} 1,$$

i.e. Z_0 denotes the counting measure generated by the set $Q(M) \cap \mathscr{Q}$. For any Lebesgue measurable set $E \subset \mathbb{R}^K$, let

$$\mu_0(E) = \mu(E \cap Q(M)),$$

i.e. μ_0 denotes the restriction of the usual K-dimensional volume to the cube $Q(M)$.

Let $\chi_{\tau,\lambda}$ denote the characteristic function of the set

$$A(\lambda, \mathbf{0}, \tau) = \{\tau(\lambda \mathbf{y}) : \mathbf{y} \in A\},$$

where A is the given compact convex body, τ is a proper orthogonal transformation in $\mathbb{R}^K$, and λ is a real number satisfying $0 < \lambda \leqslant 1$.

Consider the function

$$F_{\tau,\lambda} = \chi_{\tau,\lambda} * (\mathrm{d}Z_0 - \mathrm{d}\mu_0), \tag{94}$$

where $*$ denotes the convolution operation. More explicitly,

$$\begin{aligned} F_{\tau,\lambda}(\mathbf{x}) &= \int_{\mathbb{R}^K} \chi_{\tau,\lambda}(\mathbf{x} - \mathbf{y})(\mathrm{d}Z_0 - \mathrm{d}\mu_0)(\mathbf{y}) \\ &= \sum_{\mathbf{q}_j \in (-A(\lambda, \mathbf{x}, \tau)) \cap Q(M)} 1 - \mu((-A(\lambda, \mathbf{x}, \tau)) \cap Q(M)), \end{aligned} \tag{95}$$

where $-A(\lambda, \mathbf{x}, \tau) = \{\tau(\lambda \mathbf{y}) + \mathbf{x} : -\mathbf{y} \in A\}$. It follows that if $(-A(\lambda, \mathbf{x}, \tau)) \subset Q(M)$, then

$$F_{\tau,\lambda}(\mathbf{x}) = \sum_{\mathbf{q}_j \in (-A(\lambda, \mathbf{x}, \tau))} 1 - \mu(-A(\lambda, \mathbf{x}, \tau)); \tag{96}$$

in other words, $F_{\tau,\lambda}(\mathbf{x})$ is essentially a discrepancy function of $-A(\lambda, \mathbf{x}, \tau)$.

Our plan is to estimate the L^2-norm of $F_{\tau,\lambda}(\mathbf{x})$ from below, and then, by standard averaging arguments, obtain the desired lower bound for $\Omega[-A]$. The last trivial step is to apply the substitution $A \to (-A)$ (we prefer this (at-first-sight) artificial way for convenience of notation).

By the Parseval–Plancherel identity (7),

$$\int_{\mathbb{R}^K} |F_{\tau,\lambda}(\mathbf{x})|^2 \, \mathrm{d}\mathbf{x} = \int_{\mathbb{R}^K} |\hat{F}_{\tau,\lambda}(\mathbf{t})|^2 \, \mathrm{d}\mathbf{t}, \tag{97}$$

where $\hat{F}_{\tau,\lambda}$ denotes the Fourier transform of $F_{\tau,\lambda}$. By (6) and (94),

$$\hat{F}_{\tau,\lambda} = \hat{\chi}_{\tau,\lambda} \widehat{(\mathrm{d}Z_0 - \mathrm{d}\mu_0)}, \tag{98}$$

and so by (97),

$$\int_T \int_0^1 \int_{\mathbb{R}^K} |F_{\tau,\lambda}(\mathbf{x})|^2 \, \mathrm{d}\mathbf{x} \, \mathrm{d}\lambda \, \mathrm{d}\tau$$

$$= \int_{\mathbb{R}^K} \left(\int_T \int_0^1 |\hat{\chi}_{\tau,\lambda}(\mathbf{t})|^2 \, \mathrm{d}\lambda \, \mathrm{d}\tau \right) |(\widehat{\mathrm{d}Z_0 - \mathrm{d}\mu_0})(\mathbf{t})|^2 \, \mathrm{d}\mathbf{t}, \qquad (99)$$

where T is the group of proper orthogonal transformations in $\mathbb{R}^K$ and $\mathrm{d}\tau$ is the volume element of the invariant measure on T, normalized such that $\int_T \mathrm{d}\tau = 1$.

We mention in advance that $M \geqslant 100d(A)$, where $d(A)$ denotes the diameter of A. Thus we may assume that

$$\tfrac{1}{2}(2M)^K < \sum_{\mathbf{q}_j \in Q(M)} 1 < 2(2M)^K. \qquad (100)$$

Indeed, if (100) does not hold, then the result follows immediately from standard averaging arguments.

In what follows, we shall use both $\ll$ and the O-notation with constants which may depend on the dimension K only.

Let $\phi = (\widehat{\mathrm{d}Z_0 - \mathrm{d}\mu_0})$. We need

Lemma 6.9. *We have*

$$\int_{Q(100)} |\phi(\mathbf{t})|^2 \, \mathrm{d}\mathbf{t} \gg M^K.$$

Proof. By Lemma 6.7′ ($i = 0$), we have

$$\int_{Q(100)} |\phi(\mathbf{t})|^2 \, \mathrm{d}\mathbf{t} \gg \int_{\mathbb{R}^K} (W_M(\mathbf{x}))^2 \, \mathrm{d}\mathbf{x},$$

where $W_M(\mathbf{x}) = \#((U^K + \mathbf{x}) \cap Q(M) \cap \mathscr{Q})$. Moreover, by (100),

$$\int_{\mathbb{R}^K} (W_M(\mathbf{x}))^2 \, \mathrm{d}\mathbf{x} \geqslant \int_{\mathbb{R}^K} W_M(\mathbf{x}) \, \mathrm{d}\mathbf{x} = \sum_{\mathbf{q}_j \in Q(M)} 1 > \tfrac{1}{2}(2M)^K.$$

Summarizing, we have

$$\int_{Q(100)} |\phi(\mathbf{t})|^2 \, \mathrm{d}\mathbf{t} \gg \tfrac{1}{2}(2M)^K. \qquad \blacksquare$$

Clearly (see also (100))

$$|\phi(\mathbf{t})| = (2\pi)^{-\frac{1}{2}K} \left| \int_{\mathbb{R}^K} \mathrm{e}^{-\mathrm{i}\mathbf{x}\cdot\mathbf{t}} (\mathrm{d}Z_0 - \mathrm{d}\mu_0)(\mathbf{x}) \right|$$

$$\leqslant (2\pi)^{-\frac{1}{2}K} \left(\sum_{\mathbf{q}_j \in Q(M)} 1 + \mu(Q(M)) \right) \ll M^K. \qquad (101)$$

Therefore, if $c_{26}(K)$ is a sufficiently small positive constant, it follows from Lemma 6.9 and (101) that

$$\int_{Q(c_{26}(K)/M)} |\phi(\mathbf{t})|^2\,d\mathbf{t} < \tfrac{1}{2}\int_{Q(100)} |\phi(\mathbf{t})|^2\,d\mathbf{t}. \tag{102}$$

It follows from Lemma 6.9 and (102) that there exists an integer m, satisfying $1 \leqslant m = O(\log M)$, such that

$$\int_{Q_m} |\phi(\mathbf{t})|^2\,d\mathbf{t} \gg M^K m^{-2}, \tag{103}$$

where $Q_m = Q(100\cdot 2^{-m+1})\backslash Q(100\cdot 2^{-m})$.

We give, first of all, a rough outline of the proof. By (99) and (103), it suffices to establish a suitable lower bound for

$$\inf_{\mathbf{t}\in Q_m} \int_T \int_0^1 |\hat{\chi}_{\tau,\lambda}(\mathbf{t})|^2\,d\lambda\,d\tau.$$

Clearly

$$\hat{\chi}_{\tau,\lambda}(\mathbf{t}) = \lambda^K \hat{\chi}_A(\tau^{-1}(\lambda\mathbf{t})),$$

where χ_A denotes the characteristic function of the given compact and convex body $A \subset \mathbb{R}^K$. Let $G(\mathbf{x})$, where $\mathbf{x}\in\mathbb{R}^K$, be a function satisfying

$$|\hat{G}(\mathbf{t})| \begin{cases} = 0 & (\mathbf{t}\in Q(\delta 2^{-m})), \\ \leqslant 2 & (\mathbf{t}\in Q(100K^{-\frac{1}{2}}2^{-m})\backslash Q(\delta 2^{-m}) = \tilde{Q}), \\ = 0 & (\mathbf{t}\notin Q(100K^{-\frac{1}{2}}2^{-m})), \end{cases} \tag{104}$$

where $\delta > 0$ is a sufficiently small but fixed constant depending only on K. Then for every $\mathbf{t}^*\in Q_m$,

$$\begin{aligned}
\int_T\int_0^1 |\hat{\chi}_{\tau,\lambda}(\mathbf{t}^*)|^2\,d\lambda\,d\tau &= \int_0^1 \lambda^{2K}\left(\int_T |\hat{\chi}_A(\tau^{-1}(\lambda\mathbf{t}^*))|^2\,d\tau\right)d\lambda \\
&= \int_0^1 \lambda^{2K}\sigma^{-1}(C(\lambda|\mathbf{t}^*|,0))\left(\int_{|\mathbf{t}|=\lambda|\mathbf{t}^*|} |\hat{\chi}_A(\mathbf{t})|^2\,d\sigma(\mathbf{t})\right)d\lambda \\
&= \frac{1}{|\mathbf{t}^*|}\int_0^{|\mathbf{t}^*|}\left(\frac{y}{|\mathbf{t}^*|}\right)^{2K}\sigma^{-1}(C(y,0))\left(\int_{|\mathbf{t}|=y} |\hat{\chi}_A(\mathbf{t})|^2\,d\sigma(\mathbf{t})\right)dy \\
&= \frac{1}{|\mathbf{t}^*|}\int_{|\mathbf{t}|\leqslant|\mathbf{t}^*|}\left(\frac{|\mathbf{t}|}{|\mathbf{t}^*|}\right)^{2K}\sigma^{-1}(C(|\mathbf{t}|,0))|\hat{\chi}_A(\mathbf{t})|^2\,d\mathbf{t} \\
&\geqslant \frac{1}{|\mathbf{t}^*|}\int_{\tilde{Q}}\left(\frac{|\mathbf{t}|}{|\mathbf{t}^*|}\right)^{2K}\sigma^{-1}(C(|\mathbf{t}|,0))|\hat{\chi}_A(\mathbf{t})|^2\,d\mathbf{t} \\
&\gg_\delta \mu^{-1}(\tilde{Q})\int_{\tilde{Q}} |\hat{\chi}_A(\mathbf{t})|^2\,d\mathbf{t} \gg \mu^{-1}(Q_m)\int_{\mathbb{R}^K} |\hat{\chi}_A(\mathbf{t})|^2|\hat{G}(\mathbf{t})|^2\,d\mathbf{t},
\end{aligned} \tag{105}$$

where $\tilde{Q}$ is defined in (104), σ denotes the $(K-1)$-dimensional surface area and $C(r,\mathbf{0})$ is the ball $\{\mathbf{x}\in\mathbb{R}^K:|\mathbf{x}|\leqslant r\}$. Choosing $f=\chi_A * G$, we have, by (6) and the Parseval–Plancherel identity (7), that

$$\int_{\mathbf{R}^K}|\hat{\chi}_A(\mathbf{t})|^2|\hat{G}(\mathbf{t})|^2\,\mathrm{d}\mathbf{t}=\int_{\mathbf{R}^K}\left(\int_{\mathbf{R}^K}\chi_A(\mathbf{x}-\mathbf{y})G(\mathbf{y})\,\mathrm{d}\mathbf{y}\right)^2\mathrm{d}\mathbf{x}. \tag{106}$$

Therefore, in order to give a lower bound to the left-hand side of (105), it is sufficient to investigate the right-hand side of (106).

We shall construct the desired function G in the form of a difference $G=h-H$. Besides (104), the functions h and H will satisfy the following properties:

(i) We have

$$\int_{\mathbf{R}^K}h(\mathbf{y})\,\mathrm{d}\mathbf{y}=\int_{\mathbf{R}^K}H(\mathbf{y})\,\mathrm{d}\mathbf{y}=1.$$

(ii) Both functions h and H are 'predominantly' positive.

Let $r(h)$ be the smallest radius such that

$$\int_{|\mathbf{y}|\leqslant r(h)}h(\mathbf{y})\,\mathrm{d}\mathbf{y}\geqslant\frac{99}{100},$$

and similarly, let $r(H)$ be the smallest radius such that

$$\int_{|\mathbf{y}|\leqslant r(H)}H(\mathbf{y})\,\mathrm{d}\mathbf{y}\geqslant\frac{99}{100}.$$

We also need:

(iii) $r(H)$ is 'much smaller' than $r(h)$. In other words, the main contribution to the integral of $H(\mathbf{y})$ $(\mathbf{y}\in\mathbb{R}^K)$ comes from values of $\mathbf{y}$ in a much smaller ball centred at the origin than that in the case of $h(\mathbf{y})$.

We briefly describe the geometric heuristics of the proof. Assume that $\mathbf{x}_0\in A$ and that the euclidean distance of $\mathbf{x}_0$ and the boundary ∂A of A is in the interval $[r(H),2r(H)]$. Since the ball $C(r(H),\mathbf{x}_0)$ is contained in A, it is expected that

$$\int_{\mathbf{R}^K}\chi_A(\mathbf{x}_0-\mathbf{y})H(\mathbf{y})\,\mathrm{d}\mathbf{y}>\frac{9}{10}$$

(note that H is not necessarily positive everywhere on the set $\{\mathbf{x}_0-\mathbf{z}:\mathbf{z}\in A\}$, but predominantly positive). On the other hand, the intersection $C(r(h),\mathbf{x}_0)\cap A$ forms, roughly speaking, a 'half-ball', so it is rightly expected that the integral

$$\int_{\mathbf{R}^K}\chi_A(\mathbf{x}_0-\mathbf{y})h(\mathbf{y})\,\mathrm{d}\mathbf{y}$$

is about half of the integral $\int_{|y|\leqslant r(h)} h(\mathbf{y})\,d\mathbf{y}$, i.e. about $\frac{1}{2}$. Summarizing, we expect, for these values of $\mathbf{x}_0$, that the integral

$$\left|\int_{\mathbb{R}^K} \chi_A(\mathbf{x}_0-\mathbf{y})G(\mathbf{y})\,d\mathbf{y}\right| = \left|\int_{\mathbb{R}^K} \chi_A(\mathbf{x}_0-\mathbf{y})(H-h)(\mathbf{y})\,d\mathbf{y}\right|$$

is greater than a positive absolute constant, which yields the desired lower bound for the right-hand side of (106).

We now give the explicit form of the functions h and H as follows. For $\mathbf{x}\in\mathbb{R}^K$, let

$$h(\mathbf{x}) = \frac{1}{(2\varepsilon_1)^K}\left(\prod_{j=1}^{K}\frac{2\sin(\varepsilon_1 x_j)}{(2\pi)^{\frac{1}{2}}x_j}\right)\left(\prod_{j=1}^{K}\frac{2\sin(\varepsilon_2 x_j)}{(2\pi)^{\frac{1}{2}}x_j}\right), \tag{107}$$

and

$$H(\mathbf{x}) = \frac{1}{(2\varepsilon_3)^K}\left(\prod_{j=1}^{K}\frac{2\sin(\varepsilon_3 x_j)}{(2\pi)^{\frac{1}{2}}x_j}\right)\left(\prod_{j=1}^{K}\frac{2\sin(\varepsilon_4 x_j)}{(2\pi)^{\frac{1}{2}}x_j}\right), \tag{108}$$

where $0<\varepsilon_1<\varepsilon_2<\varepsilon_3<\varepsilon_4$ and $\varepsilon_2-\varepsilon_1=\varepsilon_4-\varepsilon_3$. These four parameters ε_i, where $1\leqslant i\leqslant 4$, will be specified later.

In the next five lemmas, we list the basic properties of h and H for later use.

Lemma 6.10. *We have*

$$(\hat{h}-\hat{H})(\mathbf{t})\begin{cases}=0 & (\mathbf{t}\in Q(\varepsilon_4-\varepsilon_3)=Q(\varepsilon_2-\varepsilon_1)),\\ \leqslant 2 & (\mathbf{t}\in Q(\varepsilon_3+\varepsilon_4)\backslash Q(\varepsilon_4-\varepsilon_3)),\\ =0 & (\mathbf{t}\notin Q(\varepsilon_3+\varepsilon_4)).\end{cases} \tag{109}$$

Proof. Since

$$\widehat{\frac{2\sin(bx)}{(2\pi)^{\frac{1}{2}}x}} = \chi_{[-b,b]}, \tag{110}$$

where $\chi_{[-b,b]}$ denotes the characteristic function of the interval $[-b,b]$, we have, by (40), that

$$\hat{h}(\mathbf{t}) = \frac{1}{(2\varepsilon_1)^K}\prod_{j=1}^{K}(\chi_{[-\varepsilon_1,\varepsilon_1]}*\chi_{[-\varepsilon_2,\varepsilon_2]})(t_j) \tag{111}$$

and

$$\hat{H}(\mathbf{t}) = \frac{1}{(2\varepsilon_3)^K}\prod_{j=1}^{K}(\chi_{[-\varepsilon_3,\varepsilon_3]}*\chi_{[-\varepsilon_4,\varepsilon_4]})(t_j), \tag{112}$$

where $*$ denotes the convolution operation. If $0<a<b$, then obviously

$$(\chi_{[-a,a]}*\chi_{[-b,b]})(t)\begin{cases}=2a & (|t|\leqslant b-a),\\ \leqslant 2a & (b-a\leqslant|t|\leqslant b+a),\\ =0 & (|t|\geqslant b+a).\end{cases} \tag{113}$$

(109) follows on combining (111)–(113). ■

We define ε_i, where $1 \leqslant i \leqslant 4$, by

$$\begin{cases} \varepsilon_3 + \varepsilon_4 = 100K^{-\frac{1}{2}}2^{-m}, \\ \varepsilon_3 = (1 - c^{-4})\varepsilon_4, \\ \varepsilon_2 = c^{-2}\varepsilon_4, \\ \varepsilon_1 = \varepsilon_2 - c^{-4}\varepsilon_4 = (1 - c^{-2})\varepsilon_2, \end{cases} \tag{114}$$

where the parameter $c \geqslant 4$ will be specified later as a sufficiently large constant depending only on K. Observe that $\varepsilon_2 - \varepsilon_1 = \varepsilon_4 - \varepsilon_3$.

Combining the well-known identity (see any textbook on harmonic analysis)

$$\int_{\mathbb{R}^K} f(\mathbf{x})\overline{g(\mathbf{x})}\, d\mathbf{x} = \int_{\mathbb{R}^K} \hat{f}(\mathbf{t})\overline{\hat{g}(\mathbf{t})}\, d\mathbf{t},$$

where $f, g \in L^2(\mathbb{R}^K)$, and (107), (108) and (110), we conclude that

Lemma 6.11. *We have*

$$\int_{\mathbb{R}^K} h(\mathbf{y})\, d\mathbf{y} = \int_{\mathbb{R}^K} (2\varepsilon_1)^{-K} \chi_{Q(\varepsilon_1)}(\mathbf{t})\chi_{Q(\varepsilon_2)}(\mathbf{t})\, d\mathbf{t} = 1 \tag{115}$$

and

$$\int_{\mathbb{R}^K} H(\mathbf{y})\, d\mathbf{y} = \int_{\mathbb{R}^K} (2\varepsilon_3)^{-K} \chi_{Q(\varepsilon_3)}(\mathbf{t})\chi_{Q(\varepsilon_4)}(\mathbf{t})\, d\mathbf{t} = 1. \tag{116}$$

Elementary estimates combined with (107), (108) and (114) give

Lemma 6.12. *We have*

$$\mathbf{y} \in Q(1/\varepsilon_2) \Rightarrow (h(\mathbf{y}) \geqslant 0 \textit{ and } h(\mathbf{y}) \gg \varepsilon_2^K) \tag{117}$$

and

$$\mathbf{y} \in Q(1/\varepsilon_4) \Rightarrow (H(\mathbf{y}) \geqslant 0 \textit{ and } H(\mathbf{y}) \gg \varepsilon_4^K). \tag{118}$$

We also have

Lemma 6.13. *For any $\beta \geqslant 1$, we have*

$$\int_{\mathbb{R}^K \setminus Q(\beta/\varepsilon_2)} |h(\mathbf{y})|\, d\mathbf{y} \ll 1/\beta \tag{119}$$

and

$$\int_{\mathbb{R}^K \setminus Q(\beta/\varepsilon_4)} |H(\mathbf{y})|\, d\mathbf{y} \ll 1/\beta. \tag{120}$$

Proof. Writing

$$Q\left(\frac{1}{2\varepsilon_i}; \mathbf{l}\right) = \prod_{j=1}^{K} \left[\frac{l_j - \frac{1}{2}}{\varepsilon_i}, \frac{l_j + \frac{1}{2}}{\varepsilon_i}\right],$$

where $\mathbf{l}=(l_1,\ldots,l_K)\in\mathbb{Z}^K$ and $1\leqslant i\leqslant 4$, we have

$$\mathbf{y}\in Q\left(\frac{1}{2\varepsilon_2};\mathbf{l}\right)\Rightarrow h(\mathbf{y})\ll \varepsilon_2^K\prod_{j=1}^K\frac{1}{(1+|l_j|)^2} \tag{121}$$

and

$$\mathbf{y}\in Q\left(\frac{1}{2\varepsilon_4};\mathbf{l}\right)\Rightarrow H(\mathbf{y})\ll \varepsilon_4^K\prod_{j=1}^K\frac{1}{(1+|l_j|)^2}. \tag{122}$$

It follows that for any $\beta\geqslant 1$,

$$\int_{\mathbb{R}^K\setminus Q(\beta/\varepsilon_2)}|h(\mathbf{y})|\,d\mathbf{y}\leqslant \sum_{\substack{\mathbf{l}\in\mathbb{Z}^K\\ \max_{1\leqslant j\leqslant K}|l_j|\geqslant\beta-\frac{1}{2}}}\int_{Q(1/2\varepsilon_2;\mathbf{l})}|h(\mathbf{y})|\,d\mathbf{y}$$

$$\ll \sum_{\substack{\mathbf{l}\in\mathbb{Z}^K\\ \max_{1\leqslant j\leqslant K}|l_j|\geqslant\beta-\frac{1}{2}}}\prod_{j=1}^K\frac{1}{(1+|l_j|)^2}$$

$$=\left(\sum_{l\in\mathbb{Z}}\frac{1}{(1+|l|)^2}\right)^K-\left(\sum_{\substack{l\in\mathbb{Z}\\ |l|<\beta-\frac{1}{2}}}\frac{1}{(1+|l|)^2}\right)^K\ll\sum_{\substack{l\in\mathbb{Z}\\ |l|\geqslant\beta-\frac{1}{2}}}\frac{1}{(1+|l|)^2}\ll\frac{1}{\beta}.$$

This gives (119). Repeating the same argument, we get (120). ∎

Let

$$h^-(\mathbf{y})=\begin{cases}h(\mathbf{y}) & (h(\mathbf{y})<0),\\ 0 & (h(\mathbf{y})\geqslant 0);\end{cases}$$

and

$$H^-(\mathbf{y})=\begin{cases}H(\mathbf{y}) & (H(\mathbf{y})<0),\\ 0 & (H(\mathbf{y})\geqslant 0).\end{cases}$$

Lemma 6.14. *Let* $\mathbf{l}=(l_1,\ldots,l_K)\in\mathbb{Z}^K$, *where* $|l_j|\leqslant c+\frac{1}{2}$ *for all* j *satisfying* $1\leqslant j\leqslant K$.

(i) *If* $\mathbf{y}\in Q(1/2\varepsilon_2;\mathbf{l})$, *then*

$$|h^-(\mathbf{y})|\ll c^{-2}\varepsilon_2^K\prod_{j=1}^K\frac{1}{(1+|l_j|)^2}. \tag{123}$$

(ii) *If* $\mathbf{y}\in Q(1/2\varepsilon_4;\mathbf{l})$, *then*

$$|H^-(\mathbf{y})|\ll c^{-6}\varepsilon_4^K\prod_{j=1}^K\frac{1}{(1+|l_j|)^2}. \tag{124}$$

Proof. We need the elementary observation that if $0<a<b$ and $\sin(ax)\sin(bx)<0$, then

$$|\sin(ax)\sin(bx)|\leqslant\tfrac{1}{4}(b-a)^2x^2. \tag{125}$$

Indeed, for some d satisfying $a < d < b$, we have $\sin(dx) = 0$, and so $|\sin(ax)\sin(bx)| = |\sin((d-a)x)\sin((b-d)x)| \leqslant (d-a)(b-d)x^2 \leqslant (b-a)^2x^2/4$. Consider (i). If $h(\mathbf{y}) < 0$, then for some v satisfying $1 \leqslant v \leqslant K$, we have $\sin(\varepsilon_1 y_v)\sin(\varepsilon_2 y_v) < 0$. It follows from (107), (114) and (125) that

$$|h^-(\mathbf{y})| \ll \frac{|\sin(\varepsilon_1 y_v)\sin(\varepsilon_2 y_v)|}{\varepsilon_1 y_v \varepsilon_2 y_v} \varepsilon_2^K \prod_{\substack{j=1 \\ j \neq v}}^{K} \frac{1}{(1+|l_j|)^2}$$

$$\leqslant \frac{(\varepsilon_2 - \varepsilon_1)^2}{4\varepsilon_1 \varepsilon_2} \varepsilon_2^K \prod_{\substack{j=1 \\ j \neq v}}^{K} \frac{1}{(1+|l_j|)^2}$$

$$\ll \left(\frac{1}{c^2}\right)^2 \varepsilon_2^K \prod_{\substack{j=1 \\ j \neq v}}^{K} \frac{1}{(1+|l_j|)^2} \ll \frac{1}{c^2} \varepsilon_2^K \prod_{j=1}^{K} \frac{1}{(1+|l_j|)^2}.$$

Similarly for (ii): If $H(\mathbf{y}) < 0$, then for some v satisfying $1 \leqslant v \leqslant K$, we have $\sin(\varepsilon_3 y_v)\sin(\varepsilon_4 y_v) < 0$. It follows from (108), (114) and (125) that (similar to the argument above)

$$|H^-(\mathbf{y})| \ll \left(\frac{1}{c^4}\right)^2 \varepsilon_4^K \prod_{\substack{j=1 \\ j \neq v}}^{K} \frac{1}{(1+|l_j|)^2} \ll \frac{1}{c^6} \varepsilon_4^K \prod_{j=1}^{K} \frac{1}{(1+|l_j|)^2}. \qquad \blacksquare$$

For any compact convex body $P \subset \mathbb{R}^K$ and real number ρ satisfying $0 < \rho \leqslant r(P)$, where $r(P)$ denotes the radius of the largest inscribed ball in P, let $P^-[\rho]$ denote the set of all centres of balls contained in P and of radius ρ. It is obvious that $P^-[\rho]$ is also compact and convex.

We shall apply the following simple result in discrete geometry (see Hadwiger (1955)): For any compact and convex body $A \subset \mathbb{R}^K$, there exist boxes B and D in arbitrary position but with parallel edges such that

$$B \subset A \subset D \quad \text{and} \quad \frac{\mu(D)}{K!} \leqslant \mu(A) \leqslant K^K \mu(B). \tag{126}$$

Let b_j and d_j $(j = 1, 2, \ldots, K)$ denote the length of the parallel edges of B and D respectively. Without loss of generality, we may assume that

$$b_1 \leqslant b_2 \leqslant \cdots \leqslant b_K. \tag{127}$$

It follows immediately from (126) that for $j = 1, \ldots, K$,

$$b_j \leqslant d_j \leqslant K^K K! b_j. \tag{128}$$

Furthermore, we shall use without proof the following well-known fact in geometry: If P_1 and P_2 are compact convex bodies in $\mathbb{R}^K$, then

$$P_1 \subset P_2 \Rightarrow \sigma(\partial P_1) \leqslant \sigma(\partial P_2), \tag{129}$$

where ∂P denotes the boundary surface of P, and σ is the $(K-1)$-dimensional surface area.

Let $\varepsilon = c\varepsilon_2 = c^{-1}\varepsilon_4$. We are going to estimate from below the right-hand side of (106). We distinguish two cases: (I) $1/\varepsilon \leqslant b_1/8$; and (II) $1/\varepsilon > b_1/8$.

For case (I), we prove

Lemma 6.15. *Suppose that $1/\varepsilon \leqslant b_1/8$. Then if $c \geqslant c_{27}(K)$, we have*

$$\int_{\mathbb{R}^K}\left(\int_{\mathbb{R}^K} \chi_A(\mathbf{x}-\mathbf{y})(h-H)(\mathbf{y})\,d\mathbf{y}\right)^2 d\mathbf{x} \gg \varepsilon^{-1}\sigma(\partial A).$$

Let $A_1 = A^-[1/\varepsilon]$ and $A_2 = A_1 \backslash A_1^-[1/\varepsilon]$. The idea of the proof of Lemma 6.15 is to consider the inner integral when $\mathbf{x} \in A_2$ and obtain a good lower bound. We therefore start by giving a lower bound for the volume of A_2. It is sufficient to use the following very crude estimate.

Lemma 6.16. *If $P \subset \mathbb{R}^K$ is a compact convex body and $0 < \rho \leqslant r(P)$, then*

$$\mu(P \backslash P^-[\rho]) > c_{28}(K)\rho\sigma(\partial P^-[\rho]),$$

where the positive constant $c_{28}(K)$ depends only on the dimension K.

Proof. Let

$$E = \partial P^-[\rho], \quad F = C(\rho, \mathbf{0}) = \{\mathbf{y} \in \mathbb{R}^K : |\mathbf{y}| \leqslant \rho\} \quad \text{and} \quad G = P \backslash P^-[\rho].$$

Simple double integration gives

$$\begin{aligned}
&\int_E \mu((F+\mathbf{x}) \cap G)\,d\sigma(\mathbf{x}) \\
&= \int_E \int_G \chi_\rho(\mathbf{y}-\mathbf{x})\,d\mathbf{y}\,d\sigma(\mathbf{x}) = \int_G \int_E \chi_\rho(\mathbf{y}-\mathbf{x})\,d\sigma(\mathbf{x})\,d\mathbf{y} \\
&= \int_G \int_E \chi_\rho(\mathbf{x}-\mathbf{y})\,d\sigma(\mathbf{x})\,d\mathbf{y} = \int_G \sigma((F+\mathbf{y}) \cap E)\,d\mathbf{y},
\end{aligned} \tag{130}$$

where χ_ρ denotes the characteristic function of the ball F. Since $P^-[\rho]$ is convex, for any $\mathbf{x} \in E$, the intersection $(F+\mathbf{x}) \cap G$ certainly contains a half-ball of radius ρ. Hence

$$\mu((F+\mathbf{x}) \cap G) \geqslant \tfrac{1}{2}\mu(F) \tag{131}$$

for any $\mathbf{x} \in E$. On the other hand, by (129), for every $\mathbf{y} \in \mathbb{R}^K$,

$$\sigma((F+\mathbf{y}) \cap E) \leqslant \sigma(\partial F). \tag{132}$$

Combining (130)–(132), we obtain

$$\tfrac{1}{2}\sigma(E)\mu(F) \leqslant \int_E \mu((F+\mathbf{x})\cap G)\,\mathrm{d}\sigma(\mathbf{x})$$
$$= \int_G \sigma((F+\mathbf{y})\cap E)\,\mathrm{d}\mathbf{y} \leqslant \mu(G)\sigma(\partial F),$$

and so

$$\mu(G) \geqslant \frac{\sigma(E)\mu(F)}{2\sigma(\partial F)} = c_{28}(K)\rho\sigma(E). \qquad \blacksquare$$

By definition (see (126))

$$B = \tau_1\left(\prod_{j=1}^{K}[-\tfrac{1}{2}b_j, \tfrac{1}{2}b_j]\right) + \mathbf{x}_1$$

with some appropriate orthogonal transformation τ_1 and translation $\mathbf{x}_1$. By hypothesis, $2/\varepsilon \leqslant b_1/4 \leqslant b_2/4 \leqslant \cdots \leqslant b_K/4$, and so we have (the reader is advised to draw a diagram for the case $K=2$)

$$A_1^-[1/\varepsilon] \supset \tau_1\left(\prod_{j=1}^{K}[-b_j/4, b_j/4]\right) + \mathbf{x}_1. \tag{133}$$

By (129) and (133),

$$\sigma(\partial A_1^-[1/\varepsilon]) \gg \sigma(\partial B), \tag{134}$$

and so by (128) and (129),

$$\sigma(\partial B) \gg \sigma(\partial D) \geqslant \sigma(\partial A). \tag{135}$$

It follows from (134) and (135) that

$$\sigma(\partial A_1^-[1/\varepsilon]) \gg \sigma(\partial A). \tag{136}$$

Combining Lemma 6.16 and (136), we conclude that

$$\mu(A_2) = \mu(A_1 \backslash A_1^-[1/\varepsilon]) \gg \varepsilon^{-1}\sigma(\partial A). \tag{137}$$

Let $\mathbf{x}_0 \in A_2$. We estimate the integral

$$\left|\int_{\mathbf{R}^K} \chi_A(\mathbf{x}_0 - \mathbf{y})(h-H)(\mathbf{y})\,\mathrm{d}\mathbf{y}\right|$$

from below. Since $\mathbf{x}_0 \in A_1 = A^-[1/\varepsilon]$, we have

$$(Q(1/K\varepsilon) + \mathbf{x}_0) \subset A.$$

Therefore

$$\int_{\mathbf{R}^K} \chi_A(\mathbf{x}_0 - \mathbf{y})H(\mathbf{y})\,\mathrm{d}\mathbf{y} \geqslant \int_{Q(1/K\varepsilon)} H(\mathbf{y})\,\mathrm{d}\mathbf{y} - \int_{\mathbf{R}^K \backslash Q(1/K\varepsilon)} |H(\mathbf{y})|\,\mathrm{d}\mathbf{y}. \tag{138}$$

By Lemma 6.11,

$$\int_{Q(1/K\varepsilon)} H(\mathbf{y})\,d\mathbf{y} = 1 - \int_{\mathbb{R}^K\setminus Q(1/K\varepsilon)} H(\mathbf{y})\,d\mathbf{y} \geqslant 1 - \int_{\mathbb{R}^K\setminus Q(1/K\varepsilon)} |H(\mathbf{y})|\,d\mathbf{y}. \tag{139}$$

Combining (138) and (139), we have

$$\int_{\mathbb{R}^K} \chi_A(\mathbf{x}_0 - \mathbf{y})H(\mathbf{y})\,d\mathbf{y} \geqslant 1 - 2\int_{\mathbb{R}^K\setminus Q(1/K\varepsilon)} |H(\mathbf{y})|\,d\mathbf{y}. \tag{140}$$

Since $\varepsilon = c^{-1}\varepsilon_4$, it follows from Lemma 6.13 that

$$\int_{\mathbb{R}^K\setminus Q(1/K\varepsilon)} |H(\mathbf{y})|\,d\mathbf{y} \ll c^{-1}.$$

Hence for any $\mathbf{x}_0 \in A_2$, we have

$$\int_{\mathbb{R}^K} \chi_A(\mathbf{x}_0 - \mathbf{y})H(\mathbf{y})\,d\mathbf{y} \geqslant 1 - O(c^{-1}). \tag{141}$$

On the other hand, since $\mathbf{x}_0 \in A_2 = A_1 \setminus A_1^-[1/\varepsilon]$, where $A_1 = A^-[1/\varepsilon]$, we see that the usual euclidean distance of $\mathbf{x}_0$ and the complement $\mathbb{R}^K\setminus A$ of A does not exceed $2/\varepsilon$. Using this observation and the convexity of A, it is easily seen that

$$(Q(1/\varepsilon_2) + \mathbf{x}_0)\setminus A \text{ contains a ball of radius } 1/4\varepsilon_2. \tag{142}$$

Here we also used the hypotheses that $\varepsilon = c\varepsilon_2$ and $c \geqslant 4$. Clearly

$$\int_{\mathbb{R}^K} \chi_A(\mathbf{x}_0 - \mathbf{y})h(\mathbf{y})\,d\mathbf{y} \leqslant \int_{Q(1/\varepsilon_2)} \chi_A(\mathbf{x}_0 - \mathbf{y})h(\mathbf{y})\,d\mathbf{y} + \int_{\mathbb{R}^K\setminus Q(1/\varepsilon_2)} h(\mathbf{y})\,d\mathbf{y} + \left|\int_{\mathbb{R}^K\setminus Q(1/\varepsilon_2)} h^-(\mathbf{y})\,d\mathbf{y}\right|. \tag{143}$$

By (115), (117) and (142),

$$\begin{aligned}\int_{Q(1/\varepsilon_2)} &\chi_A(\mathbf{x}_0 - \mathbf{y})h(\mathbf{y})\,d\mathbf{y} + \int_{\mathbb{R}^K\setminus Q(1/\varepsilon_2)} h(\mathbf{y})\,d\mathbf{y} \\ &= \int_{\mathbb{R}^K} h(\mathbf{y})\,d\mathbf{y} + \int_{Q(1/\varepsilon_2)} (\chi_A(\mathbf{x}_0 - \mathbf{y}) - 1)h(\mathbf{y})\,d\mathbf{y} \\ &\leqslant 1 - c_{29}(K)\varepsilon_2^K \mu(\{\mathbf{y}\in\mathbb{R}^K : |\mathbf{y}| \leqslant 1/4\varepsilon_2\}) = 1 - c_{30}(K),\end{aligned} \tag{144}$$

where $c_{30}(K) > 0$. Also,

$$\left|\int_{\mathbb{R}^K\setminus Q(1/\varepsilon_2)} h^-(\mathbf{y})\,d\mathbf{y}\right| \leqslant \int_{\mathbb{R}^K\setminus Q(c/\varepsilon_2)} |h(\mathbf{y})|\,d\mathbf{y} + \sum_{\substack{\mathbf{l}\in\mathbb{Z}^K \\ \max_{1\leqslant j\leqslant K} |l_j| \leqslant c+\frac{1}{2}}} \left|\int_{Q(1/2\varepsilon_2;\mathbf{l})} h^-(\mathbf{y})\,d\mathbf{y}\right|, \tag{145}$$

where $Q(1/2\varepsilon_2;\mathbf{l})$ is defined in the proof of Lemma 6.13. By Lemma 6.13,

$$\int_{\mathbf{R}^K\setminus Q(c/\varepsilon_2)} |h(\mathbf{y})|\,d\mathbf{y} \ll c^{-1}. \tag{146}$$

By Lemma 6.14,

$$\begin{aligned}
&\sum_{\substack{\mathbf{l}\in\mathbf{Z}^K\\ \max_{1\leqslant j\leqslant K}|l_j|\leqslant c+\frac{1}{2}}} \left|\int_{Q(1/2\varepsilon_2;\mathbf{l})} h^-(\mathbf{y})\,d\mathbf{y}\right| \\
&\ll \sum_{\substack{\mathbf{l}\in\mathbf{Z}^K\\ \max_{1\leqslant j\leqslant K}|l_j|\leqslant c+\frac{1}{2}}} c^{-2}\varepsilon_2^K\mu(Q(1/2\varepsilon_2;\mathbf{l}))\prod_{j=1}^K \frac{1}{(1+|l_j|)^2} \\
&= c^{-2} \sum_{\substack{\mathbf{l}\in\mathbf{Z}^K\\ \max_{1\leqslant j\leqslant K}|l_j|\leqslant c+\frac{1}{2}}} \prod_{j=1}^K \frac{1}{(1+|l_j|)^2} \\
&\leqslant c^{-2}\left(\sum_{l\in\mathbf{Z}}\frac{1}{(1+|l|)^2}\right)^K \ll c^{-2}.
\end{aligned} \tag{147}$$

Combining (143)–(147), we have, for any $\mathbf{x}_0\in A_2$, that

$$\int_{\mathbf{R}^K} \chi_A(\mathbf{x}_0-\mathbf{y})h(\mathbf{y})\,d\mathbf{y} \leqslant 1 - c_{30}(K) + O(c^{-1}) + O(c^{-2}). \tag{148}$$

It therefore follows from (141) and (148) that

$$\int_{\mathbf{R}^K} \chi_A(\mathbf{x}_0-\mathbf{y})(H-h)(\mathbf{y})\,d\mathbf{y} \geqslant c_{30}(K) - O(c^{-1}) > \tfrac{1}{2}c_{30}(K) > 0 \tag{149}$$

if $c\geqslant c_{27}(K)$ and $\mathbf{x}_0\in A_2$. Combining (137) and (149), we see that if $c\geqslant c_{27}(K)$, then

$$\begin{aligned}
&\int_{\mathbf{R}^K}\left(\int_{\mathbf{R}^K}\chi_A(\mathbf{x}-\mathbf{y})(h-H)(\mathbf{y})\,d\mathbf{y}\right)^2 d\mathbf{x} \\
&\geqslant \int_{A_2}\left(\int_{\mathbf{R}^K}\chi_A(\mathbf{x}-\mathbf{y})(h-H)(\mathbf{y})\,d\mathbf{y}\right)^2 d\mathbf{x} \gg \mu(A_2) \gg \varepsilon^{-1}\sigma(\partial A).
\end{aligned} \tag{150}$$

This completes the proof of Lemma 6.15.

For case (II), we prove

Lemma 6.17. *Suppose that $1/\varepsilon > b_1/8$. Let j_0 be the largest index j, satisfying $1\leqslant j\leqslant K$, such that $\varepsilon_4 b_j\leqslant 1$. Then if $c\geqslant c_{31}(K)$, we have*

$$\int_{\mathbf{R}^K}\left(\int_{\mathbf{R}^K}\chi_A(\mathbf{x}-\mathbf{y})(h-H)(\mathbf{y})\,d\mathbf{y}\right)^2 d\mathbf{x} \gg \left(\prod_{1\leqslant i\leqslant j_0}\varepsilon_4 b_i^2\right)\left(\prod_{j_0<j\leqslant K} b_j\right).$$

Since the boxes B and D have parallel edges, we have

$$B=\tau_1\left(\prod_{j=1}^{K}[-\tfrac{1}{2}b_j,\tfrac{1}{2}b_j]\right)+\mathbf{x}_1 \quad \text{and} \quad D=\tau_1\left(\prod_{j=1}^{K}[-\tfrac{1}{2}d_j,\tfrac{1}{2}d_j]\right)+\mathbf{x}_2,$$

where τ_1 is an orthogonal transformation and $\mathbf{x}_1,\mathbf{x}_2\in\mathbb{R}^K$. Let

$$B_1=\tau_1\left(\prod_{j=1}^{K}[-\tfrac{1}{2}b_j-(2K\varepsilon_4)^{-1},\tfrac{1}{2}b_j+(2K\varepsilon_4)^{-1}]\right)+\mathbf{x}_1.$$

Let $\mathbf{x}_0\in B_1$. We shall estimate

$$\left|\int_{\mathbf{R}^K}\chi_A(\mathbf{x}_0-\mathbf{y})(h-H)(\mathbf{y})\,d\mathbf{y}\right|^2$$

from below. Clearly

$$\int_{\mathbf{R}^K}\chi_A(\mathbf{x}_0-\mathbf{y})H(\mathbf{y})\,d\mathbf{y}\geqslant\int_{Q(1/\varepsilon_4)}\chi_A(\mathbf{x}_0-\mathbf{y})H(\mathbf{y})\,d\mathbf{y}$$
$$-\left|\int_{\mathbf{R}^K\setminus Q(1/\varepsilon_4)}\chi_A(\mathbf{x}_0-\mathbf{y})H^-(\mathbf{y})\,d\mathbf{y}\right|. \quad (151)$$

Since $\mathbf{x}_0\in B_1$, we see that the usual euclidean distance between $\mathbf{x}_0$ and B is less than $1/2\varepsilon_4$. Hence

$$\mu((Q(1/\varepsilon_4)+\mathbf{x}_0)\cap B)\gg\prod_{j=1}^{K}\min\{1/\varepsilon_4,b_j\}. \quad (152)$$

By (118) and (152),

$$\int_{Q(1/\varepsilon_4)}\chi_A(\mathbf{x}_0-\mathbf{y})H(\mathbf{y})\,d\mathbf{y}\geqslant c_{32}(K)\varepsilon_4^K\mu((Q(1/\varepsilon_4)+\mathbf{x}_0)\cap A)$$
$$\geqslant c_{32}(K)\varepsilon_4^K\mu((Q(1/\varepsilon_4)+\mathbf{x}_0)\cap B)\geqslant c_{33}(K)\varepsilon_4^K\prod_{j=1}^{K}\min\{1/\varepsilon_4,b_j\}$$
$$=c_{33}(K)\prod_{j=1}^{K}\min\{1,\varepsilon_4b_j\}. \quad (153)$$

Also

$$\left|\int_{\mathbf{R}^K\setminus Q(1/\varepsilon_4)}\chi_A(\mathbf{x}_0-\mathbf{y})H^-(\mathbf{y})\,d\mathbf{y}\right|$$
$$\leqslant\int_{\mathbf{R}^K\setminus Q(c/\varepsilon_4)}\chi_A(\mathbf{x}_0-\mathbf{y})|H(\mathbf{y})|\,d\mathbf{y}$$
$$+\sum_{\substack{\mathbf{l}\in\mathbf{Z}^K\\ \max_{1\leqslant j\leqslant K}|l_j|\leqslant c+\frac{1}{2}}}\left|\int_{Q(1/2\varepsilon_4;\mathbf{l})}\chi_A(\mathbf{x}_0-\mathbf{y})H^-(\mathbf{y})\,d\mathbf{y}\right|, \quad (154)$$

where $Q(1/2\varepsilon_4;\mathbf{l})$ is defined in the proof of Lemma 6.13. By (122),

$$\begin{aligned}\int_{\mathbf{R}^K\setminus Q(c/\varepsilon_4)} &\chi_A(\mathbf{x}_0-\mathbf{y})|H(\mathbf{y})|\,\mathrm{d}\mathbf{y}\\ &\leqslant \sum_{\substack{\mathbf{l}\in\mathbf{Z}^K\\ \max_{1\leqslant j\leqslant K}|l_j|\geqslant c-\frac{1}{2}}} \int_{Q(1/2\varepsilon_4;\mathbf{l})} \chi_A(\mathbf{x}_0-\mathbf{y})|H(\mathbf{y})|\,\mathrm{d}\mathbf{y}\\ &\ll \sum_{\substack{\mathbf{l}\in\mathbf{Z}^K\\ \max_{1\leqslant j\leqslant K}|l_j|\geqslant c-\frac{1}{2}}} \varepsilon_4^K\mu((Q(1/2\varepsilon_4;\mathbf{l})+\mathbf{x}_0)\cap A)\prod_{j=1}^K\frac{1}{(1+|l_j|)^2}. \end{aligned}\tag{155}$$

Since

$$\mu((Q(1/2\varepsilon_4;\mathbf{l})+\mathbf{x}_0)\cap A)\leqslant \mu((Q(1/2\varepsilon_4;\mathbf{l})+\mathbf{x}_0)\cap D) \ll \prod_{j=1}^K \min\{1/\varepsilon_4, d_j\},\tag{156}$$

we have, by (155), that

$$\begin{aligned}\int_{\mathbf{R}^K\setminus Q(c/\varepsilon_4)} &\chi_A(\mathbf{x}_0-\mathbf{y})|H(\mathbf{y})|\,\mathrm{d}\mathbf{y}\\ &\ll\left(\prod_{j=1}^K\min\{1/\varepsilon_4,d_j\}\right)\varepsilon_4^K \sum_{\substack{\mathbf{l}\in\mathbf{Z}^K\\ \max_{1\leqslant j\leqslant K}|l_j|\geqslant c-\frac{1}{2}}}\prod_{j=1}^K\frac{1}{(1+|l_j|)^2}\\ &\ll\left(\prod_{j=1}^K\min\{1,\varepsilon_4 d_j\}\right)\left(\left(\sum_{l\in\mathbf{Z}}\frac{1}{(1+|l|)^2}\right)^K-\left(\sum_{\substack{l\in\mathbf{Z}\\|l|<c-\frac{1}{2}}}\frac{1}{(1+|l|)^2}\right)^K\right)\\ &\ll\left(\prod_{j=1}^K\min\{1,\varepsilon_4 d_j\}\right)\left(\sum_{\substack{l\in\mathbf{Z}\\|l|\geqslant c-\frac{1}{2}}}\frac{1}{(1+|l|)^2}\right)\ll c^{-1}\prod_{j=1}^K\min\{1,\varepsilon_4 d_j\}. \end{aligned}\tag{157}$$

By Lemma 6.14 and (156),

$$\begin{aligned}\sum_{\substack{\mathbf{l}\in\mathbf{Z}^K\\ \max_{1\leqslant j\leqslant K}|l_j|\leqslant c+\frac{1}{2}}} &\left|\int_{Q(1/2\varepsilon_4;\mathbf{l})}\chi_A(\mathbf{x}_0-\mathbf{y})H^-(\mathbf{y})\,\mathrm{d}\mathbf{y}\right|\\ &\ll \sum_{\substack{\mathbf{l}\in\mathbf{Z}^K\\ \max_{1\leqslant j\leqslant K}|l_j|\leqslant c+\frac{1}{2}}} c^{-6}\varepsilon_4^K\mu((Q(1/2\varepsilon_4;\mathbf{l})+\mathbf{x}_0)\cap A)\prod_{j=1}^K\frac{1}{(1+|l_j|)^2}\\ &\leqslant c^{-6}\varepsilon_4^K\left(\prod_{j=1}^K\min\{1/\varepsilon_4,d_j\}\right)\sum_{\mathbf{l}\in\mathbf{Z}^K}\prod_{j=1}^K\frac{1}{(1+|l_j|)^2}\\ &=c^{-6}\left(\prod_{j=1}^K\min\{1,\varepsilon_4 d_j\}\right)\left(\sum_{l\in\mathbf{Z}}\frac{1}{(1+|l|)^2}\right)^K\ll c^{-6}\prod_{j=1}^K\min\{1,\varepsilon_4 d_j\}. \end{aligned}\tag{158}$$

Combining (154), (157) and (158), we have that

$$\left|\int_{\mathbf{R}^K\setminus Q(1/\varepsilon_4)} \chi_A(\mathbf{x}_0-\mathbf{y})H^-(\mathbf{y})\,d\mathbf{y}\right| \ll c^{-1}\prod_{j=1}^{K}\min\{1,\varepsilon_4 d_j\}. \tag{159}$$

It follows from the inequality $d_j \leqslant K^K K!b_j$ (see (128)), (151), (153) and (159) that

$$\begin{aligned}
&\int_{\mathbf{R}^K} \chi_A(\mathbf{x}_0-\mathbf{y})H(\mathbf{y})\,d\mathbf{y}\\
&\quad\geqslant c_{33}(K)\prod_{j=1}^{K}\min\{1,\varepsilon_4 b_j\} - c_{34}(K)c^{-1}\prod_{j=1}^{K}\min\{1,\varepsilon_4 d_j\}\\
&\quad\geqslant (c_{33}(K)-c_{34}(K)K^K K!c^{-1})\prod_{j=1}^{K}\min\{1,\varepsilon_4 b_j\}\\
&\quad\geqslant \tfrac{1}{2}c_{33}(K)\prod_{j=1}^{K}\min\{1,\varepsilon_4 b_j\}
\end{aligned} \tag{160}$$

if $c \geqslant c_{35}(K)$. On the other hand, it follows from (121) that

$$\begin{aligned}
\left|\int_{\mathbf{R}^K} \chi_A(\mathbf{x}_0-\mathbf{y})h(\mathbf{y})\,d\mathbf{y}\right| &\leqslant \sum_{\mathbf{l}\in\mathbf{Z}^K}\int_{Q(1/2\varepsilon_2;\mathbf{l})} \chi_A(\mathbf{x}_0-\mathbf{y})|h(\mathbf{y})|\,d\mathbf{y}\\
&\ll \sum_{\mathbf{l}\in\mathbf{Z}^K} \varepsilon_2^K \mu((Q(1/2\varepsilon_2;\mathbf{l})+\mathbf{x}_0)\cap A)\prod_{j=1}^{K}\frac{1}{(1+|l_j|)^2}.
\end{aligned} \tag{161}$$

Since

$$\begin{aligned}
\mu((Q(1/2\varepsilon_2;\mathbf{l})+\mathbf{x}_0)\cap A) &\leqslant \mu((Q(1/2\varepsilon_2;\mathbf{l})+\mathbf{x}_0)\cap D)\\
&\ll \prod_{j=1}^{K}\min\{1/\varepsilon_2, d_j\},
\end{aligned}$$

we have, by (161), that

$$\begin{aligned}
&\left|\int_{\mathbf{R}^K} \chi_A(\mathbf{x}_0-\mathbf{y})h(\mathbf{y})\,d\mathbf{y}\right|\\
&\quad\ll \left(\prod_{j=1}^{K}\min\{1/\varepsilon_2, d_j\}\right)\varepsilon_2^K \sum_{\mathbf{l}\in\mathbf{Z}^K}\prod_{j=1}^{K}\frac{1}{(1+|l_j|)^2}\\
&\quad= \left(\prod_{j=1}^{K}\min\{1,\varepsilon_2 d_j\}\right)\left(\sum_{l\in\mathbf{Z}}\frac{1}{(1+|l|)^2}\right)^K \ll \prod_{j=1}^{K}\min\{1,\varepsilon_2 d_j\}.
\end{aligned} \tag{162}$$

By hypothesis, $1/\varepsilon > b_1/8$, and so $1 > \varepsilon b_1/8$. Furthermore, $d_j \leqslant K^K K!b_j$ and $\varepsilon = c^{-1}\varepsilon_4 = c\varepsilon_2$. It follows from (160) and (162) that

$$\int_{\mathbf{R}^K} \chi_A(\mathbf{x}_0-\mathbf{y})(H-h)(\mathbf{y})\,d\mathbf{y}$$

$$\geqslant \tfrac{1}{2}c_{33}(K)\prod_{j=1}^{K}\min\{1,\varepsilon_4 b_j\} - c_{36}(K)\prod_{j=1}^{K}\min\{1,\varepsilon_2 d_j\}$$

$$\geqslant \tfrac{1}{2}c_{33}(K)\min\{1,c\varepsilon b_1\}\prod_{j=2}^{K}\min\{1,\varepsilon_4 b_j\}$$

$$- c_{36}(K)(K^K K!)^{K-1}\min\{1,K^K K!\varepsilon b_1 c^{-1}\}\prod_{j=2}^{K}\min\{1,\varepsilon_4 b_j\}$$

$$= (\tfrac{1}{2}c_{33}(K)\min\{1,c\varepsilon b_1\} - c_{37}(K)\min\{1,K^K K!\varepsilon b_1 c^{-1}\})\prod_{j=2}^{K}\min\{1,\varepsilon_4 b_j\}$$

$$\geqslant \tfrac{1}{4}c_{33}(K)\min\{1,c\varepsilon b_1\}\prod_{j=2}^{K}\min\{1,\varepsilon_4 b_j\} = \tfrac{1}{4}c_{33}(K)\prod_{j=1}^{K}\min\{1,\varepsilon_4 b_j\} \tag{163}$$

whenever $c \geqslant c_{31}(K) \geqslant c_{35}(K)$ (we recall that $\mathbf{x}_0$ is an arbitrary point in B_1). Obviously

$$\mu(B_1) \gg \prod_{j=1}^{K}\max\{b_j, 1/\varepsilon_4\}. \tag{164}$$

Therefore, combining (163) and (164), we have

$$\int_{\mathbb{R}^K}\left(\int_{\mathbb{R}^K}\chi_A(\mathbf{x}-\mathbf{y})(h-H)(\mathbf{y})\,\mathrm{d}\mathbf{y}\right)^2\mathrm{d}\mathbf{x}$$

$$\geqslant \int_{B_1}\left(\int_{\mathbb{R}^K}\chi_A(\mathbf{x}-\mathbf{y})(h-H)(\mathbf{y})\,\mathrm{d}\mathbf{y}\right)^2\mathrm{d}\mathbf{x}$$

$$\gg \left(\prod_{j=1}^{K}\min\{1,\varepsilon_4 b_j\}\right)^2\left(\prod_{j=1}^{K}\max\{b_j,1/\varepsilon_4\}\right) \tag{165}$$

for $c \geqslant c_{31}(K)$. Lemma 6.17 follows immediately.

We now deduce Theorem 17A from Lemmas 6.15 and 6.17. Let

$$c = \max\{4, c_{27}(K), c_{31}(K)\}.$$

Combining (106) and Lemmas 6.15 and 6.17, we have

$$\int_{\mathbb{R}^K}|\hat{\chi}_A(\mathbf{t})|^2|(\hat{h}-\hat{H})(\mathbf{t})|^2\,\mathrm{d}\mathbf{t}$$

$$= \int_{\mathbb{R}^K}\left(\int_{\mathbb{R}^K}\chi_A(\mathbf{x}-\mathbf{y})(h-H)(\mathbf{y})\,\mathrm{d}\mathbf{y}\right)^2\mathrm{d}\mathbf{x}$$

$$\gg \begin{cases} \varepsilon^{-1}\sigma(\partial A) & \text{(case (I))},\\ \left(\prod_{1\leqslant i\leqslant j_0}\varepsilon_4 b_i^2\right)\left(\prod_{j_0<j\leqslant K} b_j\right) & \text{(case (II))}, \end{cases} \tag{166}$$

where j_0 is the largest index j such that $\varepsilon_4 b_j \leqslant 1$.

Now

$$\hat{\chi}_{\tau,\lambda}(\mathbf{t}) = \lambda^K \hat{\chi}_A(\tau^{-1}(\lambda\mathbf{t})),$$

where $\chi_{\tau,\lambda}$ denotes the characteristic function of $A(\lambda, \mathbf{0}, \tau)$. Note that (see (103) and (114))

$$Q_m = Q(2K^{\frac{1}{2}}(\varepsilon_3 + \varepsilon_4)) \backslash Q(K^{\frac{1}{2}}(\varepsilon_3 + \varepsilon_4)). \tag{167}$$

It follows from (105) and (167) that for every $\mathbf{t}^* \in Q_m$,

$$\begin{aligned}\int_T \int_0^1 |\hat{\chi}_{\tau,\lambda}(\mathbf{t}^*)|^2 \,\mathrm{d}\lambda\,\mathrm{d}\tau &\gg \mu^{-1}(Q_m) \int_{\mathbf{R}^K} |\hat{\chi}_A(\mathbf{t})|^2 |(\hat{h} - \hat{H})(\mathbf{t})|^2 \,\mathrm{d}\mathbf{t} \\ &\gg \varepsilon_4^{-K} \int_{\mathbf{R}^K} |\hat{\chi}_A(\mathbf{t})|^2 |(\hat{h} - \hat{H})(\mathbf{t})|^2 \,\mathrm{d}\mathbf{t},\end{aligned} \tag{168}$$

and so, on combining (166) and (168), we have, for every $\mathbf{t}^* \in Q_m$,

$$\int_T \int_0^1 |\hat{\chi}_{\tau,\lambda}(\mathbf{t}^*)|^2 \,\mathrm{d}\lambda\,\mathrm{d}\tau \gg \begin{cases} \varepsilon_4^{-K-1}\sigma(\partial A) & \text{(case (I))}, \\ \left(\prod_{1 \leqslant i \leqslant j_0} b_i^2\right)\left(\prod_{j_0 < j \leqslant K} b_j \varepsilon_4^{-1}\right) & \text{(case (II))}. \end{cases} \tag{169}$$

It follows from (99), (169), (103) and (114) that

$$\begin{aligned}&\int_T \int_0^1 \int_{\mathbf{R}^K} |F_{\tau,\lambda}(\mathbf{x})|^2 \,\mathrm{d}\mathbf{x}\,\mathrm{d}\lambda\,\mathrm{d}\tau \\ &= \int_{\mathbf{R}^K} \left(\int_T \int_0^1 |\hat{\chi}_{\tau,\lambda}(\mathbf{t})|^2 \,\mathrm{d}\lambda\,\mathrm{d}\tau\right) |\phi(\mathbf{t})|^2 \,\mathrm{d}\mathbf{t} \\ &\geqslant \int_{Q_m} \left(\int_T \int_0^1 |\hat{\chi}_{\tau,\lambda}(\mathbf{t})|^2 \,\mathrm{d}\lambda\,\mathrm{d}\tau\right) |\phi(\mathbf{t})|^2 \,\mathrm{d}\mathbf{t} \\ &\geqslant \left(\min_{\mathbf{t} \in Q_m} \int_T \int_0^1 |\hat{\chi}_{\tau,\lambda}(\mathbf{t})|^2 \,\mathrm{d}\lambda\,\mathrm{d}\tau\right) \int_{Q_m} |\phi(\mathbf{t})|^2 \,\mathrm{d}\mathbf{t} \\ &\gg \begin{cases} M^K m^{-2} \varepsilon_4^{-K-1} \sigma(\partial A) \gg M^K m^{-2} 2^{(K+1)m} \sigma(\partial A) & \text{(case (I))}, \\ M^K m^{-2} \left(\prod_{1 \leqslant i \leqslant j_0} b_i^2\right)\left(\prod_{j_0 < j \leqslant K} 2^m b_j\right) & \text{(case (II))}, \end{cases}\end{aligned} \tag{170}$$

where $\phi = (\widehat{\mathrm{d}Z_0 - \mathrm{d}\mu_0})$, j_0 is the largest index j with $\varepsilon_4 b_j \leqslant 1$ and m is an integer with $1 \leqslant m = \mathrm{O}(\log M)$.

Clearly

$$\prod_{j=1}^K b_j = \mu(B) \gg \mu(A) \quad \text{and} \quad \prod_{j=2}^K b_j \gg \sigma(\partial B) \gg \sigma(\partial D) \geqslant \sigma(\partial A)$$

by (126)–(129), and

$$b_1 \gg d_1 \geqslant r(A) \geqslant 1$$

by (128) and the hypothesis of Theorem 17A. Combining these with (170), we have

$$\int_T \int_0^1 \int_{\mathbb{R}^K} |F_{\tau,\lambda}(\mathbf{x})|^2 \, d\mathbf{x} \, d\lambda \, d\tau$$

$$\gg \begin{cases} (\mu(A))^2 M^K (\log M)^{-2} & \text{(case (II) with } j_0 = K\text{)}, \\ \sigma(\partial A) M^K & \text{(all other cases)}. \end{cases} \tag{171}$$

We now complete the proof by standard averaging arguments. It is clear from (95) that $F_{\tau,\lambda}(\mathbf{x}) = 0$ whenever $\mathbf{x} \notin Q(M + d(A))$, where $d(A)$ denotes the diameter of A. It follows from (171) that

$$\int_T \int_0^1 \int_{Q(M+d(A))} |F_{\tau,\lambda}(\mathbf{x})|^2 \, d\mathbf{x} \, d\lambda \, d\tau \gg \min\left\{\sigma(\partial A), \frac{(\mu(A))^2}{(\log M)^2}\right\} M^K.$$

Hence either

$$\int_{Q^*} \int_T \int_0^1 |F_{\tau,\lambda}(\mathbf{x})|^2 \, d\lambda \, d\tau \, d\mathbf{x} \gg \min\left\{\sigma(\partial A), \frac{(\mu(A))^2}{(\log M)^2}\right\} M^K, \tag{172}$$

where $Q^* = Q(M + d(A)) \backslash Q(M - d(A))$, or

$$\int_{Q^{**}} \int_T \int_0^1 |F_{\tau,\lambda}(\mathbf{x})|^2 \, d\lambda \, d\tau \, d\mathbf{x} \gg \min\left\{\sigma(\partial A), \frac{(\mu(A))^2}{(\log M)^2}\right\} M^K, \tag{173}$$

where $Q^{**} = Q(M - d(A))$.

We now specify the value of the parameter M. Let

$$M = (d(A))^{2K+2}.$$

Suppose that (172) holds. Then there exist $\lambda_0 \in (0, 1]$, $\mathbf{x}_0 \in \mathbb{R}^K$ and $\tau_0 \in T$ such that

$$|F_{\tau_0,\lambda_0}(\mathbf{x}_0)|^2 \gg \frac{M}{d(A)} \min\left\{\sigma(\partial A), \frac{(\mu(A))^2}{(\log M)^2}\right\}. \tag{174}$$

Since $r(A) \geqslant 1$ by hypothesis, we have

$$\mu(A) \gg \sigma(\partial A) \gg d(A) = M^{1/(2K+2)}. \tag{175}$$

Combining (174) and (175), we have

$$|F_{\tau_0,\lambda_0}(\mathbf{x}_0)|^2 \gg \frac{M}{d(A)} = (d(A))^{2K+1} \gg d(A)(\mu(A))^2,$$

and so

$$|F_{\tau_0,\lambda_0}(\mathbf{x}_0)| > 2\mu(A) \quad \text{if} \quad d(A) \geqslant c_{38}(K). \tag{176}$$

It follows from (95) and (176) that the cardinality of

$$(-A(\lambda_0, \mathbf{x}_0, \tau_0)) \cap Q(M) \cap \{\mathbf{q}_1, \mathbf{q}_2, \mathbf{q}_3, \ldots\}$$

certainly exceeds $2\mu(A)$. Consequently

$$\sum_{\mathbf{q}_j \in (-A(\lambda_0,\mathbf{x}_0,\tau_0))} 1 - \mu(-A(\lambda_0, \mathbf{x}_0, \tau_0)) > \mu(A) \gg \sigma(\partial A) \geqslant (\sigma(\partial A))^{\frac{1}{2}},$$

and the proof is complete (we apply the trivial substitution $A \to (-A)$).

Suppose now that (173) holds. Then there exist $\lambda_1 \in (0, 1]$, $\mathbf{x}_1 \in \mathbb{R}^K$ and $\tau_1 \in T$ such that $A(\lambda_1, \mathbf{x}_1, \tau_1) \subset Q(M)$ and

$$|F_{\tau_1,\lambda_1}(\mathbf{x}_1)|^2 \gg \min\left\{\sigma(\partial A), \frac{(\mu(A))^2}{(\log M)^2}\right\}. \tag{177}$$

By (96), (175) and (177), we conclude that if $d(A) \geqslant c_{38}(K)$, then

$$\begin{aligned}\left|\sum_{\mathbf{q}_j \in (-A(\lambda_1,\mathbf{x}_1,\tau_1))} 1 - \mu(-A(\lambda_1, \mathbf{x}_1, \tau_1))\right|^2 &= |F_{\tau_1,\lambda_1}(\mathbf{x}_1)|^2 \\ &\gg \min\{\sigma(\partial A), \mu(A) M^{1/(2K+2)} (\log M)^{-2}\} \\ &\gg \min\{\sigma(\partial A), \mu(A)\} \gg \sigma(\partial A),\end{aligned}$$

and again the proof is complete (we apply the substitution $A \to (-A)$).

Finally, if $d(A) < c_{38}(K)$, then we apply the following trivial argument. Choosing $\lambda_2 = (2\mu(A))^{-1/K} \in (0, 1]$, we have $\mu(\lambda_2 A) = \frac{1}{2}$, and so for any $\mathbf{x}$ and τ,

$$\left|\sum_{\mathbf{q}_j \in A(\lambda_2,\mathbf{x},\tau)} 1 - \mu(A(\lambda_2, \mathbf{x}, \tau))\right| \geqslant \tfrac{1}{2}.$$

The proof of Theorem 17A is now complete.

Remark. The proof actually gives the existence of a set $A(\lambda_0, \mathbf{v}_0, \tau_0)$ such that

$$\left|\sum_{\mathbf{q}_j \in A(\lambda_0,\mathbf{v}_0,\tau_0)} 1 - \mu(A(\lambda_0, \mathbf{v}_0, \tau_0))\right| \gg (\sigma(\partial A))^{\frac{1}{2}}$$

and $1 \geqslant \lambda_0 \geqslant c_{39}(K) > 0$. In other words, the contraction factor λ_0 is greater than a positive constant depending at most on K.

7
Further applications of the Fourier transform method

In the first part of this chapter, we study the (traditional) case when rotation is forbidden. To avoid the considerable technical difficulties caused by higher dimensions, we restrict ourselves to the 2-dimensional case (i.e. $K = 2$). For a theorem with arbitrary K, see Theorem 19A in Chapter 6.

Again, let $\mathcal{Q} = \{\mathbf{q}_1, \mathbf{q}_2, \mathbf{q}_3, \ldots\}$ be a completely arbitrary infinite discrete set in $\mathbb{R}^2$. Given a compact region $A \subset \mathbb{R}^2$, write

$$\mathcal{D}[\mathcal{Q}; A] = \sum_{\mathbf{q}_j \in A} 1 - \mu(A),$$

where μ denotes the usual area. For any real number λ satisfying $-1 \leqslant \lambda \leqslant 1$ and any vector $\mathbf{v} \in \mathbb{R}^2$, set

$$A(\lambda, \mathbf{v}) = \{\lambda \mathbf{x} + \mathbf{v} : \mathbf{x} \in A\}.$$

Clearly $A(\lambda, \mathbf{v})$ is a homothetic image of A (note that reflection across the origin is allowed, as $-1 \leqslant \lambda \leqslant 1$). Let

$$\Delta[\mathcal{Q}; A] = \sup_{|\lambda| \leqslant 1, \mathbf{v}} |\mathcal{D}[\mathcal{Q}; A(\lambda, \mathbf{v})]|, \tag{1}$$

and define the usual-discrepancy of A by

$$\Delta[A] = \inf \Delta[\mathcal{Q}; A],$$

where the infimum is taken over all infinite discrete sets $\mathcal{Q} \subset \mathbb{R}^2$.

Now assume that A is convex. In contrast to the rotation-discrepancy $\Omega[A]$, the usual-discrepancy $\Delta[A]$ of a convex region A depends mainly on the 'smoothness' of its boundary curve.

Theorem 19B (Beck (TAc)). *Let $A \subset \mathbb{R}^2$ be a compact convex region with $\mu(A) \geqslant 2$. If the boundary curve Γ of A is twice continuously differentiable and if, for some real number $\gamma > 0$, the ratio*

$$\frac{\text{minimum curvature of } \Gamma}{\text{maximum curvature of } \Gamma}$$

is greater than γ, *then*

$$\Delta[A] > c_1(\gamma)\frac{(\mu(A))^{\frac{1}{4}}}{(\log \mu(A))^{\frac{1}{2}}}.$$

This is essentially the same lower bound as for circular discs (see $K = 2$ of Theorem 17A).

In the general case (i.e. when we have no assumption on the smoothness of Γ), we can guarantee only a much smaller usual-discrepancy.

Theorem 19C (Beck (TAc)). *Let* $A \subset \mathbb{R}^2$ *be a compact convex region with* $\mu(A) \geqslant 2$. *Then*

$$\Delta[A] > c_2 (\log \mu(A))^{\frac{1}{2}}.$$

Theorem 19C probably remains true if we replace the exponent $\frac{1}{2}$ of $(\log \mu(A))$ by the exponent 1. If true, this is best possible (apart from constant factors). The important particular case of squares was proved by Halász (see Theorem 3C).

If A is a polygon, then Theorem 19C is not very far from the truth in the sense that we cannot expect larger usual-discrepancy than a power of $(\log \mu(A))$. We shall prove the following more general result.

Theorem 20A (Beck). *Let* $S_1, S_2, \ldots, S_l$ *be* $l \geqslant 2$ *straight lines on the plane* $\mathbb{R}^2$. *Denote by* $POL(S_1, \ldots, S_l)$ *the family of convex polygons* $A \subset \mathbb{R}^2$ *such that every side of* A *is parallel to one of the given lines* $S_i\,(1 \leqslant i \leqslant l)$. *Then there exists an infinite discrete set* $\mathcal{Q} = \{\mathbf{q}_1, \mathbf{q}_2, \mathbf{q}_3, \ldots\}$ *on the plane such that for any* $A \in POL(S_1, \ldots, S_l)$,

$$|\mathcal{D}[\mathcal{Q}; A]| < c_3(l, \varepsilon)(\log (d(A) + 2))^{5+\varepsilon},$$

where $d(A)$ *denotes the diameter of* A.

It is worthwhile to mention here the following surprising result of Schmidt. If we consider the family of right-angled triangles whose short sides are parallel to the coordinate axes (i.e. two sides have fixed directions and only the third side has freedom of direction), then the supremum of the discrepancy has much larger order of magnitude (see Theorem A6 in Schmidt (1969*b*)).

If we consider only the family of homothetic images of a fixed convex polygon, then in Theorem 20A one can replace $(\log d(A))$ by $(\log \mu(A))$. Indeed, given any arbitrary compact convex region A, one can always find a linear transformation of determinant 1 such that for the image $\tilde{A}$ of A we have $d(\tilde{A}) \ll (\mu(\tilde{A}))^{\frac{1}{2}}$, where the implicit constant in $\ll$ is universal. Clearly $\Delta[\tilde{\mathcal{Q}}; \tilde{A}]] = \Delta[\mathcal{Q}; A]$, and the result follows.

We now return to Theorem 19B and 19C. To get further information on the intermediate cases, we introduce the concept of an approximability number which describes how well a convex region can be approximated by an inscribed polygon of few sides.

Definition. For any compact convex region $A \subset \mathbb{R}^2$ and integer $l \geqslant 3$, let $A_l \subset A$ denote an inscribed l-gon (i.e. polygon of l sides) of largest area. Denote by $\xi(A)$ the smallest integer $l \geqslant 3$ such that $\mu(A \backslash A_l) \leqslant l^2$. We call $\xi(A)$ the approximability number of A.

We remark that $\xi(A)$ depends on both the shape and the size of A. For instance, if $C(r)$ is a circular disc of radius $r \geqslant 1$, then as elementary calculation shows,

$$\xi(C(r)) \gg \ll r^{\frac{1}{2}}.$$

If P_l is a polygon of l sides, then $\xi(P_l) \leqslant l$ and $\xi(\lambda P_l) = l$ if the real number λ is sufficiently large depending on P_l. In general, if we know the equation of the boundary curve of A, then the determination (or at least the estimation) of $\xi(A)$ is an easy elementary problem.

By a well-known theorem of Sas in discrete geometry, we know that

$$\mu(A_l) \geqslant \mu(A) \frac{l}{2\pi} \sin\left(\frac{2\pi}{l}\right) \tag{2}$$

and equality holds only for the ellipse (see Sas (1939) and Fejes-Tóth (1953)). Using the elementary inequality

$$1 - \frac{\sin x}{x} \leqslant \frac{x^2}{6},$$

it follows from (2) that

$$3 \leqslant \xi(A) \ll (\mu(A))^{\frac{1}{4}} \tag{3}$$

whenever $\mu(A) \geqslant 1$.

The following results justify the introduction of the concept of approximability number.

Theorem 19D (Beck (TA*c*)). *Let $A \subset \mathbb{R}^2$ be a compact convex region with $\mu(A) \geqslant 2$. Then*

$$\Delta[A] \geqslant c_4 \frac{(\xi(A))^{\frac{1}{2}}}{(\log \mu(A))^{\frac{1}{4}}}.$$

Perhaps the exponent $\frac{1}{2}$ of $\xi(A)$ can be replaced by the exponent 1. If true, this is best possible, as the following result shows.

Theorem 20B (Beck (TAc)). *Let $A \subset \mathbb{R}^2$ be a compact convex region with $\mu(A) \geqslant 2$. Then for arbitrary $\varepsilon > 0$,*

$$\Delta[A] < c_5(\varepsilon)\xi(A)(\log \mu(A))^{4+\varepsilon}.$$

Comparing Theorems 19C, 19D and 20B, we conclude that the usual-discrepancy $\Delta[A]$ of a convex region A lies between two fixed powers of $\xi(A) + \log \mu(A)$.

Combining Theorem 19B, (3) and Theorem 20B, we see that for very smooth regions, the lower and upper bounds are rather close to each other. The very same can be said for the family of regular polygons.

Theorem 19E (Beck (TAc)). *If $A \subset \mathbb{R}^2$ is a regular polygon with $\mu(A) \geqslant 2$, then*

$$\Delta[A] > c_6 \frac{\xi(A)}{(\log \mu(A))^{\frac{1}{2}}}.$$

Now, as in Chapter 6, we shall deduce some results about point sets in the unit torus. We recall that given any N-element subset $\mathscr{P}$ of the unit square U_0^2,

$$\mathscr{P}^* = \{\mathbf{x} + \mathbf{l} : \mathbf{x} \in \mathscr{P}, \mathbf{l} \in \mathbb{Z}^2\}.$$

If $A \subset \mathbb{R}^2$ is a compact set, then

$$D^{\text{tor}}[\mathscr{P}; A] = \sum_{\mathbf{x} \in \mathscr{P}^* \cap A} 1 - N\mu(A).$$

We are going to define $\Delta^{\text{tor}}[\mathscr{P}; A]$ and $\Delta_N^{\text{tor}}[A]$ along the same lines as for $\Omega^{\text{tor}}[\mathscr{P}; A]$ and $\Omega_N^{\text{tor}}[A]$ in Chapter 6, the only differences being that rotation is now forbidden and, for technical reasons, the contraction coefficient runs from -1 to 1 (see also (1)). In other words, let

$$\Delta^{\text{tor}}[\mathscr{P}; A] = \sup_{|\lambda| \leqslant 1, \mathbf{v}} |D^{\text{tor}}[\mathscr{P}; A(\lambda, \mathbf{v})]|$$

and

$$\Delta_N^{\text{tor}}[A] = \inf \Delta^{\text{tor}}[\mathscr{P}; A],$$

where the infimum is extended over all N-element sets $\mathscr{P} \subset U_0^2$.

It follows immediately from Theorems 19B and 19C respectively that

Corollary 19F (Beck). *Let $A \subset \mathbb{R}^2$ be a compact convex region such that the boundary curve of A is twice continuously differentiable and has strictly positive curvature. Then ($N \geqslant 2$)*

$$\liminf_{N \to \infty} \frac{\Delta_N^{\text{tor}}[A]}{N^{\frac{1}{4}}(\log N)^{-\frac{1}{2}}} > 0.$$

Corollary 19G (Beck). *Let $A \subset \mathbb{R}^2$ be a compact convex region with $\mu(A) > 0$. Then $(N \geqslant 2)$*

$$\liminf_{N\to\infty} \frac{\Delta_N^{\mathrm{tor}}[A]}{(\log N)^{\frac12}} > 0.$$

Given any integer $N \geqslant 1$, denote by $\xi_N(A)$ the smallest integer $l \geqslant 3$ such that $\mu(A \backslash A_l) \leqslant l^2/N$. Then from Theorem 19D we obtain

Corollary 19H (Beck). *Let $A \subset \mathbb{R}^2$ be a compact convex region with $\mu(A) > 0$. Then for $N \geqslant 2$,*

$$\Delta_N^{\mathrm{tor}}[A] > c_7(A)\frac{(\xi_N(A))^{\frac12}}{(\log N)^{\frac14}}.$$

As an analogue of Theorem 20B, we state

Theorem 20C (Beck (TAc)). *Let $A \subset \mathbb{R}^2$ be a compact convex region. Then for $N \geqslant 2$,*

$$\Delta_N^{\mathrm{tor}}[A] < c_8(A, \varepsilon)\xi_N(A)(\log N)^{4+\varepsilon}.$$

If A is a polygon, then $\xi_N(A)$ is certainly not greater than the number of sides of A. Thus, as a particular case of Theorem 20C, we have

Corollary 20D (Beck). *Let $A \subset \mathbb{R}^2$ be a convex polygon. Then for any $\varepsilon > 0$,*

$$\lim_{N\to\infty} \frac{\Delta_N^{\mathrm{tor}}[A]}{(\log N)^{4+\varepsilon}} = 0.$$

Comparing Corollaries 19F and 20D, we see that the torus discrepancy $\Delta_N^{\mathrm{tor}}[A]$ is 'large' or 'small' according as A is 'smooth' or 'cornered' respectively. The next result demonstrates the existence of a compact convex region with very irregular discrepancy behaviour. We shall actually prove that 'most' (in the sense of category) convex regions have this property

Let $CONV(2)$ be the metric space of all compact convex regions in $\mathbb{R}^2$ endowed with the Hausdorff metric, defined by

$$\mathrm{dist}_{\mathscr{H}}(A, B) = \max\left(\max_{\mathbf{x}\in A}\min_{\mathbf{y}\in B}|\mathbf{x}-\mathbf{y}|, \max_{\mathbf{y}\in B}\min_{\mathbf{x}\in A}|\mathbf{x}-\mathbf{y}|\right)$$

for $A, B \in CONV(2)$. Here $|\mathbf{x}-\mathbf{y}|$ denotes, as usual, the euclidean distance. By the Blaschke selection theorem (see, for example, Hadwiger (1957), p. 154 and p. 201), $CONV(2)$ is locally compact. It follows from the Baire category theorem that the sets of first category in $CONV(2)$ are 'small' compared to their complements.

We have

Theorem 21 (Beck). *Let $f:\mathbb{N}\to\mathbb{R}^+$ and $g:\mathbb{N}\to\mathbb{R}^+$ satisfy* ($N \geqslant 2$)

$$\lim_{N\to\infty} \frac{f(N)}{(\log N)^{4+\varepsilon}} > 0$$

for some $\varepsilon > 0$ and

$$\lim_{N\to\infty} \frac{g(N)}{N^{\frac{1}{4}}(\log N)^{-\frac{1}{4}}} = 0.$$

Then for all $A \in CONV(2)$, except those in a set of first category,

(*i*) $\Delta_N^{\text{tor}}[A] < f(N)$ *for infinitely many N, and*
(*ii*) $\Delta_N^{\text{tor}}[A] > g(N)$ *for infinitely many N.*

Note that Theorem 21 was motivated by Gruber and Kenderov (1982) (see especially Theorem 2).

We now consider very briefly the case when both contraction and rotation are forbidden; in other words, we study the supremum of the discrepancy over the family $\{A + \mathbf{x}:\mathbf{x}\in\mathbb{R}^2\}$, where $A \subset \mathbb{R}^2$ is a convex region. For the sake of simplicity, consider the unit torus only. In contrast to Corollary 19G, this supremum does not necessarily tend to infinity as the cardinality N of the point distribution tends to infinity. For let $\alpha\in(0,1)$ be an irrational number, and consider the N-element set

$$\{(\{i\alpha\}, i/N):i = 1, 2, \ldots, N\}.$$

Let Q be an aligned square of side length α. By a classical observation of Ostrowski (1927), the supremum of the discrepancy over the family $\{Q + \mathbf{x}:\mathbf{x}\in\mathbb{R}^2\}$ is $O(1)$, and the constant is independent of α. Actually, it is true for any aligned rectangle of size $\alpha \times \beta$, where $0 \leqslant \beta \leqslant 1$.

Ostrowski's argument goes as follows. Because of the second coordinate i/N ($1 \leqslant i \leqslant N$), it is sufficient to verify the following statement: Let $1 \leqslant n \leqslant N$ and $\gamma\in\mathbb{R}$, and write

$$Z(n, \alpha, \gamma) = \#\{k:1 \leqslant k \leqslant n \text{ and } \{k\alpha + \gamma\} < \alpha\}.$$

Then $|Z(n, \alpha, \gamma) - n\alpha| \leqslant 1$ for all $n = 1, \ldots, N$ and $\gamma\in\mathbb{R}$.

In order to prove this, observe firstly that

$$Z(n, \alpha, \gamma) = \#\{(k, m):m\in\mathbb{Z}, 1 \leqslant k \leqslant n \text{ and } 0 \leqslant k\alpha + \gamma - m < \alpha\}.$$

All $m\in\mathbb{Z}$ that satisfy the inequality $0 \leqslant k\alpha + \gamma - m < \alpha$ for some k satisfying

$1 \leqslant k \leqslant n$ must satisfy $\gamma < m \leqslant n\alpha + \gamma$. On the other hand, for every $m \in \mathbb{Z}$, there exists a unique integer k satisfying $0 \leqslant k\alpha + \gamma - m < \alpha$. Furthermore, if m satisfies $\gamma < m \leqslant n\alpha + \gamma$, then this integer k satisfies $0 \leqslant k\alpha < (n+1)\alpha$ and hence $1 \leqslant k \leqslant n$. It follows that $Z(n, \alpha, \gamma)$ is equal to the number of integers m satisfying $\gamma < m \leqslant n\alpha + \gamma$. The inequality

$$|Z(n, \alpha, \gamma) - n\alpha| \leqslant 1$$

follows immediately.

We now consider the case when only contraction is forbidden. In the case of rotated squares, we can guarantee large discrepancy for any single value of the side length.

Theorem 22A (Beck). *Suppose that $r > 0$. Let $\mathscr{P}$ be an arbitrary distribution of N points in U_0^2. Then there is a rotated square A of side length r such that*

$$\left| \sum_{\mathbf{p} \in \mathscr{P}^* \cap A} 1 - N\mu(A) \right| > c_9(r) N^{\frac{1}{8}}.$$

Probably the exponent $\frac{1}{8}$ of N can be replaced by the exponent $\frac{1}{4}$ (or at least by the exponent $(\frac{1}{4} - \varepsilon)$).

There are many open problems in this area. For example, we are unable to prove the analogue of Theorem 22A for circular discs (i.e. we cannot prove the existence of a lower bound $f(r, N)$ such that $\lim_{N \to \infty} f(r, N) = +\infty$ for any single value of the radius $r > 0$). Note that Montgomery (1985) proved the following partial result: There is a disc D of radius $\frac{1}{2}$ or $\frac{1}{4}$ such that

$$\left| \sum_{\mathbf{p} \in \mathscr{P}^* \cap D} 1 - N\mu(D) \right| \gg N^{\frac{1}{4}}.$$

What we can prove is that the intersection $D \cap U_0^2$ has large discrepancy for a disc D of radius r.

Theorem 22B (Beck). *Suppose that $r > 0$. Let $\mathscr{P}$ be an arbitrary distribution of N points in U_0^2. Then there is a disc D of radius r such that*

$$\left| \sum_{\mathbf{p} \in \mathscr{P} \cap D} 1 - N\mu(D \cap U_0^2) \right| > c_{10}(r) N^{\frac{1}{4}}.$$

We discuss next Roth's problem concerning disc-segments (the problem was formulated in, for example, the last paragraph of §1 of Schmidt (1969*b*)).

Let $\mathscr{P}$ be an arbitrary distribution of N points in the circular disc D of unit area. For each disc-segment A (i.e. an intersection of D with a half-plane), let $Z[\mathscr{P}: A]$ denote the number of elements of $\mathscr{P}$ which lie in A, and write

$$\Delta_0[\mathscr{P}; A] = |Z[\mathscr{P}; A] - N\mu(A)|,$$

where μ denotes the usual area. Set

$$\Delta_0[N] = \inf_{\mathscr{P}} \sup_{A} \Delta_0[\mathscr{P}; A],$$

where the supremum is extended over all disc-segments A of D, and the infimum is taken over all N-element subsets $\mathscr{P}$ of D. Roth suspected that $\Delta_0[N]$ cannot be bounded as $N \to \infty$. The answer is positive.

Theorem 23A (Beck). *We have*

$$\Delta_0[N] > c_{11} N^{\frac{1}{4}} (\log N)^{-\frac{7}{2}}.$$

In the opposite direction, we state

Theorem 23B (Beck). *We have*

$$\Delta_0[N] < c_{12} N^{\frac{1}{4}} (\log N)^{\frac{1}{2}}.$$

The last results in this chapter are concerned with the following interesting geometrical problem: For what set of N points on the unit sphere is the sum of all euclidean distances between points maximal, and what is the maximum?

Our starting point is the following surprising 'invariance principle' due to Stolarsky: The sum of distances between points plus the quadratic average of a discrepancy type quantity is constant. Thus the sum of distances is maximized by a well-distributed set of points. We now introduce some notation to make this statement more precise.

Let S^K denote the surface of the unit sphere in $\mathbb{R}^{K+1}$. Let $\mathscr{P} = \{\mathbf{z}_1, \ldots, \mathbf{z}_N\}$ be a distribution of N points on S^K. We define

$$L(N, K, \mathscr{P}) = \sum_{1 \leqslant i < j \leqslant N} |\mathbf{z}_i - \mathbf{z}_j| \tag{4}$$

and

$$L(N, K) = \max_{\mathscr{P}} L(N, K, \mathscr{P}), \tag{5}$$

where the maximum is taken over all N-element subsets $\mathscr{P}$ of S^K.

The determination of $L(N, K)$ is a long-standing open problem in discrete geometry. For $K = 1$, the solution is given by the regular N-gon (see Fejes-Tóth (1956)). It is also known that for $N = K + 2$, the regular simplex is optimal. For $N > K + 2$ and $K \geqslant 2$, the exact value of $L(N, K)$ is unknown. The reason for this is that if N is sufficiently large compared to K, then there are no 'regular' configurations on the sphere, so the extremal point system(s) is (are), as expected, quite complicated and 'ad hoc'.

It is natural to compare the discrete sum $L(N,K,\mathscr{P})$ with the integral

$$\frac{N^2}{2\sigma(S^K)}\int_{S^K}|\mathbf{x}_0-\mathbf{x}|\,\mathrm{d}\sigma(\mathbf{x}),$$

where σ denotes the surface area, $\mathrm{d}\sigma(\mathbf{x})$ represents an element of the surface area on S^K and $\mathbf{x}_0=(1,0,\ldots,0)\in\mathbb{R}^{K+1}$. Note that here the correct coefficient is $\frac{1}{2}N^2$ rather than $\binom{N}{2}$, since in (4), we can write $1\leqslant i\leqslant j\leqslant N$ in place of $1\leqslant i<j\leqslant N$ without changing the value of $L(N,K,\mathscr{P})$.

Stolarsky's beautiful invariance principle mentioned above is as follows (see Stolarsky (1973) and note that Stolarsky dealt with a general class of metrics, not just the euclidean distance function):

$$L(N,K,\mathscr{P})+\int_{-1}^{1}\left(\frac{1}{\sigma(S^K)}\int_{S^K}|Z[\mathscr{P},\mathbf{x},t]-N\sigma^*(t)|^2\mathrm{d}\sigma(\mathbf{x})\right)\mathrm{d}t$$

$$=\frac{N^2}{2\sigma(S^K)}\int_{S^K}|\mathbf{x}_0-\mathbf{x}|\,\mathrm{d}\sigma(\mathbf{x}). \tag{6}$$

Here $Z[\mathscr{P},\mathbf{x},t]$ denotes the number of points of $\mathscr{P}$ in the spherical cap $C(\mathbf{x},t)=\{\mathbf{y}\in S^K:\mathbf{x}\cdot\mathbf{y}\leqslant t\}$, where $\mathbf{x}\cdot\mathbf{y}$ denotes the standard inner product and $\sigma^*(t)=\sigma^*(C(\mathbf{x},t))$ denotes the normalized surface area of $C(\mathbf{x},t)$ (so that $\sigma^*(S^K)=1$).

The second term on the left-hand side of (6) is clearly a measure of the irregularity of $\mathscr{P}$ relative to the spherical caps.

Set

$$c_0(K)=\frac{1}{2\sigma(S^K)}\int_{S^K}|\mathbf{x}_0-\mathbf{x}|\,\mathrm{d}\sigma(\mathbf{x}).$$

The quantity $c_0(K)N^2$ can be considered as the solution of the 'continuous relaxation' of the distance problem above (i.e. on the right-hand side of (5), the maximum is extended over all non-negative Borel measures on S^K with total measure N).

Note that the constants $c_0(K)$ can be calculated explicitly (e.g. $c_0(1)=2/\pi$, $c_0(2)=\frac{2}{3}$).

Identity (6) immediately yields $c_0(K)N^2-L(N,K,\mathscr{P})>0$. The problem is to determine the correct order of magnitude of $c_0(K)N^2-L(N,K)$.

Theorem 24A (Stolarsky (1973)). *We have*

$$c_0(K)N^2-L(N,K)<c_{13}(K)N^{1-1/K}.$$

Improving on the earlier results of Stolarsky (1973) and Harman (1982) in the opposite direction, we state

Theorem 24B (Beck (1984*b*)). *We have*

$$c_0(K)N^2 - L(N,K) > c_{14}(K)N^{1-1/K}.$$

We emphasize the qualitative difference between dimensions $K = 1$ and $K \geqslant 2$. By Theorems 24A and 24B, $c_0(1)N^2 - L(N,1)$ remains bounded as $N \to \infty$, but for $K \geqslant 2$, $c_0(K)N^2 - L(N,K)$ tends to infinity as $N \to \infty$ with polynomial speed in N.

Combining Theorem 24B and (6), we obtain

Corollary 24C (Beck (1984*b*)). *Let $\mathscr{P}$ be a distribution of N points on the unit sphere S^K. Then there exists a spherical cap $A \subset S^K$ with discrepancy*

$$\left| \sum_{\mathbf{x} \in \mathscr{P} \cap A} 1 - N\sigma^*(A) \right| > c_{15}(K)N^{\frac{1}{2}-1/2K}.$$

We should mention here the earlier result of Schmidt (1969*b*): There exists a spherical cap on S^K with discrepancy $N^{\frac{1}{2}-1/2K-\varepsilon}$.

In the opposite direction, we have

Theorem 24D (Beck). *For arbitrary integer $N \geqslant 2$, there exists an N-element set $\{\mathbf{z}_1, \ldots, \mathbf{z}_N\} \subset S^K$ such that for any spherical cap $A \subset S^K$,*

$$\left| \sum_{\mathbf{z}_i \in A} 1 - N\sigma^*(A) \right| < c_{16}(K)N^{\frac{1}{2}-1/2K}(\log N)^{\frac{1}{2}}.$$

We shall discuss Theorems 19B, 19C, 19D, 19E and 21 in §7.1, Theorems 22A and 22B in §7.2, Theorem 23A in §7.3 and Theorem 24B in §7.4. The proofs of all these lower bounds are based on the Fourier transform method introducted in Chapter 6. This method is suitable for families which are closed under translation. Although the families of disc-segments and of spherical caps are not closed under translation, we are able to employ the method indirectly by introducing some auxiliary 'translation invariant' families (namely rectangles and $(K + 1)$-dimensional balls respectively).

The proofs of the upper bounds are postponed to Chapter 8.

7.1 Homothetic convex sets

The main part of this section is concerned with the proofs of Theorems 19D, 19C, 19E and 19B (in this order), and we conclude the section by proving Theorem 21.

Let $A \subset \mathbb{R}^2$ be a compact convex region with $\mu(A) \geqslant 2$, and let $\mathscr{Q} = \{\mathbf{q}_1, \mathbf{q}_2, \mathbf{q}_3, \ldots\} \subset \mathbb{R}^2$ be an infinite discrete set. Let M be a real parameter to be fixed later, but sufficiently large compared to the diameter $d(A)$ of A.

For any Lebesgue measurable set $H \subset \mathbb{R}^2$, let

$$\mu_0(H) = \mu(H \cap [-M, M]^2).$$

For any $H \subset \mathbb{R}^2$, let

$$Z_0(H) = \sum_{\mathbf{q}_j \in H \cap [-M,M]^2} 1.$$

For the sake of brevity, let χ_λ, where $\lambda \in \mathbb{R}$, denote the characteristic function of the set $\lambda A = A(\lambda, \mathbf{0}) = \{\lambda \mathbf{x} : \mathbf{x} \in A\}$.

Consider now the function

$$F_\lambda = \chi_\lambda * (\mathrm{d}Z_0 - \mathrm{d}\mu_0), \tag{7}$$

where $*$ denotes the convolution operation. More explicitly,

$$\begin{aligned} F_\lambda(\mathbf{x}) &= \int_{\mathbb{R}^2} \chi_\lambda(\mathbf{x} - \mathbf{y})(\mathrm{d}Z_0(\mathbf{y}) - \mathrm{d}\mu_0(\mathbf{y})) \\ &= \sum_{\mathbf{q}_j \in A(-\lambda, \mathbf{x}) \cap [-M,M]^2} 1 - \mu(A(-\lambda, \mathbf{x}) \cap [-M, M]^2). \end{aligned} \tag{8}$$

Note that if $A(-\lambda, \mathbf{x}) \subset [-M, M]^2$, then

$$F_\lambda(\mathbf{x}) = \sum_{\mathbf{q}_j \in A(-\lambda, \mathbf{x})} 1 - \mu(A(-\lambda, \mathbf{x})); \tag{9}$$

in other words, $F_\lambda(\mathbf{x})$ is essentially a discrepancy function.

Our plan is to estimate a certain L^2-norm of (9) from below and, by standard averaging arguments, obtain the desired lower bound for $\Delta[\mathscr{Q}; A]$.

We may clearly assume that

$$\tfrac{1}{2}(2M)^2 < \sum_{\mathbf{q}_j \in [-M,M]^2} 1 < 2(2M)^2; \tag{10}$$

otherwise, by standard averaging arguments, we immediately obtain a discrepancy of size $\gg \mu(A)$ for the family $A + \mathbf{x} \subset [-M, M]^2$ (note that M is large compared to $d(A)$).

Applying a certain linear transformation of determinant 1, we may further assume that

$$\frac{r(A)}{d(A)} \geqslant c_{17} > 0, \tag{11}$$

where $r(A)$ denotes the radius of the largest inscribed circle in A and $d(A)$ denotes, as usual, the diameter of A.

Since the discrepancy function F_λ is defined as a convolution (see (7)), we shall use Fourier analysis. We recall that if

$$\hat{f}(\mathbf{t}) = \frac{1}{2\pi}\int_{\mathbf{R}^2} e^{-i\mathbf{x}\cdot\mathbf{t}} f(\mathbf{x})\,d\mathbf{x}$$

denotes the Fourier transform of f, then

$$\widehat{f * g} = \hat{f}\hat{g} \tag{12}$$

and

$$\widehat{fg} = \hat{f} * \hat{g} \tag{13}$$

for any $f, g \in L^2(\mathbb{R}^2)$. Furthermore, we have (Parseval–Plancherel identity)

$$\int_{\mathbf{R}^2} |f(\mathbf{x})|^2 d\mathbf{x} = \int_{\mathbf{R}^2} |\hat{f}(\mathbf{t})|^2 d\mathbf{t} \tag{14}$$

and (isometry)

$$\int_{\mathbf{R}^2} f(\mathbf{x})\overline{g(\mathbf{x})}\,d\mathbf{x} = \int_{\mathbf{R}^2} \hat{f}(\mathbf{t})\overline{\hat{g}(\mathbf{t})}\,d\mathbf{t}. \tag{15}$$

We are going to estimate the integral

$$\int_{-\infty}^{\infty}\left(\int_{\mathbf{R}^2} |F_\lambda(\mathbf{x})|^2 d\mathbf{x}\right) e^{-\lambda^2} d\lambda$$

from below. In view of (14), we have

$$\int_{-\infty}^{\infty}\left(\int_{\mathbf{R}^2} |F_\lambda(\mathbf{x})|^2 d\mathbf{x}\right) e^{-\lambda^2} d\lambda = \int_{-\infty}^{\infty}\left(\int_{\mathbf{R}^2} |\hat{F}_\lambda(\mathbf{t})|^2 d\mathbf{t}\right) e^{-\lambda^2} d\lambda. \tag{16}$$

From (7) and (12), we have

$$\hat{F}_\lambda = \hat{\chi}_\lambda \phi,$$

where

$$\phi(\mathbf{t}) = (\widehat{dZ_0 - d\mu_0})(\mathbf{t}) = \frac{1}{2\pi}\int_{\mathbf{R}^2} e^{-i\mathbf{x}\cdot\mathbf{t}}(dZ_0(\mathbf{x}) - d\mu_0(\mathbf{x})).$$

It follows from (16) that

$$\int_{-\infty}^{\infty}\left(\int_{\mathbf{R}^2} |F_\lambda(\mathbf{x})|^2 d\mathbf{x}\right) e^{-\lambda^2} d\lambda = \int_{\mathbf{R}^2}\left(\int_{-\infty}^{\infty} |\hat{\chi}_\lambda(\mathbf{t})|^2 e^{-\lambda^2} d\lambda\right) |\phi(\mathbf{t})|^2 d\mathbf{t}. \tag{17}$$

We shall study the right-hand side of (17), and begin by investigating the term $\phi(\mathbf{t})$.

For any real number $b>0$, let

$$T(b)=\{\mathbf{t}=(t_1,t_2)\in\mathbb{R}^2:|t_1|\leqslant 100b,|t_2|\leqslant 100/b\}, \tag{18}$$

and denote by $T(b,\tau)$ the image of $T(b)$ after rotation τ, where $0\leqslant\tau<\pi$ (since $K=2$, we identify the orthogonal transformation τ with the angle τ of rotation).

Lemma 7.1. *For any real number $b>0$ and angle τ, where $0\leqslant\tau<\pi$,*

$$\int_{T(b,\tau)}|\phi(\mathbf{t})|^2\,\mathrm{d}\mathbf{t}\gg M^2.$$

Remark. Here and in what follows, the implicit constants in Vinogradov's notation $\ll$ are positive and absolute.

Proof of Lemma 7.1. We essentially repeat the proof of Lemma 6.7′. For convenience of notation, we prove the lemma for $T(b)=T(b,0)$ only. The general case follows exactly the same lines. Let

$$f(\mathbf{x})=\left(\frac{2\sin(50bx_1)}{(2\pi)^{\frac{1}{2}}x_1}\right)^2\left(\frac{2\sin(50x_2/b)}{(2\pi)^{\frac{1}{2}}x_2}\right)^2.$$

We see from (13) that the Fourier transform $\hat{f}$ of f is equal to the convolution of the characteristic function of the rectangle $\frac{1}{2}T(b)$ with itself, i.e.

$$\hat{f}(\mathbf{t})=(100b-|t_1|)^+(100/b-|t_2|)^+,$$

where $y^+=y$ if $y>0$ and 0 otherwise. It follows from (12) that the Fourier transform of the convolution

$$g(\mathbf{x})=\int_{\mathbb{R}^2}f(\mathbf{x}-\mathbf{y})(\mathrm{d}Z_0-\mathrm{d}\mu_0)(\mathbf{y})$$

is $\hat{f}\phi$. Thus by (14),

$$\begin{aligned}\int_{\mathbb{R}^2}g^2(\mathbf{x})\,\mathrm{d}\mathbf{x}&=\int_{\mathbb{R}^2}|\hat{f}(\mathbf{t})|^2|\phi(\mathbf{t})|^2\,\mathrm{d}\mathbf{t}\\&=\int_{\mathbb{R}^2}((100b-|t_1|)^+(100/b-|t_2|)^+)^2|\phi(\mathbf{t})|^2\,\mathrm{d}\mathbf{t}\\&\leqslant 10^8\int_{T(b)}|\phi(\mathbf{t})|^2\,\mathrm{d}\mathbf{t}.\end{aligned}\tag{19}$$

On the other hand, elementary calculation gives

$$f(\mathbf{x}-\mathbf{q}_j)-\int_{\mathbb{R}^2}f(\mathbf{x}-\mathbf{y})\,\mathrm{d}\mathbf{y}\geqslant c_{18}>0 \tag{20}$$

whenever $\mathbf{q}_j \in (10^{-4}T(1/b) + \mathbf{x})$. Since $f(\mathbf{x})$ is a non-negative function, it follows from (20) that

$$|g(\mathbf{x})| = \left| \int_{\mathbf{R}^2} f(\mathbf{x} - \mathbf{y})(\mathrm{d}Z_0 - \mathrm{d}\mu_0)(\mathbf{y}) \right|$$

$$= \left| \sum_{\mathbf{q}_j \in [-M,M]^2} f(\mathbf{x} - \mathbf{q}_j) - \int_{[-M,M]^2} f(\mathbf{x} - \mathbf{y})\,\mathrm{d}\mathbf{y} \right| \geqslant c_{18} S(\mathbf{x}), \quad (21)$$

where $S(\mathbf{x})$ denotes the number of points $\mathbf{q}_j$ which lie in

$$(10^{-4}T(1/b) + \mathbf{x}) \cap [-M, M]^2.$$

Combining (21) and (10), we have

$$\int_{\mathbf{R}^2} g^2(\mathbf{x})\,\mathrm{d}\mathbf{x} \gg \int_{\mathbf{R}^2} S^2(\mathbf{x})\,\mathrm{d}\mathbf{x} \geqslant \int_{\mathbf{R}^2} S(\mathbf{x})\,\mathrm{d}\mathbf{x} = \mu(10^{-4}T(1/b)) \sum_{\mathbf{q}_j \in [-M,M]^2} 1$$

$$= 10^{-4} \sum_{\mathbf{q}_j \in [-M,M]^2} 1 \gg M^2. \quad (22)$$

Lemma 7.1 follows on combining (19) and (22). ■

Next, we deal with the first factor

$$\int_{-\infty}^{\infty} |\hat{\chi}_\lambda(\mathbf{t})|^2 \mathrm{e}^{-\lambda^2}\,\mathrm{d}\lambda$$

on the right-hand side of (17).

We define a partition

$$T(b, \tau) = T_1(b, \tau) \cup T_2(b, \tau) \cup T_3(b, \tau) \cup \cdots$$

as follows: For every integer $j \geqslant 1$, let

$$T_j(b, \tau) = \left\{ \mathbf{t} \in T(b, \tau) : \frac{d(T(b, \tau))}{2^{j+1}} < |\mathbf{t}| \leqslant \frac{d(T(b, \tau))}{2^j} \right\}, \quad (23)$$

where $d(T(b, \tau)) = 200(b^2 + b^{-2})^{\frac{1}{2}}$ denotes the diameter of $T(b, \tau)$.

Let $[\mathbf{w}', \mathbf{w}'']$ be a chord of A, i.e. both endpoints $\mathbf{w}'$ and $\mathbf{w}''$ of the straight line segment $[\mathbf{w}', \mathbf{w}''] = \{\alpha\mathbf{w}' + (1 - \alpha)\mathbf{w}'' : 0 \leqslant \alpha \leqslant 1\}$ lie on the boundary curve Γ of the convex region A. Let $\Gamma(\mathbf{w}', \mathbf{w}'')$ denote the shorter arc on Γ with endpoints $\mathbf{w}'$ and $\mathbf{w}''$. For any $\mathbf{v} \in \Gamma(\mathbf{w}', \mathbf{w}'')$, let $\operatorname{dist}(\mathbf{v}, [\mathbf{w}', \mathbf{w}''])$ denote the usual euclidean (perpendicular) distance between $\mathbf{v}$ and the line passing through $\mathbf{w}'$ and $\mathbf{w}''$, and write

$$\delta_A(\mathbf{w}', \mathbf{w}'') = \max_{\mathbf{v} \in \Gamma(\mathbf{w}', \mathbf{w}'')} \operatorname{dist}(\mathbf{v}, [\mathbf{w}', \mathbf{w}'']).$$

Also, let $\tau(\mathbf{w}', \mathbf{w}'')$ be the angle between the x_1-axis and the perpendicular bisector of $[\mathbf{w}', \mathbf{w}'']$.

Lemma 7.2. *Let A be a convex region satisfying $r(A) \geqslant 1$ and*

$$\frac{r(A)}{d(A)} \geqslant c_{17} > 0. \tag{24}$$

Let $[\mathbf{w}', \mathbf{w}'']$ be a chord of A such that for some real number b with $1 \leqslant b \leqslant r(A)$, $|\mathbf{w}' - \mathbf{w}''| \geqslant b$ and $\delta_A(\mathbf{w}', \mathbf{w}'') \leqslant 1/b$. Then for every integer $j \geqslant 1$ and for every $\mathbf{t}^* \in T_j(b, \tau)$ with $\tau = \tau(\mathbf{w}', \mathbf{w}'')$,

$$\int_{-\infty}^{\infty} |\hat{\chi}_\lambda(\mathbf{t}^*)|^2 \mathrm{e}^{-\lambda^2} \, \mathrm{d}\lambda \gg \min\{4^j, (r(A))^4\}.$$

We now prove Lemma 7.2. Since $\hat{\chi}_\lambda(\mathbf{t}) = \lambda^2 \hat{\chi}_A(\lambda \mathbf{t})$, where χ_A denotes the characteristic function of A, it suffices to investigate $\hat{\chi}_A(\mathbf{t})$ only.

Let $S(\mathbf{0}, \beta)$ denote the straight line passing through the origin $\mathbf{0}$ of the $t_1 t_2$-plane such that the angle between $S(\mathbf{0}, \beta)$ and the t_1-axis is β. Let $B(j, b, \tau)$ be the set of angles β such that the straight line $S(\mathbf{0}, \beta)$ intersects the domain $T_j(b, \tau)$. Clearly

$$|\tau - \beta| < \min\{c_{19} 2^j b^{-2}, \pi\} \tag{25}$$

for all $\beta \in B(j, b, \tau)$.

Let $\beta \in B(j, b, \tau)$ be fixed, and write $\mathbf{e}(\beta) = (\cos\beta, \sin\beta)$. Then for any $\mathbf{t} \in S(\mathbf{0}, \beta)$, we have

$$\begin{aligned} \hat{\chi}_A(\mathbf{t}) &= \frac{1}{2\pi} \int_A \mathrm{e}^{-\mathrm{i}\mathbf{x}\cdot\mathbf{t}} \, \mathrm{d}\mathbf{x} \\ &= \frac{1}{(2\pi)^{\frac{1}{2}}} \frac{1}{(2\pi)^{\frac{1}{2}}} \int_{-\infty}^{\infty} \mathrm{e}^{-\mathrm{i}xt} g_\beta(x) \, \mathrm{d}x = \frac{1}{(2\pi)^{\frac{1}{2}}} \hat{g}_\beta(t), \end{aligned} \tag{26}$$

where the real number t is defined by the equation $\mathbf{t} = t\mathbf{e}(\beta)$, $g_\beta(x)$ is the euclidean length of the chord

$$\{\mathbf{x} \in A : \mathbf{x} \cdot \mathbf{t} = xt\},$$

and $\hat{g}_\beta(t)$ denotes the Fourier transform of the 'chord function' $g_\beta(x)$ (of one variable $x \in \mathbb{R}$).

From (9), (17) and (26), we see that the 'maximum discrepancy' $\Delta[\mathcal{Q}; A]$ depends heavily on the behaviour of the chord function $g_\beta(x)$.

We introduce two auxiliary functions. Let

$$h(x) = \varepsilon \mathrm{e}^{-\frac{1}{2}(\varepsilon x)^2} \quad \text{and} \quad H(x) = 4\varepsilon \mathrm{e}^{-\frac{1}{2}(4\varepsilon x)^2}, \tag{27}$$

where the real parameter $\varepsilon > 0$ will be fixed later.

Remark. These functions play the same roles as $h(\mathbf{x})$ and $H(\mathbf{x})$ in the proof of Theorem 17A (see (107) and (108) in Chapter 6), but this case is technically much simpler.

We recall the well-known result that if $f(x)=\mathrm{e}^{-\frac{1}{2}a^2x^2}$, where $a>0$, then $\hat{f}(t)=a^{-1}\mathrm{e}^{-t^2/2a^2}$. It follows from (27) that

$$\hat{h}(t)=\mathrm{e}^{-t^2/(2\varepsilon^2)} \quad \text{and} \quad \hat{H}(t)=\mathrm{e}^{-t^2/(32\varepsilon^2)}. \tag{28}$$

We now specify the value of ε. Let

$$\varepsilon=\tfrac{1}{8}\operatorname{dist}(\mathbf{0},T_j(b,\tau)), \tag{29}$$

where dist denotes the usual euclidean distance. It follows from (23) and (29) that

$$\varepsilon=\frac{25(b^2+b^{-2})^{\frac{1}{2}}}{2^{j+1}} \tag{30}$$

and for every $\mathbf{t}^*\in T_j(b,\tau)$,

$$8\varepsilon<|\mathbf{t}^*|\leqslant 16\varepsilon. \tag{31}$$

Using the trivial identity $\hat{\chi}_\lambda(\mathbf{t})=\lambda^2\hat{\chi}_A(\lambda\mathbf{t})$ and (26), we have that for any $\mathbf{t}^*\in S(\mathbf{0},\beta)$,

$$\begin{aligned}\int_{-\infty}^{\infty}\mathrm{e}^{-\lambda^2}|\hat{\chi}_\lambda(\mathbf{t}^*)|^2\,\mathrm{d}\lambda&=\int_{-\infty}^{\infty}\mathrm{e}^{-\lambda^2}\lambda^4|\hat{\chi}_A(\lambda\mathbf{t}^*)|^2\,\mathrm{d}\lambda\\&=\frac{1}{2\pi}\int_{-\infty}^{\infty}\mathrm{e}^{-\lambda^2}\lambda^4|\hat{g}_\beta(\lambda|\mathbf{t}^*|)|^2\,\mathrm{d}\lambda\\&=\frac{1}{2\pi|\mathbf{t}^*|}\int_{-\infty}^{\infty}\mathrm{e}^{-(t/|\mathbf{t}^*|)^2}\left(\frac{t}{|\mathbf{t}^*|}\right)^4|\hat{g}_\beta(t)|^2\,\mathrm{d}t.\end{aligned} \tag{32}$$

If $\mathbf{t}^*\in T_j(b,\tau)$, then $|\mathbf{t}^*|>8\varepsilon$ by (31). It follows from (28) that

$$|(\hat{h}-\hat{H})(t)|=|\mathrm{e}^{-t^2/2\varepsilon^2}-\mathrm{e}^{-t^2/32\varepsilon^2}|\leqslant\mathrm{e}^{-t^2/32\varepsilon^2}\leqslant\mathrm{e}^{-(t/|\mathbf{t}^*|)^2}. \tag{33}$$

On the other hand, $|\mathbf{t}^*|\leqslant 16\varepsilon$ by (31), and so

$$\begin{aligned}(\hat{h}-\hat{H})(t)&=|(1-\mathrm{e}^{-t^2/32\varepsilon^2})-(1-\mathrm{e}^{-t^2/2\varepsilon^2})|\leqslant(1-\mathrm{e}^{-t^2/32\varepsilon^2})+(1-\mathrm{e}^{-t^2/2\varepsilon^2})\\&\leqslant\frac{t^2}{32\varepsilon^2}+\frac{t^2}{2\varepsilon^2}\leqslant\frac{t^2}{\varepsilon^2}\ll\left(\frac{t}{|\mathbf{t}^*|}\right)^2.\end{aligned} \tag{34}$$

Hence

$$|(\hat{h}-\hat{H})(t)|^2\ll\min\left\{\left(\frac{t}{|\mathbf{t}^*|}\right)^4,\mathrm{e}^{-2(t/|\mathbf{t}^*|)^2}\right\}\ll\mathrm{e}^{-(t/|\mathbf{t}^*|)^2}\left(\frac{t}{|\mathbf{t}^*|}\right)^4, \tag{35}$$

and so, by (31), (32) and (35), we conclude that for every $\mathbf{t}^*\in S(\mathbf{0},\beta)\cap T_j(b,\tau)$,

$$\int_{-\infty}^{\infty}\mathrm{e}^{-\lambda^2}|\hat{\chi}_\lambda(\mathbf{t}^*)|^2\,\mathrm{d}\lambda\gg\frac{1}{\varepsilon}\int_{-\infty}^{\infty}|(\hat{h}-\hat{H})(t)|^2|\hat{g}_\beta(t)|^2\,\mathrm{d}t. \tag{36}$$

By (12) and (14), we have

$$\int_{-\infty}^{\infty}\left(\int_{-\infty}^{\infty} g_\beta(y)(h-H)(x-y)\,\mathrm{d}y\right)^2 \mathrm{d}x=\int_{-\infty}^{\infty} |\hat{g}_\beta(t)|^2|(\hat{h}-\hat{H})(t)|^2\,\mathrm{d}t,$$

and so it follows from (36) that

$$\int_{-\infty}^{\infty} \mathrm{e}^{-\lambda^2}|\hat{\chi}_\lambda(\mathbf{t}^*)|^2\,\mathrm{d}\lambda \gg \frac{1}{\varepsilon}\int_{-\infty}^{\infty}\left(\int_{-\infty}^{\infty} g_\beta(y)(h-H)(x-y)\,\mathrm{d}y\right)^2 \mathrm{d}x \quad (37)$$

for every $\mathbf{t}^* \in S(\mathbf{0},\beta)\cap T_j(b,\tau)$.

We are going to study the right-hand side of (37).

For any arbitrary angle α, let z_α^+ (resp. z_α^-) be that real number for which $g_\alpha(z_\alpha^+ + \delta)=0$ and $g_\alpha(z_\alpha^+ - \delta)>0$ (resp. $g_\alpha(z_\alpha^- - \delta)=0$ and $g_\alpha(z_\alpha^- + \delta)>0$) for all sufficiently small positive δ. Here, we have used the convention that $g_\beta(x)=0$ if $\{\mathbf{x}\in A:\mathbf{x}\cdot\mathbf{t}=xt\}$ is empty. The reader may wish, at this point, to draw a diagram.

We need the following elementary lemma which describes the behaviour of the chord function of a convex region.

Lemma 7.3. *Let A be a convex region such that $r(A)\geqslant 1$ and (24) holds. Assume further that for some angle τ and real number b satisfying $1\leqslant b\leqslant r(A)$, $g_\tau(z_\tau^+ - 1/b)\geqslant \frac{1}{2}b$. Then for any y with $4/b\leqslant y\leqslant r(A)$ and for any angle β satisfying $|\tau-\beta|\leqslant y/b$, we have*

$$g_\beta(z_\beta^+ - y)\gg \max\{b,y\}.$$

We postpone the proof of this lemma.

For later use, we mention here the following two apparently trivial consequences of the convexity of A: Let the angle α be given. If $0<y\leqslant\delta$, then

$$g_\alpha(z_\alpha^+ - y)\geqslant \frac{y}{\delta}g_\alpha(z_\alpha^+ - \delta); \quad (38)$$

on the other hand, if $0<\delta\leqslant y\leqslant r(A)$, then

$$g_\alpha(z_\alpha^+ - y)\geqslant \frac{|z_\alpha^+ - z_\alpha^-|-y}{|z_\alpha^+ - z_\alpha^-|-\delta}g_\alpha(z_\alpha^+ - \delta)\geqslant \frac{2r(A)-y}{2r(A)-\delta}g_\alpha(z_\alpha^+ - \delta)\geqslant \tfrac{1}{2}g_\alpha(z_\alpha^+ - \delta). \quad (39)$$

By the hypotheses of Lemma 7.2, for some real number b with $1\leqslant b\leqslant r(A)$,

$$\max\{g_\tau(z_\tau^+ - \delta), g_\tau(z_\tau^- + \delta)\}\geqslant |\mathbf{w}'-\mathbf{w}''|\geqslant b,$$

where $\delta=\delta_A(\mathbf{w}',\mathbf{w}'')\leqslant 1/b$. Suppose that

$$g_\tau(z_\tau^+ - \delta)\geqslant b, \quad (40)$$

where $\delta \leqslant 1/b$. Using $1 \leqslant b \leqslant r(A)$, (39) and (40), we have

$$g_\tau(z_\tau^+ - 1/b) \geqslant \tfrac{1}{2}b. \tag{41}$$

Combining (41) and Lemma 7.3, we obtain

$$g_\beta(z_\beta^+ - y) \gg \max\{b, y\} \tag{42}$$

whenever $4/b \leqslant y \leqslant r(A)$ and $|\tau - \beta| \leqslant y/b$.

On the other hand, by (25) and (30), we have

$$|\tau - \beta| \leqslant \min\{c_{19}2^j b^{-2}, \pi\} < b^{-1}\min\{c_{20}/\varepsilon, r(A)\} \tag{43}$$

for every $\beta \in B(j, b, \tau)$. Choosing $y = \min\{c_{20}/\varepsilon, r(A)\}$, it follows from (38), (39), (42) and (43) that

$$g_\beta(z_\beta^+ - \min\{\varepsilon^{-1}, r\}) \gg \begin{cases} b & (\varepsilon^{-1} \leqslant r), \\ r & (\varepsilon^{-1} > r), \end{cases} \tag{44}$$

uniformly for all $\beta \in B(j, b, \tau)$. Here and from now on, $r = r(A)$.

We are now ready to estimate the right-hand side of (37). Let x_0 be an arbitrary real number in the interval $[z_\beta^+ + \varepsilon^{-1}, z_\beta^+ + 2\varepsilon^{-1}]$. Then it follows from (27) and elementary calculation that

$$0 < H(x_0 - y) \leqslant \tfrac{1}{2}h(x_0 - y) \tag{45}$$

for all $y \leqslant z_\beta^+$. Thus by (27), (38), (44) and (45), for all $\beta \in B(j, b, \tau)$ and $x_0 \in [z_\beta^+ + \varepsilon^{-1}, z_\beta^+ + 2\varepsilon^{-1}]$,

$$\begin{aligned} \int_{-\infty}^{\infty} g_\beta(y)(h - H)(x_0 - y)\,dy &\geqslant \tfrac{1}{2}\int_{-\infty}^{\infty} g_\beta(y)h(x_0 - y)\,dy \\ &\geqslant \tfrac{1}{2}\int_{z'}^{z''} g_\beta(y)h(x_0 - y)\,dy \\ &\gg \begin{cases} b & (\varepsilon^{-1} \leqslant r), \\ r^2\varepsilon & (\varepsilon^{-1} > r), \end{cases} \end{aligned} \tag{46}$$

where $z' = z_\beta^+ - \min\{\varepsilon^{-1}, r\}$ and $z'' = z_\beta^+ - \tfrac{1}{2}\min\{\varepsilon^{-1}, r\}$. Hence, for all $\beta \in B(j, b, \tau)$,

$$\begin{aligned} &\int_{-\infty}^{\infty}\left(\int_{-\infty}^{\infty} g_\beta(y)(h - H)(x - y)\,dy\right)^2 dx \\ &\geqslant \int_{z_\beta^+ + \varepsilon^{-1}}^{z_\beta^+ + 2\varepsilon^{-1}}\left(\int_{-\infty}^{\infty} g_\beta(y)(h - H)(x - y)\,dy\right)^2 dx \\ &\gg \begin{cases} b^2\varepsilon^{-1} & (\varepsilon^{-1} \leqslant r), \\ r^4\varepsilon & (\varepsilon^{-1} > r). \end{cases} \end{aligned} \tag{47}$$

Since

$$T_j(b,\tau) = \bigcup_{\beta \in B(j,b,\tau)} (S(\mathbf{0},\beta) \cap T_j(b,\tau)),$$

it follows from (37) and (47) that for every $\mathbf{t}^* \in T_j(b,\tau)$,

$$\int_{-\infty}^{\infty} \mathrm{e}^{-\lambda^2} |\hat{\chi}_\lambda(\mathbf{t}^*)|^2 \, \mathrm{d}\lambda \gg \begin{cases} b^2 \varepsilon^{-2} & (\varepsilon^{-1} \leqslant r), \\ r^4 & (\varepsilon^{-1} > r). \end{cases} \tag{48}$$

Lemma 7.2 follows from (30) and (48).

In order to apply Lemma 7.2, we require the following lemma.

Lemma 7.4. *Let $A \subset \mathbb{R}^2$ be an arbitrary compact convex region. For every integer $l \geqslant 3$, let $A_l \subset A$ denote an inscribed l-gon (i.e. polygon of l sides) of largest area. Denote by $k = k(A) \geqslant 3$ the smallest integer such that $\mu(A \backslash A_k) \leqslant k$. Let $\mathbf{v}_1, \mathbf{v}_2, \ldots, \mathbf{v}_k$ be the vertices of A_k in their order on the boundary Γ of A. Suppose that $k = k(A) \geqslant 6$. Then there exists an index set $J \subset \{1, 2, \ldots, k\}$ such that*

(a_1) $\#J \gg k$;
(a_2) $|\mathbf{v}_i - \mathbf{v}_{i+1}| \gg k^{\frac{1}{2}}$ for every $i \in J$;
(a_3) $|\mathbf{v}_i - \mathbf{v}_{i+1}| \delta_A(\mathbf{v}_i, \mathbf{v}_{i+1}) \ll 1$ for every $i \in J$; and
(a_4) if $i, j \in J$, $i \neq j$ and $q \leqslant |\mathbf{v}_i - \mathbf{v}_{i+1}|, |\mathbf{v}_j - \mathbf{v}_{j+1}| < 2q$ for some real number q, then

$$|\tau(\mathbf{v}_i, \mathbf{v}_{i+1}) - \tau(\mathbf{v}_j, \mathbf{v}_{j+1})| \gg q^{-2} \pmod{\pi}.$$

Again, as for Lemma 7.3, we postpone the proof of this lemma.

It is useful at this point to note (the trivial estimate) that if $\mu(\lambda A) = \frac{1}{2}$, then for every $x \in \mathbb{R}^2$, we have

$$\left| \sum_{\mathbf{q}_j \in A(\lambda, \mathbf{x})} 1 - \mu(A(\lambda, \mathbf{x})) \right| \geqslant \tfrac{1}{2}. \tag{49}$$

We now investigate Theorem 19D. This can be deduced from the following intermediate result by averaging arguments.

Lemma 7.5. *Writing $d = d(A)$, we have*

$$(\log d) \int_{-\infty}^{\infty} \left(\int_{\mathbb{R}^2} |F_\lambda(\mathbf{x})|^2 \, \mathrm{d}\mathbf{x} \right) \mathrm{e}^{-\lambda^2 (\log d)^2} \, \mathrm{d}\lambda \gg \frac{\xi(A)}{(\log d)^{\frac{1}{2}}} M^2.$$

To prove Lemma 7.5, note first of all that we may assume that the approximability number $\xi(A)$ is sufficiently large, otherwise Theorem 19D follows immediately from the trivial estimate (49).

Let $c_{21} \geqslant 6$ be so large that if $k(A) \geqslant c_{21}$, then for every $i \in J$,

$$|\mathbf{v}_i - \mathbf{v}_{i+1}| \geqslant 1 \quad \text{and} \quad \delta_A(\mathbf{v}_i, \mathbf{v}_{i+1}) \leqslant 1, \tag{50}$$

where the index set J is defined in Lemma 7.4. The existence of such a universal constant c_{21} easily follows from (a_2) and (a_3) of Lemma 7.4.

Now let $\xi(A) \geqslant \max\{c_{21}, c_{17}^{-1} + 1\}$. From the definition of $\xi(A)$, we clearly have that for any l with $3 \leqslant l < \xi(A)$,

$$\mu(A \setminus A_l) > l^2.$$

Since $l^2 > l$, we conclude that

$$k = k(A) \geqslant \xi(A) \geqslant c_{21} \geqslant 6. \tag{51}$$

It follows from Lemma 7.4 that there exists an index set $J = J(A) \subset \{1, 2, \ldots, k\}$ satisfying properties (a_1)–(a_4) and (50). On the other hand, using the trivial inequality $\mu(A) > (\xi(A) - 1)^2 \geqslant c_{17}^{-2}$, we have

$$r(A) \geqslant c_{17} d(A) \geqslant c_{17}(\mu(A))^{\frac{1}{2}} \geqslant 1.$$

For any $i \in J$, let

$$b_i = \min\{|\mathbf{v}_i - \mathbf{v}_{i+1}|, 1/\delta_A(\mathbf{v}_i, \mathbf{v}_{i+1}), r(A)\}. \tag{52}$$

Clearly

$$|\mathbf{v}_i - \mathbf{v}_{i+1}| \geqslant b_i \quad \text{and} \quad \delta_A(\mathbf{v}_i, \mathbf{v}_{i+1}) \leqslant 1/b_i;$$

and so by (50) and (52), $1 \leqslant b_i \leqslant r(A)$. We can therefore apply Lemma 7.2 to every chord $[\mathbf{v}_i, \mathbf{v}_{i+1}]$, where $i \in J$.

For any integer $n \geqslant 1$, let

$$J(n) = \{i \in J : 2^{n-1} \leqslant b_i < 2^n\}.$$

Obviously

$$J(n) = \varnothing \quad \text{if } 2^{n-1} > r(A). \tag{53}$$

Also, by Lemma 7.2, we have that for every $n \geqslant 1$, $m \geqslant 1$ and $i \in J(n)$,

$$\int_{-\infty}^{\infty} e^{-\lambda^2} |\hat{\chi}_\lambda(\mathbf{t})|^2 \, d\lambda \gg \min\{4^m, (r(A))^4\} \tag{54}$$

for all $\mathbf{t} \in T_m(b_i, \tau(\mathbf{v}_i, \mathbf{v}_{i+1}))$.

In order to deal with the 'overlapping' of sets of the type $T_m(b_i, \tau(\mathbf{v}_i, \mathbf{v}_{i+1}))$, we introduce the following: For every integer $h \geqslant 1$, let

$$S(h) = \{\mathbf{t} \in \mathbb{R}^2 : \mathbf{t} \text{ belongs to exactly } h \text{ sets of type } T_m(b_i, \tau(\mathbf{v}_i, \mathbf{v}_{i+1})), \text{ where } n \geqslant 1, m \geqslant 1 \text{ and } i \in J(n)\}.$$

For convenience of notation, write $\tau_i = \tau(\mathbf{v}_i, \mathbf{v}_{i+1})$ for every $i \in J$.

Let $\mathbf{t}^*\in S(h)$ be arbitrary but fixed. Obviously

$$h=\sum_{n=1}^{\infty}\sum_{m=1}^{\infty}\#J(\mathbf{t}^*,n,m)=\sum_{(n,m)\in\mathbb{Z}^2(\mathbf{t}^*)}\#J(\mathbf{t}^*,n,m), \tag{55}$$

where

$$J(\mathbf{t}^*,n,m)=\{i\in J(n):\mathbf{t}^*\in T_m(b_i,\tau_i)\}$$

and

$$\mathbb{Z}^2(\mathbf{t}^*)=\{(n,m):n\geqslant 1,m\geqslant 1 \text{ and } \mathbf{t}^*\in T_m(b_i,\tau_i) \text{ for some } i\in J(n)\}.$$

By (a_3) of Lemma 7.4 and (11), for every $i\in J$,

$$|\mathbf{v}_i-\mathbf{v}_{i+1}|\ll 1/\delta_A(\mathbf{v}_i,\mathbf{v}_{i+1}) \quad\text{and}\quad |\mathbf{v}_i-\mathbf{v}_{i+1}|\leqslant d(A)\ll r(A),$$

and so it follows from (52) that

$$|\mathbf{v}_i-\mathbf{v}_{i+1}|\gg\ll b_i. \tag{56}$$

By definition,

$$2^{n-1}\leqslant b_i<2^n \quad\text{for every } i\in J(n), \tag{57}$$

and so by (56),

$$|\mathbf{v}_i-\mathbf{v}_{i+1}|\gg\ll 2^n \quad\text{for every } i\in J(n). \tag{58}$$

Moreover, by (25), we have that if $\mathbf{t}^*\in T_m(b,\tau)$, then

$$|\tau-\beta(\mathbf{t}^*)|\ll\min\{2^m b^{-2},1\}, \tag{59}$$

where $\beta=\beta(\mathbf{t}^*)$ is the angle of $\mathbf{t}^*$, i.e. $\mathbf{t}^*\in S(\mathbf{0},\beta)\cap T_m(b,\tau)$. Combining ($a_4$) of Lemma 7.4 and (57)–(59), we conclude that

$$\#J(\mathbf{t}^*,n,m)\leqslant 2\frac{\max_{i\in J(n)}|\tau_i-\beta(\mathbf{t}^*)|}{\min_{\substack{i,j\in J(n)\\ i\neq j}}|\tau_i-\tau_j|}+1\ll\frac{\min\{2^m 4^{-n},1\}}{4^{-n}}=\min\{2^m,4^n\}, \tag{60}$$

and so by (53),

$$\#J(\mathbf{t}^*,n,m)\ll\min\{2^m,(r(A))^2\}. \tag{61}$$

We have, from (30), (31) and (57), that if $\mathbf{t}^*\in T_m(b_i,\tau_i)$ for some $i\in J(n)$, then

$$|\mathbf{t}^*|\gg\ll 2^{n-m}. \tag{62}$$

Let m_0 be the largest index $m\geqslant 1$ such that $\mathbf{t}^*\in T_m(b_i,\tau_i)$ for some $i\in J=\bigcup_{n=1}^{\infty}J(n)$. Then by (53), (55), (61) and (62), we have that for any $\mathbf{t}^*\in S(h)$,

$$h\ll\sum_{(n,m)\in\mathbb{Z}^2(\mathbf{t}^*)}\min\{2^m,(r(A))^2\}\ll\min\{2^{m_0},(r(A))^2\log(r(A)+1)\}. \tag{63}$$

Combining this with (54), we see that for any $\mathbf{t}^* \in S(h)$, where $h \geqslant 1$,

$$\int_{-\infty}^{\infty} e^{-\lambda^2} |\hat{\chi}_\lambda(\mathbf{t}^*)|^2 \, d\lambda \gg h. \tag{64}$$

On the other hand, by Lemma 7.1, (a_1) of Lemma 7.4 and (51),

$$\sum_{h \geqslant 1} h \int_{S(h)} |\phi(\mathbf{t})|^2 \, d\mathbf{t} = \sum_{n \geqslant 1} \sum_{i \in J(n)} \sum_{m \geqslant 1} \int_{T_m(b_i, \tau_i)} |\phi(\mathbf{t})|^2 \, d\mathbf{t}$$

$$= \sum_{n \geqslant 1} \sum_{i \in J(n)} \int_{T(b_i, \tau_i)} |\phi(\mathbf{t})|^2 \, d\mathbf{t} \gg \sum_{n \geqslant 1} \sum_{i \in J(n)} M^2 = (\# J) M^2$$

$$\gg k(A) M^2 \geqslant \xi(A) M^2. \tag{65}$$

Combining (17), (64) and (65), we obtain the key inequality

$$\int_{-\infty}^{\infty} \left(\int_{\mathbf{R}^2} |F_\lambda(\mathbf{x})|^2 \, d\mathbf{x} \right) e^{-\lambda^2} \, d\lambda = \int_{\mathbf{R}^2} \left(\int_{-\infty}^{\infty} e^{-\lambda^2} |\hat{\chi}_\lambda(\mathbf{t})|^2 \, d\lambda \right) |\phi(\mathbf{t})|^2 \, d\mathbf{t}$$

$$\geqslant \sum_{h \geqslant 1} \int_{S(h)} \left(\int_{-\infty}^{\infty} e^{-\lambda^2} |\hat{\chi}_\lambda(\mathbf{t})|^2 \, d\lambda \right) |\phi(\mathbf{t})|^2 \, d\mathbf{t}$$

$$\gg \sum_{h \geqslant 1} \int_{S(h)} h |\phi(\mathbf{t})|^2 \, d\mathbf{t} \gg \xi(A) M^2. \tag{66}$$

Let $\delta \in (0, 1)$ be arbitrary. Applying the substitution $A \to (\delta A)$ in (66), we have

$$\delta^{-1} \int_{-\infty}^{\infty} \left(\int_{\mathbf{R}^2} |F_\lambda(\mathbf{x})|^2 \, d\mathbf{x} \right) e^{-(\lambda/\delta)^2} \, d\lambda \gg \xi(\delta A) M^2. \tag{67}$$

We need the following simple lemma.

Lemma 7.6. *Let B be an arbitrary compact convex region. For every $l \geqslant 3$, denote by $B_l \subset B$ an inscribed l-gon of largest area.*

(*i*) *For arbitrary $m \geqslant 3$,*

$$\mu(B \backslash B_m) \leqslant \frac{2\pi^2}{3m^2} \mu(B).$$

(*ii*) *For arbitrary $m \geqslant 3$ and $n \geqslant 3$,*

$$\mu(B \backslash B_{m(n+1)}) \leqslant \frac{2\pi^2}{3n^2} \mu(B \backslash B_m).$$

Proof. The first inequality follows immediately from Sas's theorem (2) and from the elementary result that for every $x \in \mathbb{R}$,

$$1 - \frac{\sin x}{x} \leqslant \frac{x^2}{6}. \tag{68}$$

We now prove (ii). Let $\mathbf{v}_1, \mathbf{v}_2, \ldots, \mathbf{v}_m$ denote the vertices of B_m in their order on the boundary Γ of B. For $i = 1, \ldots, m$, let $D^{(i)}$ denote the convex region bordered by the line segment $[\mathbf{v}_i, \mathbf{v}_{i+1}]$ and the short arc $\Gamma(\mathbf{v}_i, \mathbf{v}_{i+1})$ (convention: $\mathbf{v}_{m+1} = \mathbf{v}_1$). Applying Sas's theorem (2) to $D^{(i)}$ with $l = n$, we establish the existence of an inscribed polygon $D_n^{(i)} \subset D^{(i)}$ of n sides such that

$$\mu(D^{(i)} \backslash D_n^{(i)}) \leqslant \left(1 - \frac{n}{2\pi} \sin\left(\frac{2\pi}{n}\right)\right) \mu(D^{(i)}).$$

Let D be the convex hull of the polygons B_m and $D_n^{(i)}$ $(1 \leqslant i \leqslant m)$. Clearly D has at most $m + nm = m(n+1)$ vertices, and

$$\mu(B \backslash D) \leqslant \sum_{i=1}^{m} \mu(D^{(i)} \backslash D_n^{(i)}) \leqslant \sum_{i=1}^{m} \left(1 - \frac{n}{2\pi} \sin\left(\frac{2\pi}{n}\right)\right) \mu(D^{(i)})$$
$$= \left(1 - \frac{n}{2\pi} \sin\left(\frac{2\pi}{n}\right)\right) \mu(B \backslash B_m). \tag{69}$$

(ii) follows from (68) and (69). ∎

Let $t > 1$ be an arbitrary real number. By Lemma 7.6(ii),

$$\mu(tB \backslash tB_{m(n+1)}) \leqslant \frac{2\pi^2 t^2}{3n^2} \mu(B \backslash B_m). \tag{70}$$

Choosing $n = [ct^{\frac{1}{2}}]$ with some sufficiently large absolute constant c, we get

$$\frac{2\pi^2 t^2}{3n^2} \leqslant (n+1)^2. \tag{71}$$

Let $m = \xi(B)$. Then by (70) and (71),

$$\mu(tB \backslash tB_{m(n+1)}) \leqslant (n+1)^2 \mu(B \backslash B_m) \leqslant (n+1)^2 m^2.$$

Therefore,

$$\xi(tB) \leqslant m(n+1) \ll mt^{\frac{1}{2}} = \xi(B) t^{\frac{1}{2}}.$$

It follows immediately from this that

$$\xi(\delta A) \gg \delta^{\frac{1}{2}} \xi(A) \tag{72}$$

for arbitrary $\delta \in (0, 1)$.

By (67) and (72), for arbitrary $\delta \in (0, 1)$,

$$\delta^{-1} \int_{-\infty}^{\infty} \left(\int_{\mathbf{R}^2} |F_\lambda(\mathbf{x})|^2 \, d\mathbf{x} \right) e^{-(\lambda/\delta)^2} \, d\lambda \gg \delta^{\frac{1}{2}} \xi(A) M^2. \tag{73}$$

Now let $\delta = (\log d(A))^{-1}$. Then by (73), with $d = d(A)$,

$$(\log d) \int_{-\infty}^{\infty} \left(\int_{\mathbf{R}^2} |F_\lambda(\mathbf{x})|^2 \, d\mathbf{x} \right) e^{-\lambda^2 (\log d)^2} \, d\lambda \gg \frac{\xi(A)}{(\log d)^{\frac{1}{2}}} M^2, \tag{74}$$

and this completes the proof of Lemma 7.5.

We now deduce Theorem 19D. Let $M=(d(A))^6$. It is obvious from (8) and (10) that

$$|F_\lambda(\mathbf{x})|\leqslant 8M^2$$

for every $\lambda\in\mathbb{R}$ and $\mathbf{x}\in\mathbb{R}^2$, and $F_\lambda(\mathbf{x})=0$ whenever $A(-\lambda,\mathbf{x})\cap[-M, M]^2=\varnothing$. It follows that, writing $d=d(A)$,

$$\int_{|\lambda|>1}\left(\int_{\mathbb{R}^2}|F_\lambda(\mathbf{x})|^2\,\mathrm{d}\mathbf{x}\right)\mathrm{e}^{-\lambda^2(\log d)^2}\,\mathrm{d}\lambda$$
$$\ll\int_{|\lambda|>1} M^4(\max\{\lambda d, M\})^2 d^{-\lambda^2(\log d)}\,\mathrm{d}\lambda. \qquad (75)$$

From Lemma 7.5, (75) and $M=d^6$, it follows by elementary calculation that

$$\int_{|\lambda|>1}\left(\int_{\mathbb{R}^2}|F_\lambda(\mathbf{x})|^2\,\mathrm{d}\mathbf{x}\right)\mathrm{e}^{-\lambda^2(\log d)^2}\,\mathrm{d}\lambda$$
$$\leqslant\tfrac{1}{2}\int_{-\infty}^{\infty}\left(\int_{\mathbb{R}^2}|F_\lambda(\mathbf{x})|^2\,\mathrm{d}\mathbf{x}\right)\mathrm{e}^{-\lambda^2(\log d)^2}\,\mathrm{d}\lambda \qquad (76)$$

if $d=d(A)\geqslant c_{22}$. Hence, by (74) and (76),

$$(\log d)\int_{-1}^{1}\left(\int_{\mathbb{R}^2}|F_\lambda(\mathbf{x})|^2\,\mathrm{d}\mathbf{x}\right)\mathrm{e}^{-\lambda^2(\log d)^2}\,\mathrm{d}\lambda\gg\frac{\xi(A)}{(\log d)^{\frac{1}{2}}}M^2 \qquad (77)$$

if $d=d(A)\geqslant c_{22}$.

If for some $\lambda_0\in[-1,1]$ and $\mathbf{x}_0\in\mathbb{R}^2$,

$$\sum_{\mathbf{q}_j\in A(\lambda_0,\mathbf{x}_0)} 1\geqslant 2\mu(A) \qquad (78)$$

holds, then obviously

$$\Delta[\mathcal{Q};A]\geqslant\sum_{\mathbf{q}_j\in A(\lambda_0,\mathbf{x}_0)}1-\mu(A(\lambda_0,\mathbf{x}_0))\geqslant 2\mu(A)-\mu(A)=\mu(A)\gg(\xi(A))^4 \qquad (79)$$

by (3), and we have proved much more than we wanted.

We may therefore assume the impossibility of (78), i.e.

$$\sum_{\mathbf{q}_j\in A(\lambda,\mathbf{x})} 1<2\mu(A) \qquad (80)$$

for all $\lambda\in[-1,1]$ and $\mathbf{x}\in\mathbb{R}^2$. Using (8), $M=d^6$ and (80), we have, writing $d=d(A)$, that

$$\int_{-1}^{1}\left(\int_{\mathbb{R}^2\setminus[-M+d,M-d]^2}|F_\lambda(\mathbf{x})|^2\,\mathrm{d}\mathbf{x}\right)\mathrm{e}^{-\lambda^2(\log d)^2}\,\mathrm{d}\lambda\ll dM(\mu(A))^2$$
$$\leqslant d^5M=\frac{M^2}{d}, \qquad (81)$$

and so, combining with (77), we have

$$(\log d)\int_{-1}^{1}\left(\int_{[-M+d,M-d]^2}|F_\lambda(\mathbf{x})|^2\,\mathrm{d}\mathbf{x}\right)\mathrm{e}^{-\lambda^2(\log d)^2}\,\mathrm{d}\lambda \gg \frac{\xi(A)}{(\log d)^{\frac{1}{2}}}M^2 \tag{82}$$

if $d = d(A) \geqslant c_{23}$.

If $\mathbf{x}\in[-M+d, M-d]^2$ and $\lambda\in[-1,1]$, then obviously $A(-\lambda,\mathbf{x})\subset[-M,M]^2$. It follows from (9), (11) and (82), via standard averaging argument, that

$$\Delta[\mathcal{Q};A] \gg \frac{(\xi(A))^{\frac{1}{2}}}{(\log\mu(A))^{\frac{1}{2}}}$$

if $\mu(A)\geqslant c_{24}$. The remaining case $2\leqslant\mu(A)<c_{24}$ follows directly from the trivial estimate (49). Theorem 19D follows.

It remains to prove Lemmas 7.3 and 7.4.

Proof of Lemma 7.3. Let $[\mathbf{w}',\mathbf{w}'']$ and $[\mathbf{u}',\mathbf{u}'']$ denote the two chords of A such that $\tau(\mathbf{w}',\mathbf{w}'')=\tau$, $\delta_A(\mathbf{w}',\mathbf{w}'')=1/b$, $g_\tau(z_\tau^+-1/b)=|\mathbf{w}'-\mathbf{w}''|\geqslant\frac{1}{2}b$, $\tau(\mathbf{u}',\mathbf{u}'')=\beta$ and $\delta_A(\mathbf{u}',\mathbf{u}'')=y$. The problem is to find a good lower bound to $g_\beta(z_\beta^+-y)=|\mathbf{u}'-\mathbf{u}''|$. If $y>b$, then the hypothesis (24) immediately yields $g_\beta(z_\beta^+-y)\gg y=\max\{b,y\}$, and the proof is complete. We may therefore assume that $y\leqslant b$. There are four possibilities (the reader may wish to draw some diagrams).

Case 1: $\Gamma(\mathbf{u}',\mathbf{u}'')\supset\Gamma(\mathbf{w}',\mathbf{w}'')$, where Γ is the boundary curve of A. Assume that the endpoints of these chords are in the order $\mathbf{u}',\mathbf{w}',\mathbf{w}'',\mathbf{u}''$ on Γ. We clearly have

$$\frac{|z_\beta^+-z_\beta^-|-y}{|z_\beta^+-z_\beta^-|}\geqslant\frac{2r(A)-y}{2r(A)}\geqslant\frac{2r(A)-b}{2r(A)}\geqslant\frac{2r(A)-r(A)}{2r(A)}=\frac{1}{2}. \tag{83}$$

Let $\mathbf{u}^-$ be a point on Γ such that $\mathbf{u}^-\cdot\mathbf{e}(\beta)=z_\beta^-$, where $\mathbf{e}(\beta)=(\cos\beta,\sin\beta)$. Clearly $\mathbf{u}^-\in\Gamma\backslash\Gamma(\mathbf{u}',\mathbf{u}'')$ and $\operatorname{dist}(\mathbf{u}^-,[\mathbf{u}',\mathbf{u}''])=|z_\beta^+-z_\beta^-|-y$. We may assume, without loss of generality, that $\operatorname{dist}(\mathbf{u}',[\mathbf{w}',\mathbf{w}''])\leqslant\operatorname{dist}(\mathbf{u}'',[\mathbf{w}',\mathbf{w}''])$. Let $S(\mathbf{u}'|\mathbf{w}',\mathbf{w}'')$ denote the straight line passing through $\mathbf{u}'$ and parallel to the chord $[\mathbf{w}',\mathbf{w}'']$. Suppose that $\mathbf{w}_1=S(\mathbf{u}'|\mathbf{w}',\mathbf{w}'')\cap[\mathbf{w}',\mathbf{u}^-]$ and $\mathbf{w}_2=S(\mathbf{u}'|\mathbf{w}',\mathbf{w}'')\cap[\mathbf{w}'',\mathbf{u}^-]$. Let α denote the angle of the triangle $\mathbf{u}',\mathbf{w}_2,\mathbf{u}''$ at the vertex $\mathbf{w}_2$. It follows from (24) that

$$\alpha\gg 1, \tag{84}$$

where the implicit constant in $\gg$ depends only on the value of c_{17}. Moreover, by (83),

$$|\mathbf{u}'-\mathbf{w}_2|\geqslant|\mathbf{w}_1-\mathbf{w}_2|\geqslant\frac{|\mathbf{w}_1-\mathbf{u}^-|}{|\mathbf{w}'-\mathbf{u}^-|}|\mathbf{w}'-\mathbf{w}''|$$

$$\geqslant\frac{|z_\beta^+-z_\beta^-|-y}{|z_\beta^+-z_\beta^-|}|\mathbf{w}'-\mathbf{w}''|\geqslant\frac{|\mathbf{w}'-\mathbf{w}''|}{2}\geqslant\frac{b}{4}. \tag{85}$$

Combining (84) and (85), we conclude that

$$g_\beta(z_\beta^+ - y) = |\mathbf{u}' - \mathbf{u}''| \gg |\mathbf{u}' - \mathbf{w}_2| \gg b,$$

and this case is complete.

Case 2: $\Gamma(\mathbf{u}', \mathbf{u}'') \cap \Gamma(\mathbf{w}', \mathbf{w}'') = \emptyset$. Let $\mathbf{u}^+$ be a point of Γ such that $\mathbf{u}^+ \cdot \mathbf{e}(\beta) = z_\beta^+$. Then $\mathbf{u}^+ \in \Gamma(\mathbf{u}', \mathbf{u}'')$ and $\mathrm{dist}(\mathbf{u}^+, [\mathbf{u}', \mathbf{u}'']) = \delta_A(\mathbf{u}', \mathbf{u}'') = y$. We may clearly assume that the points $\mathbf{w}', \mathbf{w}'', \mathbf{u}', \mathbf{u}''$ are exactly in this order on Γ. Let γ denote the angle of the triangle $\mathbf{u}^+, \mathbf{u}', \mathbf{u}''$ at $\mathbf{u}'$. By hypothesis, $\gamma \leqslant |\tau - \beta|$, and so

$$\gamma \leqslant y/b \leqslant 1. \tag{86}$$

Let $\mathbf{u}_1$ be the projection of $\mathbf{u}^+$ on the line passing through $\mathbf{u}'$ and $\mathbf{u}''$. It easily follows from the assumption (24) that

$$|\mathbf{u}' - \mathbf{u}_1| \ll |\mathbf{u}' - \mathbf{u}''| = g_\beta(z_\beta^+ - y). \tag{87}$$

Therefore, by (86) and (87),

$$g_\beta(z_\beta^+ - y) \gg |\mathbf{u}' - \mathbf{u}_1| = y/(\tan\gamma) \gg y/(y/b) = b,$$

and the second case is complete.

Case 3: $[\mathbf{u}', \mathbf{u}''] \cap [\mathbf{w}', \mathbf{w}''] \neq \emptyset$. Without loss of generality, we may assume that the endpoints of these chords are exactly in the order $\mathbf{w}', \mathbf{u}', \mathbf{w}'', \mathbf{u}''$ on Γ. Again, let $\mathbf{u}^+$ be defined as in Case 2. If $\mathbf{u}^+ \notin \Gamma(\mathbf{w}', \mathbf{w}'')$, then we can repeat the argument of Case 2 without any modification. Thus we can assume that $\mathbf{u}^+ \in \Gamma(\mathbf{w}', \mathbf{w}'')$. Let $\mathbf{u}_1$ be the projection of $\mathbf{u}^+$ on the line passing through $\mathbf{u}'$ and $\mathbf{u}''$. Let

$$\mathbf{u}^* = \begin{cases} [\mathbf{u}^+, \mathbf{u}_1] \cap [\mathbf{w}', \mathbf{w}''] & \text{(if this exists),} \\ \mathbf{u}_1 & \text{(otherwise).} \end{cases}$$

Clearly

$$|\mathbf{u}^+ - \mathbf{u}^*| \cos(\tau - \beta) \leqslant \delta_A(\mathbf{w}', \mathbf{w}'') = 1/b.$$

Since

$$|\mathbf{u}^+ - \mathbf{u}_1| = \delta_A(\mathbf{u}', \mathbf{u}'') = y \geqslant 4/b$$

and

$$|\tau - \beta| \leqslant y/b \leqslant 1 < \pi/3,$$

we conclude that

$$|\mathbf{u}^* - \mathbf{u}_1| \geqslant |\mathbf{u}^+ - \mathbf{u}_1| - |\mathbf{u}^+ - \mathbf{u}^*| \geqslant y - \frac{1}{b\cos(\tau - \beta)} > y - \frac{2}{b} \geqslant \frac{y}{2}. \tag{88}$$

The assumption (24) yields

$$g_\beta(z_\beta^+ - y) = |\mathbf{u}' - \mathbf{u}''| \gg \max\{|\mathbf{u}' - \mathbf{u}_1|, |\mathbf{u}'' - \mathbf{u}_1|\}. \tag{89}$$

Let $\mathbf{v} = [\mathbf{u}', \mathbf{u}''] \cap [\mathbf{w}', \mathbf{w}'']$. Then by (88) and (89),

$$g_\beta(z_\beta^+ - y) \gg \max\{|\mathbf{u}' - \mathbf{u}_1|, |\mathbf{u}'' - \mathbf{u}_1|\} \geqslant |\mathbf{v} - \mathbf{u}_1|$$

$$= |\mathbf{u}^* - \mathbf{u}_1| \frac{1}{\tan(\tau - \beta)} \geqslant \frac{y}{2} \frac{1}{\tan(y/b)} \gg \frac{yb}{2y} = \frac{b}{2},$$

and the third case is complete.

Case 4: $\Gamma(\mathbf{w}', \mathbf{w}'') \supset \Gamma(\mathbf{u}', \mathbf{u}'')$. Clearly

$$y \cos(\tau - \beta) = \delta_A(\mathbf{u}', \mathbf{u}'') \cos(\tau - \beta) \leqslant \delta_A(\mathbf{w}', \mathbf{w}'') = 1/b. \qquad (90)$$

Since $|\tau - \beta| \leqslant y/b \leqslant 1 < \pi/3$, (90) contradicts the hypothesis $y \geqslant 4/b$. This case is therefore impossible.

Summarizing Cases 1–4, the lemma follows. ∎

To prove Lemma 7.4, we shall proceed by iterated application of the following lemma.

Lemma 7.7. *Let A be the same as in Lemma 7.4. Let $B_l \subset A$ be an inscribed l-gon ($l \geqslant 3$) of A such that $\mu(A \backslash B_l) \leqslant 100l$. Let $\mathbf{u}_1, \ldots, \mathbf{u}_l$ denote the vertices of B_l in their order on the boundary Γ of A. Then either there is an index set $L \subset \{1, 2, \ldots, l\}$ such that*

(a_1) *$\#L \gg l$;*
(a_2) *$|\mathbf{u}_i - \mathbf{u}_{i+1}| \gg l^{\frac{1}{2}}$ for every $i \in L$;*
(a_3) *$|\mathbf{u}_i - \mathbf{u}_{i+1}| \delta_A(\mathbf{u}_i, \mathbf{u}_{i+1}) \ll 1$ for every $i \in L$; and*
(a_4) *if $i, j \in L$, $i \neq j$ and $q \leqslant |\mathbf{u}_i - \mathbf{u}_{i+1}|, |\mathbf{u}_j - \mathbf{u}_{j+1}| < 2q$ for some real number q, then*

$$|\tau(\mathbf{u}_i, \mathbf{u}_{i+1}) - \tau(\mathbf{u}_j, \mathbf{u}_{j+1})| \gg q^{-2} \pmod{\pi};$$

or there is another inscribed polygon $B_t \subset A$ of t sides such that $t < l$ and $\mu(A \backslash B_t) \leqslant \mu(A \backslash B_l) + (l - t)$.

Proof. If a triangle $\mathbf{u}_i, \mathbf{u}_{i+1}, \mathbf{u}_{i+2}$ ($1 \leqslant i \leqslant l$ and convention: $\mathbf{u}_{l+1} = \mathbf{u}_1$ and $\mathbf{u}_{l+2} = \mathbf{u}_2$) has area $\leqslant 1$, then the second alternative holds. Indeed, throwing out the vertex $\mathbf{u}_{i+1}$, we get the polygon B_{l-1} of $(l-1)$ vertices $\mathbf{u}_1, \ldots, \mathbf{u}_i, \mathbf{u}_{i+2}, \ldots, \mathbf{u}_l$ such that

$$\mu(A \backslash B_{l-1}) = \mu(A \backslash B_l) + \mu(\text{triangle}(\mathbf{u}_i, \mathbf{u}_{i+1}, \mathbf{u}_{i+2})) \leqslant \mu(A \backslash B_l) + 1.$$

We may therefore assume that any triangle $\mathbf{u}_i, \mathbf{u}_{i+1}, \mathbf{u}_{i+2}$ has area greater than 1. Let

$$L_0 = \{1, 3, 5, \ldots, 2[\tfrac{1}{2}(l-1)] - 1\}.$$

Let $L_1^* = \{i \in L_0 : |\mathbf{u}_i - \mathbf{u}_{i+1}| \geqslant |\mathbf{u}_{i+1} - \mathbf{u}_{i+2}|\}$ and $L_1^{**} = L_0 \backslash L_1^*$. It is obvious that one of the sets L_1^* and L_1^{**} has cardinality $\geqslant (l-2)/4$. Without loss of generality, assume that $\#L_1^* \geqslant (l-2)/4$. Let

$$\begin{cases} L_2 = \{i \in L_1^* : |\mathbf{u}_i - \mathbf{u}_{i+1}| \leqslant l^{\frac{1}{2}}/10\}, \\ L_3 = \{i \in L_1^* : |\mathbf{u}_i - \mathbf{u}_{i+1}| \delta_A(\mathbf{u}_i, \mathbf{u}_{i+1}) \geqslant 10^4\}, \\ L_4 = \{i \in L_1^* : \pi - \alpha_i \leqslant |\mathbf{u}_i - \mathbf{u}_{i+1}|^{-2}\}, \end{cases}$$

where α_i denotes the angle of the triangle $\mathbf{u}_i, \mathbf{u}_{i+1}, \mathbf{u}_{i+2}$ at $\mathbf{u}_{i+1}$. The index set $L = L_1^* \backslash (L_2 \cup L_3 \cup L_4)$ clearly satisfies (a_2)–(a_4), so it remains to prove that $\#L \gg l$. We first observe that $L_4 = \emptyset$. Indeed, if $j \in L_4$, then the triangle $\mathbf{u}_j, \mathbf{u}_{j+1}, \mathbf{u}_{j+2}$ has area

$$\tfrac{1}{2}|\mathbf{u}_j - \mathbf{u}_{j+1}||\mathbf{u}_{j+1} - \mathbf{u}_{j+2}| \sin(\pi - \alpha_j) \leqslant \tfrac{1}{2}|\mathbf{u}_j - \mathbf{u}_{j+1}|^2(\pi - \alpha_j) \leqslant 1,$$

a contradiction. We next assume that $\#L_2 \geqslant l/20$. Since

$$2\pi = \sum_{i=1}^{l} (\pi - \alpha_i) \geqslant \sum_{i \in L_2} (\pi - \alpha_i),$$

it follows that there exists an index $j \in L_2$ such that

$$(\pi - \alpha_j) \leqslant \frac{2\pi}{\#L_2} \leqslant \frac{2\pi}{l/20} = \frac{40\pi}{l}.$$

Then the triangle $\mathbf{u}_j, \mathbf{u}_{j+1}, \mathbf{u}_{j+2}$ has area

$$\begin{aligned} &\tfrac{1}{2}|\mathbf{u}_j - \mathbf{u}_{j+1}||\mathbf{u}_{j+1} - \mathbf{u}_{j+2}| \sin(\pi - \alpha_j) \\ &\leqslant \tfrac{1}{2}|\mathbf{u}_j - \mathbf{u}_{j+1}|^2(\pi - \alpha_j) \leqslant \frac{1}{2}\left(\frac{l^{\frac{1}{2}}}{10}\right)^2 \frac{40\pi}{l} = \frac{\pi}{5} \leqslant 1, \end{aligned}$$

a contradiction again. Finally, we estimate $\#L_3$. By hypothesis,

$$\begin{aligned} 100l \geqslant \mu(A \backslash B_l) &\geqslant \sum_{i=1}^{l} \tfrac{1}{2}|\mathbf{u}_i - \mathbf{u}_{i+1}| \delta_A(\mathbf{u}_i, \mathbf{u}_{i+1}) \\ &\geqslant \sum_{i \in L_3} \tfrac{1}{2}|\mathbf{u}_i - \mathbf{u}_{i+1}| \delta_A(\mathbf{u}_i, \mathbf{u}_{i+1}) \geqslant (\#L_3)\frac{10^4}{2}. \end{aligned}$$

Therefore, $\#L_3 \leqslant l/50$. Summarizing, we have

$$\#L \geqslant \frac{l-2}{4} - \frac{l}{20} - \frac{l}{50} \gg l$$

as required. ∎

Proof of Lemma 7.4. Let $l = k = k(A)$ and $B_l = A_k$. By iterated application of Lemma 7.7, we conclude that either Lemma 7.4 holds, or (i) there is an inscribed polygon $B_t \subset A$ of t sides such that $3 \leqslant t < k$ and $100t < \mu(A \backslash B_t)$, or (ii) B_t is degenerate, i.e. $t<3$. In both cases,

$$\mu(A\backslash B_t)\leqslant \mu(A\backslash A_k)+(k-t)\leqslant k+(k-t)\leqslant 2k. \tag{91}$$

We first consider (i). In view of (91), $t<k/50$. Applying Lemma 7.6(ii) with $m=t$ and $n=[k/2t]\geqslant 25$, we have

$$\mu(A\backslash A_{m(n+1)})\leqslant \frac{2\pi^2}{3(25)^2}\mu(A\backslash A_t)\leqslant \frac{2\pi^2}{3(25)^2}\mu(A\backslash B_t).$$

Thus, by (91), we have, with $r=m(n+1)=t([k/2t]+1)$,

$$\mu(A\backslash A_r)\leqslant \frac{2\pi^2}{3(25)^2}2k<\tfrac{1}{2}k\leqslant r. \tag{92}$$

Since $r<k$, (92) contradicts the minimality of $k=k(A)$. Hence (i) is impossible. Next, we consider (ii). By (91),

$$\mu(A)=\mu(A\backslash B_t)\leqslant 2k.$$

Applying Lemma 7.6(i) with $m=k-1$, we get

$$\mu(A\backslash A_{k-1})\leqslant \frac{2\pi^2}{3(k-1)^2}\mu(A)\leqslant \frac{2\pi^2}{3(k-1)^2}2k=\frac{4\pi^2 k}{3(k-1)^2}. \tag{93}$$

By hypothesis, we have $k=k(A)\geqslant 6$, so

$$\frac{4\pi^2 k}{3(k-1)^2}<k-1, \tag{94}$$

and so (93) and (94) contradicts the minimality of $k=k(A)$. Hence (ii) is also impossible. The proof of Lemma 7.4 is now complete. ■

This also completes the proof of Theorem 19D.

Theorem 19C can be deduced from the following intermediate result by averaging arguments similar to those for the deduction of Theorem 19D from Lemma 7.5. The following lemma is the analogue of Lemma 7.5.

Lemma 7.8. *Writing $d=d(A)$, we have*

$$(\log d)\int_{-\infty}^{\infty}\left(\int_{\mathbf{R}^2}|F_\lambda(\mathbf{x})|^2\,\mathrm{d}\mathbf{x}\right)\mathrm{e}^{-\lambda^2(\log d)^2}\,\mathrm{d}\lambda \gg (\log\mu(A))M^2.$$

To prove Lemma 7.8, we again assume that

$$\frac{r(A)}{d(A)}\geqslant c_{17}>0.$$

We may further assume that $\mu(A)$ is greater than a sufficiently large absolute constant, otherwise we obtain Theorem 19C immediately on using the trivial estimate (49).

Let $m=[2(\mu(A))^{\frac{1}{3}}]$. It follows from Lemma 7.6(i) that

$$\mu(A\backslash A_m)\leqslant\frac{2\pi^2}{3m^2}\mu(A)<m. \tag{95}$$

Clearly

$$\text{perimeter}\,(A_m)\geqslant d(A_m)\gg d(A)\geqslant(\mu(A))^{\frac{1}{2}}\gg m^{\frac{3}{2}}. \tag{96}$$

Combining (95) and (96) and applying standard averaging arguments, we can show that for some side $[\mathbf{v}',\mathbf{v}'']$ of A_m,

$$|\mathbf{v}'-\mathbf{v}''|\gg m^{\frac{1}{2}}\quad\text{and}\quad\delta_A(\mathbf{v}',\mathbf{v}'')\ll m^{-\frac{1}{2}}. \tag{97}$$

Let $\mu(A)$ be so large that $|\mathbf{v}'-\mathbf{v}''|\geqslant 1$ and $\delta_A(\mathbf{v}',\mathbf{v}'')\leqslant 1$, and let

$$N=\min\{|\mathbf{v}'-\mathbf{v}''|,1/\delta_A(\mathbf{v}',\mathbf{v}'')\}.$$

Then clearly $N\geqslant 1$ and

$$(\mu(A))^{\frac{1}{6}}\leqslant m^{\frac{1}{2}}\ll N\leqslant d(A)\ll(\mu(A))^{\frac{1}{2}}. \tag{98}$$

It follows from Lemma 7.2 that for any $j\geqslant 1$ and any integer n satisfying $1\leqslant 2^n\leqslant N$,

$$\int_{-\infty}^{\infty}\mathrm{e}^{-\lambda^2}|\hat{\chi}_\lambda(\mathbf{t})|^2\,\mathrm{d}\lambda\gg\min\{4^j,(r(A))^4\}\gg\min\{4^j,(\mu(A))^2\} \tag{99}$$

for all $\mathbf{t}\in T_j(2^n,\tau(\mathbf{v}',\mathbf{v}''))$.

For every integer $h\geqslant 1$, let

$$\begin{aligned}S(h)=\{\mathbf{t}\in\mathbb{R}^2:&\,\mathbf{t}\text{ belongs to exactly }h\text{ sets of type}\\&T_j(2^n,\tau(\mathbf{v}',\mathbf{v}'')),\text{ where }j\geqslant 1\text{ and }1\leqslant 2^n\leqslant N\}.\end{aligned} \tag{100}$$

For every $\mathbf{t}\in\bigcup_{h=1}^{\infty}S(h)$, let $j(\mathbf{t})$ denote the largest index $j\geqslant 1$ such that $\mathbf{t}\in T_j(2^n,\tau(\mathbf{v}',\mathbf{v}''))$ for some integer n satisfying $1\leqslant 2^n\leqslant N$. By (23),

$$|\mathbf{t}|\gg\ll 2^{n-j}\quad\text{if}\quad\mathbf{t}\in T_j(2^n,\tau). \tag{101}$$

It easily follows that for $h\geqslant 1$,

$$j(\mathbf{t})\gg h\quad\text{if}\quad\mathbf{t}\in S(h). \tag{102}$$

Let H denote the largest integer $h\geqslant 1$ such that $S(h)\neq\varnothing$. By (98) and (100),

$$H\leqslant\#\{n\in\mathbb{Z}:1\leqslant 2^n\leqslant N\}\ll\log\mu(A). \tag{103}$$

It follows from (102) and (103) that for every integer h with $1\leqslant h\leqslant H$,

$$\min_{\mathbf{t}\in S(h)}\{4^{j(\mathbf{t})},(\mu(A))^2\}\gg\min_{\mathbf{t}\in S(h)}\{j(\mathbf{t}),\log\mu(A)\}\gg h. \tag{104}$$

By (17), (99) and (104),

$$\int_{-\infty}^{\infty}\left(\int_{\mathbf{R}^2}|F_\lambda(\mathbf{x})|^2\,\mathrm{d}\mathbf{x}\right)\mathrm{e}^{-\lambda^2}\,\mathrm{d}\lambda=\int_{\mathbf{R}^2}\left(\int_{-\infty}^{\infty}\mathrm{e}^{-\lambda^2}|\hat{\chi}_\lambda(\mathbf{t})|^2\,\mathrm{d}\lambda\right)|\phi(\mathbf{t})|^2\,\mathrm{d}\mathbf{t}$$

$$\geqslant\sum_{h=1}^{H}\int_{S(h)}\left(\int_{-\infty}^{\infty}\mathrm{e}^{-\lambda^2}|\hat{\chi}_\lambda(\mathbf{t})|^2\,\mathrm{d}\lambda\right)|\phi(\mathbf{t})|^2\,\mathrm{d}\mathbf{t}$$

$$\gg\sum_{h=1}^{H}\int_{S(h)}\left(\min_{\mathbf{t}\in S(h)}\{4^{j(\mathbf{t})},(\mu(A))^2\}\right)|\phi(\mathbf{t})|^2\,\mathrm{d}\mathbf{t}$$

$$\gg\sum_{h=1}^{H}h\int_{S(h)}|\phi(\mathbf{t})|^2\,\mathrm{d}\mathbf{t}=\sum_{\substack{n\in Z\\ 1\leqslant 2^n\leqslant N}}\int_{T(2^n,\tau(\mathbf{v}',\mathbf{v}''))}|\phi(\mathbf{t})|^2\,\mathrm{d}\mathbf{t}. \tag{105}$$

Combining Lemma 7.1, (98) and (105), we have the key inequality

$$\int_{-\infty}^{\infty}\left(\int_{\mathbf{R}^2}|F_\lambda(\mathbf{x})|^2\,\mathrm{d}\mathbf{x}\right)\mathrm{e}^{-\lambda^2}\,\mathrm{d}\lambda\gg(\log\mu(A))M^2. \tag{106}$$

Let $\delta\in(0,1)$ be arbitrary. Applying the substitution $A\to(\delta A)$ in (106), we get

$$\delta^{-1}\int_{-\infty}^{\infty}\left(\int_{\mathbf{R}^2}|F_\lambda(\mathbf{x})|^2\,\mathrm{d}\mathbf{x}\right)\mathrm{e}^{-(\lambda/\delta)^2}\,\mathrm{d}\lambda\gg(\log\mu(\delta A))M^2. \tag{107}$$

Now let $\delta=(\log d(A))^{-1}$. Then by (107), with $d=d(A)$,

$$(\log d)\int_{-\infty}^{\infty}\left(\int_{\mathbf{R}^2}|F_\lambda(\mathbf{x})|^2\,\mathrm{d}\mathbf{x}\right)\mathrm{e}^{-\lambda^2(\log d)^2}\,\mathrm{d}\lambda\gg(\log\mu(A))M^2, \tag{108}$$

and this completes the proof of Lemma 7.8.

Remark. If we follow the argument towards the end of the proof of Theorem 19A in Chapter 6, then we obtain the following 'contained in U^2' version of Theorem 19C: Let $A\subset U^2$ be a convex domain, and let $\mathscr{P}$ be a distribution of N points in U^2. Then there exists a homothetic copy $A(\lambda,\mathbf{v})$ of A such that $|\lambda|\leqslant 1$, $A(\lambda,\mathbf{v})\subset U^2$ and

$$\left|\sum_{\mathbf{x}\in\mathscr{P}\cap A(\lambda,\mathbf{v})}1-N\mu(A(\lambda,\mathbf{v}))\right|>c_{25}(A)(\log N)^{\frac{1}{2}},$$

where the positive constant $c_{25}(A)$ depends only on A.

We now turn to Theorem 19E. Again, we apply averaging arguments (similar to those for the deduction of Theorem 19D from Lemma 7.5) to the following intermediate result, which is the analogue of Lemma 7.5.

Lemma 7.9. *Suppose further that A is a regular polygon. Then, writing*

$d = d(A)$, we have

$$(\log d)\int_{-\infty}^{\infty}\left(\int_{\mathbf{R}^2}|F_\lambda(\mathbf{x})|^2\,\mathrm{d}\mathbf{x}\right)\mathrm{e}^{-\lambda^2(\log d)^2}\,\mathrm{d}\lambda \gg \frac{(\xi(A))^2}{\log d}M^2.$$

For the sake of brevity, let $T = T(1,0)$ and $T_j = T_j(1,0)$, i.e. $b = 1$ and $\tau = 0$ in (18) and (23). We need the following analogue of Lemma 7.2.

Lemma 7.10. *Suppose that A is a regular polygon of l sides, where $l \geqslant 3$. Then for every integer $j \geqslant 1$ and every $\mathbf{t}^* \in T_j$,*

$$\int_{-\infty}^{\infty}\mathrm{e}^{-\lambda^2}|\hat{\chi}_\lambda(\mathbf{t}^*)|^2\,\mathrm{d}\lambda \gg \min\{2^{4j}l^2, 2^{3j}r(A), (r(A))^4\}.$$

Proof. We recall (37): For every $\mathbf{t}^* \in S(\mathbf{0},\beta)\cap T_j$,

$$\int_{-\infty}^{\infty}\mathrm{e}^{-\lambda^2}|\hat{\chi}_\lambda(\mathbf{t}^*)|^2\,\mathrm{d}\lambda \gg \frac{1}{\varepsilon}\int_{-\infty}^{\infty}\left(\int_{-\infty}^{\infty}g_\beta(y)(h-H)(x-y)\,\mathrm{d}y\right)^2\mathrm{d}x, \quad (109)$$

where, by (30) and (31),

$$\varepsilon = \frac{25}{2^{j+\frac{1}{2}}} \quad \text{and} \quad 8\varepsilon < |\mathbf{t}^*| \leqslant 16\varepsilon. \quad (110)$$

In order to estimate the right-hand side of (109), we require some information on the chord function $g_\beta(y)$ of the regular polygon A of l sides. Using elementary calculation, we obtain, writing $r = r(A)$,

$$g_\beta(z_\beta^+ - \min\{\varepsilon^{-1}, r\}) \gg \begin{cases} l/\varepsilon & (\varepsilon^{-1} \leqslant r/l^2), \\ (r/\varepsilon)^{\frac{1}{2}} & (r/l^2 < \varepsilon^{-1} \leqslant r), \\ r & (\varepsilon^{-1} > r). \end{cases} \quad (111)$$

Repeating the same argument as in the proof of Lemma 7.2, we have the following analogue of (46): For all $\beta \in B(j,1,0)$ and $x_0 \in [z_\beta^+ + \varepsilon^{-1}, z_\beta^+ + 2\varepsilon^{-1}]$,

$$\int_{-\infty}^{\infty}g_\beta(y)(h-H)(x_0-y)\,\mathrm{d}y \geqslant \tfrac{1}{2}\int_{-\infty}^{\infty}g_\beta(y)h(x_0-y)\,\mathrm{d}y$$

$$\geqslant \tfrac{1}{2}\int_{z'}^{z''}g_\beta(y)h(x_0-y)\,\mathrm{d}y \gg \begin{cases} l/\varepsilon & (\varepsilon^{-1} \leqslant r/l^2), \\ (r/\varepsilon)^{\frac{1}{2}} & (r/l^2 < \varepsilon^{-1} \leqslant r), \\ r^2\varepsilon & (\varepsilon^{-1} > r), \end{cases} \quad (112)$$

where $z' = z_\beta^+ - \min\{\varepsilon^{-1}, r\}$ and $z'' = z_\beta^+ - \frac{1}{2}\min\{\varepsilon^{-1}, r\}$, and so

$$\int_{-\infty}^{\infty}\left(\int_{-\infty}^{\infty}g_\beta(y)(h-H)(x-y)\,\mathrm{d}y\right)^2\mathrm{d}x \gg \begin{cases} l^2/\varepsilon^3 & (\varepsilon^{-1} \leqslant r/l^2), \\ r/\varepsilon^2 & (r/l^2 < \varepsilon^{-1} \leqslant r), \\ r^4\varepsilon & (\varepsilon^{-1} > r). \end{cases} \quad (113)$$

Lemma 7.10 follows from (109), (110) and (113). ■

We now prove Lemma 7.9. Since A has l sides, clearly $\xi(A)\leqslant l$. Combining this with (3), we obtain, writing $r=r(A)$,

$$\min\{2^{4j}l^2,2^{3j}r,r^4\}\geqslant\min\{l^2,r\}\gg\min\{l^2,(\mu(A))^{\frac{1}{2}}\}\gg(\xi(A))^2. \tag{114}$$

Combining this with Lemma 7.10, we have that for every $\mathbf{t}^*\in T$,

$$\int_{-\infty}^{\infty}\mathrm{e}^{-\lambda^2}|\hat{\chi}_\lambda(\mathbf{t}^*)|^2\,\mathrm{d}\lambda\gg(\xi(A))^2. \tag{115}$$

It follows from (17) and (115) that

$$\int_{-\infty}^{\infty}\left(\int_{\mathbf{R}^2}|F_\lambda(\mathbf{x})|^2\,\mathrm{d}\mathbf{x}\right)\mathrm{e}^{-\lambda^2}\,\mathrm{d}\lambda=\int_{\mathbf{R}^2}\left(\int_{-\infty}^{\infty}\mathrm{e}^{-\lambda^2}|\hat{\chi}_\lambda(\mathbf{t})|^2\,\mathrm{d}\lambda\right)|\phi(\mathbf{t})|^2\,\mathrm{d}\mathbf{t}$$
$$\geqslant\int_T\left(\int_{-\infty}^{\infty}\mathrm{e}^{-\lambda^2}|\hat{\chi}_\lambda(\mathbf{t})|^2\,\mathrm{d}\lambda\right)|\phi(\mathbf{t})|^2\,\mathrm{d}\mathbf{t}\gg(\xi(A))^2\int_T|\phi(\mathbf{t})|^2\,\mathrm{d}\mathbf{t}, \tag{116}$$

and so, by Lemma 7.1 and (116), we obtain the following analogue of (66):

$$\int_{-\infty}^{\infty}\left(\int_{\mathbf{R}^2}|F_\lambda(\mathbf{x})|^2\,\mathrm{d}\mathbf{x}\right)\mathrm{e}^{-\lambda^2}\,\mathrm{d}\lambda\gg(\xi(A))^2M^2. \tag{117}$$

Let $\delta\in(0,1)$ be arbitrary. Applying the substitution $A\to(\delta A)$ in (117) and noting (72), we have

$$\delta^{-1}\int_{-\infty}^{\infty}\left(\int_{\mathbf{R}^2}|F_\lambda(\mathbf{x})|^2\,\mathrm{d}\mathbf{x}\right)\mathrm{e}^{-(\lambda/\delta)^2}\,\mathrm{d}\lambda\gg(\xi(\delta A))^2M^2\gg\delta(\xi(A))^2M^2. \tag{118}$$

Lemma 7.9 follows on letting $\delta=(\log d(A))^{-1}$.

We now modify the argument in the proof of Theorem 19E to obtain a proof of Theorem 19B. The only modification is that the chord function $g_\beta(y)$ of the smooth region A satisfies the following different inequality:

$$g_\beta(z_\beta^+-\min\{\varepsilon^{-1},r\})\gg_\gamma\begin{cases}(r/\varepsilon)^{\frac{1}{2}} & (\varepsilon^{-1}\leqslant r),\\ r & (\varepsilon^{-1}>r).\end{cases} \tag{119}$$

This essentially corresponds to the case $l=[(\mu(A))^{\frac{1}{4}}]$ in the proof of Theorem 19E. We deduce the following analogue of (118). For $\delta\in(0,1)$,

$$\delta^{-1}\int_{-\infty}^{\infty}\left(\int_{\mathbf{R}^2}|F_\lambda(\mathbf{x})|^2\,\mathrm{d}\mathbf{x}\right)\mathrm{e}^{-(\lambda/\delta)^2}\,\mathrm{d}\lambda\gg_\gamma\delta(\mu(A))^{\frac{1}{2}}M^2. \tag{120}$$

We conclude this section by proving Theorem 21.

The simple underlying idea of the proof of Theorem 21 is as follows: Let P^1 be a convex polygon. By 'smoothing' the corners of P^1 slightly one can obtain a convex region B^1 of differentiability class two. In B^1 one can inscribe a convex polygon P^2 which approximates B^1 very closely. By 'smoothing' the corners of P^2 slightly one can obtain a convex region B^2 of differentiability class two. And so on. If in this process, $P^1, B^1, P^2, B^2, \ldots$ differ by only very little, then $A = P^1 \cap B^1 \cap P^2 \cap B^2 \cap \cdots$ satisfies the requirements (i) and (ii) of Theorem 21.

After the heuristics, we begin the proof of (i). We shall actually deduce it from Theorem 20C.

Let $POL(n)$ denote the subspace of $CONV(2)$ consisting of all convex polygons of at most n vertices. For every $A \in CONV(2)$, let

$$v(A, n) = \inf \mu(A \backslash P),$$

where the infimum is extended over all $P \in POL(n)$ satisfying $P \subset A$.

The following lemma was independently proved by Schneider and Wieacker (1981) and by Gruber and Kenderov (1982).

Lemma 7.11. *Let $h: \mathbb{N} \to \mathbb{R}^+$ satisfy $h(n) \to 0$ as $n \to \infty$. Then for all $A \in CONV(2)$, except those in a set of first category,*

$$v(A, n) < h(n) \quad \textit{for infinitely many } n.$$

Proof. The function $A \to v(A, n)$ is clearly continuous for each $n \in \mathbb{N}$. Thus the sets $\{A \in CONV(2): v(A, n) \geqslant h(n)\}$ are closed for each $n \in \mathbb{N}$. It follows that the set

$$\mathscr{A}_k = \{A \in CONV(2): v(A, n) \geqslant h(n) \text{ for all } n \geqslant k\}$$

$$= \bigcap_{n=k}^{\infty} \{A \in CONV(2): v(A, n) \geqslant h(n)\} \tag{121}$$

is again closed for each $k \in \mathbb{N}$. We shall show that

$$\mathscr{A}_k \text{ is nowhere dense in } CONV(2) \text{ for each } k \in \mathbb{N}. \tag{122}$$

In view of (121), it is sufficient to show that $\mathscr{A}_k$ has empty interior. Suppose that for some $k \in \mathbb{N}$, the interior of $\mathscr{A}_k$ is non-empty. Since the set of convex polygons is dense in $CONV(2)$, there exists a convex polygon $P \in \mathscr{A}_k$. Then $v(P, n) = 0$ for all sufficiently large n. This contradicts the definition of $\mathscr{A}_k$. Hence (122) is established, and Lemma 7.11 follows. ∎

Let $h(n) = 2^{-2^n}$, and write

$$\mathscr{A} = \{A \in CONV(2): v(A, n) < h(n) \text{ for infinitely many } n\}.$$

Then for every $A \in \mathscr{A}$ and $N \geqslant 3$,

$$\liminf_{N \to \infty} \frac{\xi_N(A)}{\log \log N} < \infty,$$

and so, by Theorem 20C,

$$\Delta_N^{\text{tor}}[A] < f(N) \quad \text{for infinitely many } N$$

(here $f(N)$ is a function satisfying the hypothesis of Theorem 21). Theorem 21(i) follows, since by Lemma 7.11, $CONV(2)\backslash\mathscr{A}$ forms a set of first category.

Next we prove (ii). For every $A\in CONV(2)$, let

$$\rho(A, n) = \inf_{\mathscr{P}} l(A, n)\int_{-\infty}^{\infty}\left(\int_{U_0^2}|D^{\text{tor}}[\mathscr{P}; A(\lambda, \mathbf{x})]|^2\,\mathrm{d}\mathbf{x}\right)\mathrm{e}^{-\lambda^2 l^2(A,n)}\,\mathrm{d}\lambda,$$

where $l(A, n) = \log(2 + n\mu(A))$ and the infimum is taken over all n-element subsets $\mathscr{P}$ of U_0^2.

The function $A\to\rho(A, n)$ is continuous for each $n\in\mathbb{N}$. We indicate this as follows: Let $A, B\in CONV(2)$ and $dist_{\mathscr{H}}(A, B) < \delta$. Observe that if $\delta < \delta(A, \varepsilon)$, then

$$(1-\varepsilon)B + \mathbf{u}\subset A\subset(1+\varepsilon)B + \mathbf{v} \tag{123}$$

for some $\mathbf{u}, \mathbf{v}\in\mathbb{R}^2$. For the sake of brevity, let

$$D_A(\lambda, \mathbf{y}) = D^{\text{tor}}[\mathscr{P}; A(\lambda, \mathbf{y})].$$

Then by (123), for arbitrary $\mathbf{x}$,

$$\begin{aligned}&\min\{|D_B((1-\varepsilon)\lambda, \lambda\mathbf{u}+\mathbf{x})| - (\mu(A) - (1-\varepsilon)^2\mu(B)),\\&\qquad |D_B((1+\varepsilon)\lambda, \lambda\mathbf{v}+\mathbf{x})| - ((1+\varepsilon)^2\mu(B) - \mu(A))\}\\&\quad\leqslant |D_A(\lambda, \mathbf{x})|\\&\quad\leqslant\max\{|D_B((1-\varepsilon)\lambda, \lambda\mathbf{u}+\mathbf{x})| + (\mu(A) - (1-\varepsilon)^2\mu(B)),\\&\qquad\qquad |D_B((1+\varepsilon)\lambda, \lambda\mathbf{v}+\mathbf{x})| + ((1+\varepsilon)^2\mu(B) - \mu(A))\}.\end{aligned} \tag{124}$$

Moreover, let

$$U_B(\lambda) = \int_{U^2}|D_B(\lambda, \mathbf{y})|^2\,\mathrm{d}\mathbf{y}.$$

Clearly

$$|D_B(\lambda, \mathbf{y})| \ll n\lambda d(B),$$

and so we have (note that $n = \#\mathscr{P}$ is fixed)

$$\begin{aligned}&\left|\int_{-\infty}^{\infty}U_B((1+\varepsilon)\lambda)\mathrm{e}^{-\lambda^2}\,\mathrm{d}\lambda - \int_{-\infty}^{\infty}U_B((1-\varepsilon)\lambda)\mathrm{e}^{-\lambda^2}\,\mathrm{d}\lambda\right|\\&= \left|\int_{-\infty}^{\infty}U_B(\lambda)\left(\frac{1}{1+\varepsilon}\mathrm{e}^{-\lambda^2/(1+\varepsilon)^2} - \frac{1}{1-\varepsilon}\mathrm{e}^{-\lambda^2/(1-\varepsilon)^2}\right)\mathrm{d}\lambda\right|\\&\ll n^2(d(B))^2\int_{-\infty}^{\infty}\lambda^2\left|\frac{1}{1-\varepsilon}\mathrm{e}^{-\lambda^2/(1-\varepsilon)^2} - \frac{1}{1+\varepsilon}\mathrm{e}^{-\lambda^2/(1+\varepsilon)^2}\right|\mathrm{d}\lambda.\end{aligned} \tag{125}$$

Since the right-hand side of (125) tends to 0 as $B\to A$ and $\varepsilon\to 0$, the continuity of $A \to \rho(A, n)$ for fixed n easily follows from (124) and (125).

We need

Lemma 7.12. *Let* $G:\mathbb{N} \to \mathbb{R}^+$ *satisfy*

$$\lim_{n\to\infty} \frac{G(n)}{n^{\frac{1}{4}}(\log n)^{-1}} = 0.$$

Then for all $A \in CONV(2)$, *except those in a set of first category,*

$$\rho(A, n) > G(n) \quad \text{for infinitely many } n.$$

Proof. We repeat the argument of the previous lemma. Since the function $A \to \rho(A, n)$ is continuous, the sets $\{A \in CONV(2): \rho(A, n) \leqslant G(n)\}$ are closed for each $n \in \mathbb{N}$. It follows that the set

$$\mathscr{B}_k = \{A \in CONV(2): \rho(A, n) \leqslant G(n) \text{ for all } n \geqslant k\}$$
$$= \bigcap_{n=k}^{\infty} \{A \in CONV(2): \rho(A, n) \leqslant G(n)\}$$

is again closed for each $k \in \mathbb{N}$. We shall show that

$$\mathscr{B}_k \text{ is nowhere dense in } CONV(2) \text{ for each } k \in \mathbb{N}. \tag{126}$$

Suppose on the contrary that for some $k \in \mathbb{N}$, the interior of $\mathscr{B}_k$ is non-empty. Since the analytic convex sets are dense in $CONV(2)$, there exists an analytic set $B \in \mathscr{B}_k$. We now recall (120) with $\delta^{-1} = l(A, n) = \log(2 + n\mu(A))$: If A satisfies the hypotheses of Theorem 19B, then

$$l(A, n)\int_{-\infty}^{\infty}\left(\int_{\mathbb{R}^2} |F_\lambda(\mathbf{x})|^2\, d\mathbf{x}\right) e^{-\lambda^2 l^2(A,n)}\, d\lambda \gg_\gamma \frac{(\mu(A))^{\frac{1}{2}}}{l(A, n)} M^2. \tag{127}$$

By tending to infinity with M, it easily follows from (127) that

$$l(A, n)\int_{-\infty}^{\infty}\left(\int_{U^2} |D^{\mathrm{tor}}[\mathscr{P}; A(\lambda, \mathbf{x})]|^2\, d\mathbf{x}\right) e^{-\lambda^2 l^2(A,n)}\, d\lambda \gg_\gamma \frac{(n\mu(A))^{\frac{1}{2}}}{\log(2 + n\mu(A))}, \tag{128}$$

where $\mathscr{P}$ is an arbitrary n-element subset of U_0^2. Applying (128) to the analytic set $B \in \mathscr{B}_k$, we get

$$\liminf_{n\to\infty} \frac{\rho(B, n)}{n^{\frac{1}{4}}(\log n)^{-1}} > 0.$$

This contradicts the definition of $\mathscr{B}_k$, and Lemma 7.12 follows. ■

Let $\mathscr{P} \subset U_0^2$, where $\#\mathscr{P} = n$, be arbitrary but fixed. Clearly

$$|D^{\mathrm{tor}}[\mathscr{P}; A(\lambda, \mathbf{x})]| \ll n\lambda d(A).$$

Thus we have

$$l(A,n)\int_{|\lambda|>1}\left(\int_{U^2}|D^{\mathrm{tor}}[\mathscr{P};A(\lambda,\mathbf{x})]|^2\,\mathrm{d}\mathbf{x}\right)\mathrm{e}^{-\lambda^2 l^2(A,n)}\,\mathrm{d}\lambda$$

$$\ll l(A,n)\int_{|\lambda|>1}(n\lambda d(A))^2(l(A,n))^{-\lambda^2 l(A,n)}\,\mathrm{d}\lambda\leqslant 1$$

if n is sufficiently large depending on A.

It follows from Lemma 7.12 that for all $A\in CONV(2)$, except those in a set of first category,

$$l(A,n)\int_{-1}^{1}\left(\int_{U^2}|D^{\mathrm{tor}}[\mathscr{P};A(\lambda,\mathbf{x})]|^2\,\mathrm{d}\mathbf{x}\right)\mathrm{e}^{-\lambda^2 l^2(A,n)}\,\mathrm{d}\lambda>G(n)-1 \qquad (129)$$

for infinitely many n. Since

$$\int_{-\infty}^{\infty}\mathrm{e}^{-t^2}\,\mathrm{d}t=\pi^{\frac{1}{2}},$$

we obtain

$$\pi^{\frac{1}{2}}|\Delta^{\mathrm{tor}}[\mathscr{P};A]|^2\geqslant l(A,n)\int_{-1}^{1}\left(\int_{U^2}|D^{\mathrm{tor}}[\mathscr{P};A(\lambda,\mathbf{x})]|^2\,\mathrm{d}\mathbf{x}\right)\mathrm{e}^{-\lambda^2 l^2(A,n)}\,\mathrm{d}\lambda. \qquad (130)$$

Theorem 21(ii) now follows from (129) and (130).

7.2 Congruent sets

In this section, we prove Theorems 22A and 22B.

We first of all prove the following slightly more general and renormalized form of Theorem 22A.

Theorem 22C. *Let $\mathscr{Q}=\{\mathbf{q}_1,\mathbf{q}_2,\ldots,\}$ be a completely arbitrary infinite discrete subset of $\mathbb{R}^2$. Then for any positive real number l, there is a rotated square S of side length l such that*

$$\left|\sum_{\mathbf{q}_j\in S}1-\mu(S)\right|\gg l^{\frac{1}{4}}. \qquad (131)$$

We introduce a real parameter $M=M(l)$ which will be specified later. For any Lebesgue measurable set $H\subset\mathbb{R}^2$, let

$$\mu_0(H)=\mu(H\cap[-M,M]^2).$$

For any $H\subset\mathbb{R}^2$, write

$$Z_0(H)=\sum_{\mathbf{q}_j\in H\cap[-M,M]^2}1.$$

Let $\chi_{l,\theta}$ denote the characteristic function of the rotated square

$$S(l,\theta)=\{\mathbf{x}=(x_1,x_2)\in\mathbb{R}^2:|x_1\cos\theta+x_2\sin\theta|\leqslant\tfrac{1}{2}l,\ |-x_1\sin\theta+x_2\cos\theta|\leqslant\tfrac{1}{2}l\}.$$

Moreover, let $S(l,\theta,\mathbf{y})=S(l,\theta)+\mathbf{y}$.

Consider now the function

$$F_\theta=\chi_{l,\theta}*(\mathrm{d}Z_0-\mathrm{d}\mu_0),$$

where $*$ denotes the convolution operation. More explicitly,

$$\begin{aligned}F_\theta(\mathbf{x})&=\int_{\mathbb{R}^2}\chi_{l,\theta}(\mathbf{x}-\mathbf{y})(\mathrm{d}Z_0(\mathbf{y})-\mathrm{d}\mu_0(\mathbf{y}))\\&=\sum_{\mathbf{q}_j\in S(l,\theta,\mathbf{x})\cap[-M,M]^2}1-\mu(S(l,\theta,\mathbf{x})\cap[-M,M]^2).\end{aligned}$$

Note that if $S(l,\theta,\mathbf{x})\subset[-M,M]^2$, then $|F_\theta(\mathbf{x})|$ is equal to the discrepancy of the rotated square $S(l,\theta,\mathbf{x})$ of side length l. Applying identities (12) and (14), we have

$$\begin{aligned}\int_0^\pi\left(\int_{\mathbb{R}^2}|F_\theta(\mathbf{x})|^2\,\mathrm{d}\mathbf{x}\right)\mathrm{d}\theta&=\int_0^\pi\left(\int_{\mathbb{R}^2}|\hat{F}_\theta(\mathbf{t})|^2\,\mathrm{d}\mathbf{t}\right)\mathrm{d}\theta\\&=\int_{\mathbb{R}^2}\left(\int_0^\pi|\hat{\chi}_{l,\theta}(\mathbf{t})|^2\,\mathrm{d}\theta\right)|\phi(\mathbf{t})|^2\,\mathrm{d}\mathbf{t},\end{aligned}\qquad(132)$$

where, as before,

$$\phi(\mathbf{t})=(\widehat{\mathrm{d}Z_0-\mathrm{d}\mu_0})(\mathbf{t})=\frac{1}{2\pi}\int_{\mathbb{R}^2}\mathrm{e}^{-i\mathbf{x}\cdot\mathbf{t}}(\mathrm{d}Z_0(\mathbf{x})-\mathrm{d}\mu_0(\mathbf{x})).$$

We may assume that

$$\sum_{\mathbf{q}_j\in[-M,M]^2}1>\tfrac{1}{2}(2M)^2.\qquad(133)$$

Indeed, M will be specified later so that $M>4l$, and so if (133) does not hold, then by standard averaging arguments, there exists a translated square $[-\frac{1}{2}l,\frac{1}{2}l]^2+\mathbf{x}$ contained in $[-M,M]^2$ and having less than $8l^2/9$ points $\mathbf{q}_j$. Consequently, this square has discrepancy $\gg l^2$.

We may further assume that $l>\frac{1}{2}$. Indeed, if $l\leqslant\frac{1}{2}$, then let S be a translated square of the form $[-\frac{1}{2}l,\frac{1}{2}l]^2+\mathbf{x}$ such that $S\cap\mathcal{Q}\neq\varnothing$. Clearly

$$\sum_{\mathbf{q}_j\in S}1-\mu(S)\geqslant 1-l^2\geqslant\tfrac{3}{4}\gg l^{\frac{1}{2}}.$$

We now investigate the right-hand side of (132).

Applying Lemma 7.1 with $b=1$, we have

$$\int_{[-100,100]^2} |\phi(\mathbf{t})|^2 \, d\mathbf{t} \gg M^2. \tag{134}$$

Next, we estimate the integral

$$\int_0^\pi |\hat{\chi}_{l,\theta}(\mathbf{t})|^2 \, d\theta$$

from below simultaneously for all $\mathbf{t}\in[-100,100]^2$. By elementary calculation,

$$\hat{\chi}_{l,\theta}(\mathbf{t}) = \frac{2}{\pi} \frac{\sin(\frac{1}{2}l\mathbf{t}\cdot\mathbf{e}_1(\theta))}{\mathbf{t}\cdot\mathbf{e}_1(\theta)} \frac{\sin(\frac{1}{2}l\mathbf{t}\cdot\mathbf{e}_2(\theta))}{\mathbf{t}\cdot\mathbf{e}_2(\theta)},$$

where $\mathbf{e}_1(\theta)=(\cos\theta,\sin\theta)$ and $\mathbf{e}_2(\theta)=(-\sin\theta,\cos\theta)$. Hence

$$\int_0^\pi |\hat{\chi}_{l,\theta}(\mathbf{t})|^2 \, d\theta = \frac{4}{\pi^2}\int_0^\pi \left(\frac{\sin(\frac{1}{2}l\mathbf{t}\cdot\mathbf{e}_1(\theta))}{\mathbf{t}\cdot\mathbf{e}_1(\theta)}\right)^2 \left(\frac{\sin(\frac{1}{2}l\mathbf{t}\cdot\mathbf{e}_2(\theta))}{\mathbf{t}\cdot\mathbf{e}_2(\theta)}\right)^2 d\theta. \tag{135}$$

Now assume that $\mathbf{t}\in[-100,100]^2$ and let $\alpha=\alpha(\mathbf{t})$ be defined by the equations

$$\mathbf{t}\cdot\mathbf{e}_1(\alpha)=0 \quad \text{and} \quad \mathbf{t}\cdot\mathbf{e}_2(\alpha)=|\mathbf{t}|. \tag{136}$$

Consequently,

$$(\mathbf{t}\cdot\mathbf{e}_2(\theta))^2 = |\mathbf{t}|^2(1-\sin^2(\theta-\alpha)).$$

It follows from this that the value of the function

$$f(\theta)=\tfrac{1}{2}l\mathbf{t}\cdot\mathbf{e}_2(\theta)$$

varies from $(\frac{1}{2}l|\mathbf{t}|-c_{26}|\mathbf{t}|)$, where $c_{26}>0$, to $\frac{1}{2}l|\mathbf{t}|$ when the argument θ varies from $(\alpha-l^{-\frac{1}{2}})$ to α. Hence there is a subinterval $I(\mathbf{t})\subset[\alpha-l^{-\frac{1}{2}},\alpha]$ with length $\gg l^{-\frac{1}{2}}$ such that

$$|\sin(\tfrac{1}{2}l\mathbf{t}\cdot\mathbf{e}_2(\theta))| \geqslant c_{27}|\mathbf{t}|$$

for all $\theta\in I(\mathbf{t})$, where $c_{27}>0$. Hence

$$\left(\frac{\sin(\frac{1}{2}l\mathbf{t}\cdot\mathbf{e}_2(\theta))}{\mathbf{t}\cdot\mathbf{e}_2(\theta)}\right)^2 \geqslant c_{27}^2 \gg 1 \tag{137}$$

for all $\theta\in I(\mathbf{t})$. Furthermore, there is a subset $H(\mathbf{t})$ of $I(\mathbf{t})$ such that its total length is $\gg l^{-\frac{1}{2}}$ and for all $\theta\in H(\mathbf{t})$,

$$|\sin(\tfrac{1}{2}l\mathbf{t}\cdot\mathbf{e}_1(\theta))| \geqslant c_{28}>0.$$

Since $H(\mathbf{t})\subset I(\mathbf{t})\subset[\alpha-l^{-\frac{1}{2}},\alpha]$ and $\mathbf{t}\in[-100,100]^2$, it follows from (136) that for all $\theta \in H(\mathbf{t})$,

$$|\mathbf{t}\cdot\mathbf{e}_1(\theta)| \leqslant |\mathbf{t}| l^{-\frac{1}{2}} \leqslant 100 l^{-\frac{1}{2}}\sqrt{2}$$

and so

$$\left(\frac{\sin(\frac{1}{2}l\mathbf{t}\cdot\mathbf{e}_1(\theta))}{\mathbf{t}\cdot\mathbf{e}_1(\theta)}\right)^2 \gg l. \tag{138}$$

Combining (135), (137) and (138), we have that for any $\mathbf{t}\in[-100, 100]^2$,

$$\int_0^\pi |\hat{\chi}_{l,\theta}(\mathbf{t})|^2\, d\theta \geqslant \frac{4}{\pi^2}\int_{H(\mathbf{t})} \left(\frac{\sin(\frac{1}{2}l\mathbf{t}\cdot\mathbf{e}_1(\theta))}{\mathbf{t}\cdot\mathbf{e}_1(\theta)}\right)^2 \left(\frac{\sin(\frac{1}{2}l\mathbf{t}\cdot\mathbf{e}_2(\theta))}{\mathbf{t}\cdot\mathbf{e}_2(\theta)}\right)^2 d\theta$$

$$\gg l\int_{H(\mathbf{t})} d\mathbf{t} \gg l^{\frac{1}{2}}. \tag{139}$$

Combining (132), (134) and (139), we conclude that

$$\int_0^\pi\left(\int_{\mathbb{R}^2}|F_\theta(\mathbf{x})|^2\, d\mathbf{x}\right)d\theta \geqslant \int_{[-100,100]^2}\left(\int_0^\pi |\hat{\chi}_{l,\theta}(\mathbf{t})|^2\, d\theta\right)|\phi(\mathbf{t})|^2\, d\mathbf{t}$$

$$\geqslant\left(\min_{\mathbf{t}\in[-100,100]^2}\int_0^\pi|\hat{\chi}_{l,\theta}(\mathbf{t})|^2\, d\theta\right)\int_{[-100,100]^2}|\phi(\mathbf{t})|^2\, d\mathbf{t} \gg l^{\frac{1}{2}}M^2. \tag{140}$$

We are now in a position to finish the proof of Theorem 22C. If for some $\theta\in[0,\pi]$ and $\mathbf{x}\in\mathbb{R}^2$, the inequality

$$|F_\theta(\mathbf{x})| = \left|\sum_{\mathbf{q}_j\in S(l,\theta,\mathbf{x})\cap[-M,M]^2} 1 - \mu(S(l,\theta,\mathbf{x})\cap[-M,M]^2)\right| > 2l^2$$

holds, then obviously the tilted square $S(l,\theta,\mathbf{x})$ contains more than $2l^2$ points $\mathbf{q}_j$, and so $S(l,\theta,\mathbf{x})$ has a 'giant' discrepancy. Thus we may assume that

$$\max_{\theta,\mathbf{x}}|F_\theta(\mathbf{x})| \leqslant 2l^2. \tag{141}$$

Since $F_\theta(\mathbf{x})=0$ whenever $S(l,\theta,\mathbf{x})\cap[-M,M]^2$ is empty, we clearly have

$$(2M)^2 \max_{S(l,\theta,\mathbf{x})\subset[-M,M]^2}|F_\theta(\mathbf{x})|^2 + c_{29}Ml \max_{S(l,\theta,\mathbf{x})\not\subset[-M,M]^2}|F_\theta(\mathbf{x})|^2$$

$$\geqslant \frac{1}{\pi}\int_0^\pi\left(\int_{\mathbb{R}^2}|F_\theta(\mathbf{x})|^2\, d\mathbf{x}\right)d\theta. \tag{142}$$

It follows from (140)–(142) that

$$\max_{S(l,\theta,\mathbf{x})\subset[-M,M]^2}|F_\theta(\mathbf{x})|^2 \gg l^{\frac{1}{2}} - O(l^5/M) \gg l^{\frac{1}{2}}$$

if $M = M(l)$ is sufficiently large. This completes the proof of Theorem 22C.

We now investigate Theorem 22B. Again we prove a renormalized form of the theorem. Let M be a natural number.

Theorem 22D. *Let $\alpha > 0$ be a real number. Suppose that $\mathscr{P} = \{\mathbf{p}_1, \mathbf{p}_2, \dots, \mathbf{p}_N\}$ is a distribution of $N = 4M^2$ points in the square $[-M, M]^2$. Then there is a circular disc D of radius $r = \alpha M$ such that*

$$\left| \sum_{\mathbf{p}_j \in D} 1 - \mu(D \cap [-M, M]^2) \right| > c_{30}(\alpha) N^{\frac{1}{4}}. \tag{143}$$

Again, our method makes use of Fourier transforms. Besides relations (12)–(15), we also need the following identity: For $j = 1, 2$ and writing $\mathbf{x} = (x_1, x_2)$ and $\mathbf{t} = (t_1, t_2)$, we have $(\mathrm{i} = \sqrt{-1})$

$$\widehat{(-\mathrm{i}x_j f)} = \frac{\partial \hat{f}}{\partial t_j}. \tag{144}$$

For any Lebesgue measurable set $H \subset \mathbb{R}^2$, let

$$\mu_0(H) = \mu(H \cap [-M, M]^2).$$

For any $H \subset \mathbb{R}^2$, write

$$Z_0(H) = \sum_{\mathbf{p}_j \in H \cap \mathscr{P}} 1.$$

Let χ_r denote the characteristic function of the disc

$$D_r = \{\mathbf{x} = (x_1, x_2) \in \mathbb{R}^2 : x_1^2 + x_2^2 \leqslant r^2\},$$

and let

$$F = \chi_r * (\mathrm{d}Z_0 - \mathrm{d}\mu_0), \tag{145}$$

where $*$ denotes the convolution operation. More explicitly,

$$\begin{aligned} F(\mathbf{x}) &= \int_{\mathbb{R}^2} \chi_r(\mathbf{x} - \mathbf{y})(\mathrm{d}Z_0(\mathbf{y}) - \mathrm{d}\mu_0(\mathbf{y})) \\ &= \sum_{\mathbf{p}_j \in (D_r + \mathbf{x}) \cap \mathscr{P}} 1 - \mu((D_r + \mathbf{x}) \cap [-M, M^2]). \end{aligned} \tag{146}$$

Applying (12) and (14) to (145), we have

$$\int_{\mathbb{R}^2} |F(\mathbf{x})|^2 \,\mathrm{d}\mathbf{x} = \int_{\mathbb{R}^2} |\hat{F}(\mathbf{t})|^2 \,\mathrm{d}\mathbf{t} = \int_{\mathbb{R}^2} |\hat{\chi}_r(\mathbf{t})|^2 |\phi(\mathbf{t})|^2 \,\mathrm{d}\mathbf{t}, \tag{147}$$

where, as usual,

$$\phi(\mathbf{t}) = (\widehat{\mathrm{d}Z_0 - \mathrm{d}\mu_0})(\mathbf{t}) = \frac{1}{2\pi} \int_{\mathbb{R}^2} \mathrm{e}^{-\mathrm{i}\mathbf{x}\cdot\mathbf{t}} (\mathrm{d}Z_0(\mathbf{x}) - \mathrm{d}\mu_0(\mathbf{x})).$$

Since we work with only a single value of the radius r, we need a deeper study of the function $\phi(\mathbf{t})$ for $\mathbf{t}\in[-100,100]^2$. We aim to find a subset of $[-100,100]^2$ in such a way that in this subset $|\phi(\mathbf{t}+\mathbf{u})-\phi(\mathbf{t})|$ is not too 'large' compared with $|\mathbf{u}|$, while the integral of $|\phi(\mathbf{t})|^2$ over this subset is 'large'.

Write, for $k, l\in\mathbb{N}^*$, $k+l\geqslant 1$ and $\mathbf{t}=(t_1,t_2)$,

$$\phi_{k,l}(\mathbf{t})=\frac{\partial^k\partial^l}{\partial t_1^k\partial t_2^l}\phi(\mathbf{t}),$$

and consider the Taylor expansion (writing $\mathbf{u}=(u_1,u_2)$)

$$\phi(\mathbf{t}+\mathbf{u})-\phi(\mathbf{t})=\sum_{\substack{k,l\in\mathbb{N}^*\\ k+l\geqslant 1}}\phi_{k,l}(\mathbf{t})\frac{u_1^k u_2^l}{k!l!} \tag{148}$$

Let c_0 be any positive number. We consider the following subset of $[-100,100]^2$:

$$T(c_0)=\{\mathbf{t}\in[-100,100]^2:|\phi_{k,l}(\mathbf{t})|\leqslant c_0 2^{k+l}M^{k+l}|\phi(\mathbf{t})| \text{ for all } k, l\in\mathbb{N}^*,\ k+l\geqslant 1\}. \tag{149}$$

Lemma 7.13.

(i) *If* $\mathbf{t}\in T(c_0)$, *then* $|\phi(\mathbf{t}+\mathbf{u})-\phi(\mathbf{t})|\leqslant c_0|\phi(\mathbf{t})|(e^{4M|\mathbf{u}|}-1)$.

(ii) *If* c_0 *is a sufficiently large constant, independent of* M, *then*

$$\int_{T(c_0)}|\phi(\mathbf{t})|^2\,d\mathbf{t}\gg M^2.$$

Proof. We start with the proof of (i). By (148) and (149), with $\mathbf{u}=(u_1,u_2)$,

$$|\phi(\mathbf{t}+\mathbf{u})-\phi(\mathbf{t})|\leqslant c_0|\phi(\mathbf{t})|\sum_{\substack{k,l\in\mathbb{N}^*\\ k+l\geqslant 1}}\frac{(2Mu_1)^k(2Mu_2)^l}{k!l!}$$

$$=c_0|\phi(\mathbf{t})|(e^{2M(|u_1|+|u_2|)}-1)\leqslant c_0|\phi(\mathbf{t})|(e^{4M|\mathbf{u}|}-1).$$

It remains to prove (ii). The proof of Lemma 7.1 in §7.1 gives the following extra information on $\phi(\mathbf{t})$:

$$\int_{[-100,100]^2}|\phi(\mathbf{t})|^2\,d\mathbf{t}\gg\int_{\mathbf{R}^2}(S(\mathbf{x}))^2\,d\mathbf{x}, \tag{150}$$

where $S(\mathbf{x})$ denotes the number of points $\mathbf{p}_j$ which lie in $[-\frac{1}{100},\frac{1}{100}]^2+\mathbf{x}$. Moreover, it was shown that

$$\int_{\mathbf{R}^2}(S(\mathbf{x}))^2\,d\mathbf{x}\gg M^2. \tag{151}$$

We are going to give an upper bound to

$$\int_{[-100,100]^2} |\phi_{k,l}(\mathbf{t})|^2 \, \mathrm{d}\mathbf{t}$$

when $k+l \geqslant 1$. In fact, we shall show that

$$\int_{[-100,100]^2} \left|\frac{\phi_{k,l}(\mathbf{t})}{M^{k+l}}\right|^2 \mathrm{d}\mathbf{t} < c_{31} \int_{[-100,100]^2} |\phi(\mathbf{t})|^2 \, \mathrm{d}\mathbf{t}, \tag{152}$$

where the constant c_{31} is independent of M, k and l. By iterated application of (144), we see that $\phi_{k,l}$ is the Fourier transform of the signed measure

$$(-\mathrm{i})^{k+l} x_1^k x_2^l (Z_0 - \mu_0),$$

i.e.

$$\phi_{k,l}(\mathbf{t}) = \frac{1}{2\pi} \int_{\mathbf{R}^2} \mathrm{e}^{-\mathrm{i}\mathbf{x}\cdot\mathbf{t}} (-\mathrm{i})^{k+l} x_1^k x_2^l (\mathrm{d}Z_0(\mathbf{x}) - \mathrm{d}\mu_0(\mathbf{x})),$$

where $\mathbf{x} = (x_1, x_2)$. Let $h(\mathbf{x})$ denote the characteristic function of the square $[-\frac{1}{100}, \frac{1}{100}]^2$, and write

$$H_{k,l}(\mathbf{x}) = \int_{\mathbf{R}^2} h(\mathbf{x} - \mathbf{y})(-\mathrm{i})^{k+l} y_1^k y_2^l (\mathrm{d}Z_0(\mathbf{y}) - \mathrm{d}\mu_0(\mathbf{y})).$$

Using (12), we have

$$\hat{H}_{k,l}(\mathbf{t}) = \hat{h}(\mathbf{t}) \phi_{k,l}(\mathbf{t}),$$

where

$$\hat{h}(\mathbf{t}) = \left(\frac{2\sin(t_1/100)}{(2\pi)^{\frac{1}{2}} t_1}\right)\left(\frac{2\sin(t_2/100)}{(2\pi)^{\frac{1}{2}} t_2}\right). \tag{153}$$

By the Parseval–Plancherel identity (14),

$$\int_{\mathbf{R}^2} |H_{k,l}(\mathbf{x})|^2 \, \mathrm{d}\mathbf{x} = \int_{\mathbf{R}^2} |\hat{H}_{k,l}(\mathbf{t})|^2 \, \mathrm{d}\mathbf{t} = \int_{\mathbf{R}^2} |\hat{h}(\mathbf{t})|^2 |\phi_{k,l}(\mathbf{t})|^2 \, \mathrm{d}\mathbf{t}. \tag{154}$$

Clearly

$$\begin{aligned} |H_{k,l}(\mathbf{x})| &= \left| \int_{\mathbf{R}^2} h(\mathbf{x} - \mathbf{y}) y_1^k y_2^l (\mathrm{d}Z_0(\mathbf{y}) - \mathrm{d}\mu_0(\mathbf{y})) \right| \\ &\leqslant M^{k+l} \left(\int_{[-M,M]^2} h(\mathbf{x} - \mathbf{y}) \, \mathrm{d}Z_0(\mathbf{y}) + \int_{[-M,M]^2} h(\mathbf{x} - y) \, \mathrm{d}\mathbf{y} \right) \\ &= M^{k+l}(S(\mathbf{x}) + Q(\mathbf{x})), \end{aligned} \tag{155}$$

where $S(\mathbf{x})$ denotes the number of points $\mathbf{p}_j$ which lie in $[-\frac{1}{100}, \frac{1}{100}]^2 + \mathbf{x}$ and

$$Q(\mathbf{x}) = \mu(([-\tfrac{1}{100}, \tfrac{1}{100}]^2 + \mathbf{x}) \cap [-M, M]^2).$$

Since $S^2(\mathbf{x}) \geqslant S(\mathbf{x})$ and $Q(\mathbf{x}) \leqslant 1$, we have, by (151), that

$$\int_{\mathbb{R}^2} (S(\mathbf{x}) + Q(\mathbf{x}))^2 \, d\mathbf{x} = \int_{\mathbb{R}^2} (S^2(\mathbf{x}) + 2S(\mathbf{x})Q(\mathbf{x}) + Q^2(\mathbf{x})) \, d\mathbf{x}$$

$$\leqslant 3 \int_{\mathbb{R}^2} S^2(\mathbf{x}) \, d\mathbf{x} + \int_{\mathbb{R}^2} Q(\mathbf{x}) \, d\mathbf{x} \ll \int_{\mathbb{R}^2} S^2(\mathbf{x}) \, d\mathbf{x} + M^2 \ll \int_{\mathbb{R}^2} S^2(\mathbf{x}) \, d\mathbf{x}. \tag{156}$$

Summarizing (153)–(156), we obtain

$$\int_{[-100,100]^2} |\phi_{k,l}(\mathbf{t})|^2 \, d\mathbf{t} \ll \left(\min_{\mathbf{t} \in [-100,100]^2} |\hat{h}(\mathbf{t})|^2 \right) \int_{\mathbb{R}^2} |\phi_{k,l}(\mathbf{t})|^2 \, d\mathbf{t}$$

$$\leqslant \int_{\mathbb{R}^2} |\hat{h}(\mathbf{t})|^2 |\phi_{k,l}(\mathbf{t})|^2 \, d\mathbf{t} = \int_{\mathbb{R}^2} |H_{k,l}(\mathbf{x})|^2 \, d\mathbf{x}$$

$$\leqslant M^{2(k+l)} \int_{\mathbb{R}^2} (S(\mathbf{x}) + Q(\mathbf{x}))^2 \, d\mathbf{x} \ll M^{2(k+l)} \int_{\mathbb{R}^2} S^2(\mathbf{x}) \, d\mathbf{x}. \tag{157}$$

(152) follows on combining (150) and (157). Now let

$$U(c_0) = [-100, 100]^2 \setminus T(c_0)$$
$$= \{\mathbf{t} \in [-100, 100]^2 : \text{there exist } k, l \text{ such that }$$
$$k + l \geqslant 1 \text{ and } |\phi_{k,l}(\mathbf{t})| > c_0 2^{k+l} M^{k+l} |\phi(\mathbf{t})|\}.$$

If c_0 is sufficiently large depending only on the constant factor c_{31} in (152), then, by (152), we have

$$\int_{U(c_0)} |\phi(\mathbf{t})|^2 \, d\mathbf{t} \leqslant \frac{1}{c_0^2} \sum_{k+l \geqslant 1} 4^{-(k+l)} \int_{U(c_0)} \left| \frac{\phi_{k,l}(\mathbf{t})}{M^{k+l}} \right|^2 d\mathbf{t}$$

$$\leqslant \frac{1}{c_0^2} \sum_{k+l \geqslant 1} 4^{-(k+l)} \int_{[-100,100]^2} \left| \frac{\phi_{k,l}(\mathbf{t})}{M^{k+l}} \right|^2 d\mathbf{t}$$

$$\leqslant \sum_{k+l \geqslant 1} 4^{-(k+l)} \int_{[-100,100]^2} |\phi(\mathbf{t})|^2 \, d\mathbf{t} = \tfrac{7}{9} \int_{[-100,100]^2} |\phi(\mathbf{t})|^2 \, d\mathbf{t}. \tag{158}$$

It follows from (150), (151) and (158) that

$$\int_{T(c_0)} |\phi(\mathbf{t})|^2 \, d\mathbf{t} \geqslant \tfrac{2}{9} \int_{[-100,100]^2} |\phi(\mathbf{t})|^2 \, d\mathbf{t} \gg M^2$$

provided that c_0 is sufficiently large. This proves (ii). ■

Let $c_{32} = [100 c_0]$, where c_0 is specified in Lemma 7.13(ii). We now partition the square $[-100, 100]^2$ into $\omega = (200 c_{32} M)^2$ congruent squares of side length $(c_{32}M)^{-1}$, and denote these squares by $B^{(i)}$ $(i = 1, 2, \ldots, \omega)$. Let

$$\Omega = \{i \in \{1, 2, \ldots, \omega\} : B^{(i)} \cap T(c_0) \neq \emptyset\}.$$

Clearly

$$\bigcup_{i \in \Omega} B^{(i)} \supset T(c_0).$$

By Lemma 7.13(i), we have, for all $i \in \Omega$,

$$\min_{\mathbf{t} \in B^{(i)}} |\phi(\mathbf{t})| \geqslant \tfrac{1}{2} \max_{\mathbf{t} \in B^{(i)}} |\phi(\mathbf{t})|. \tag{159}$$

Next we study the Fourier transform $\hat{\chi}_r$ of the characteristic function of the disc D_r of radius $r = \alpha M$, where $\alpha > 0$. Since the disc D_r is rotation-invariant, we can write

$$g(t) = \hat{\chi}_r(\mathbf{t}),$$

where $t = |\mathbf{t}|$. Let $M_1 = 2c_{32}M$, and put

$$G(y) = \int_y^{y+1/M_1} |g(t)|^2 \, \mathrm{d}t.$$

Lemma 7.14. *We have, for all* $y \in [0, 100\sqrt{2}]$,

$$G(y) \gg_\alpha M.$$

Proof. We recall formulae (53) and (54) in Chapter 6 (noting that here $K = 2$):

$$g(t) = \hat{\chi}_r(\mathbf{t}) = c_{33} r t^{-1} J_1(rt), \tag{160}$$

where

$$J_1(x) = (2/\pi x)^{\frac{1}{2}} \cos(x - 3\pi/4) + \mathrm{O}(x^{-\frac{3}{2}}).$$

Hence

$$g(t) = c_{34} r^{\frac{1}{2}} t^{-\frac{3}{2}} (\cos(rt - 3\pi/4) + \mathrm{O}(r^{-1} t^{-1}), \tag{161}$$

and so

$$G(y) \gg_\alpha r/y^3 \gg_\alpha M/y^3, \tag{162}$$

provided that $y > c_{35}/\alpha r$, where c_{35} is a sufficiently large constant. Next, we note that the Bessel function $J_1(x)$ is continuous and has finitely many zeros in any finite interval. Hence

$$\frac{M_1}{r} \int_z^{z+r/M_1} (J_1(x))^2 \, \mathrm{d}x \gg_\alpha 1 \tag{163}$$

for all $z \in [0, c_{35}/\alpha]$. Combining (160) and (163), we see that

$$G(y) \gg_\alpha r^4 \gg_\alpha M^4 \tag{164}$$

whenever $0 \leqslant y \leqslant c_{35}/\alpha r$. Lemma 7.14 follows from (162) and (164). ∎

We are now ready to complete the proof of Theorem 22D. It easily follows from Lemma 7.14 that for all $i = 1, 2, \ldots \omega$,

$$\frac{1}{\mu(B^{(i)})}\int_{B^{(i)}} |\hat{\chi}_r(\mathbf{t})|^2 \, d\mathbf{t} \gg_\alpha M. \tag{165}$$

On the other hand, by (159), we have, for all $i \in \Omega$,

$$\left(\min_{\mathbf{t}\in B^{(i)}} |\phi(\mathbf{t})|^2\right)\mu(B^{(i)}) \geqslant \tfrac{1}{4}\int_{B^{(i)}} |\phi(\mathbf{t})|^2 \, d\mathbf{t}. \tag{166}$$

By (147),

$$\int_{\mathbb{R}^2} |F(\mathbf{x})|^2 \, d\mathbf{x} = \int_{\mathbb{R}^2} |\hat{\chi}_r(\mathbf{t})|^2 |\phi(\mathbf{t})|^2 \, d\mathbf{t} \geqslant \sum_{i\in\Omega}\int_{B^{(i)}} |\hat{\chi}_r(\mathbf{t})|^2 |\phi(\mathbf{t})|^2 \, d\mathbf{t}$$

$$\geqslant \sum_{i\in\Omega}\left(\frac{1}{\mu(B^{(i)})}\int_{B^{(i)}} |\hat{\chi}_r(\mathbf{t})|^2 \, d\mathbf{t}\right)\left(\left(\min_{\mathbf{t}\in B^{(i)}} |\phi(\mathbf{t})|^2\right)\mu(B^{(i)})\right). \tag{167}$$

Combining (165)–(167) and Lemma 7.13(ii), we have

$$\int_{\mathbb{R}^2} |F(\mathbf{x})|^2 \, d\mathbf{x} \gg_\alpha M\left(\sum_{i\in\Omega}\int_{B^{(i)}} |\phi(\mathbf{t})|^2 \, d\mathbf{t}\right) \geqslant M\int_{T(c_0)} |\phi(\mathbf{t})|^2 \, d\mathbf{t} \gg M^3. \tag{168}$$

Since $F(\mathbf{x}) = 0$ whenever the set $(D_r + \mathbf{x}) \cap [-M, M]^2$ is empty, (168) yields

$$\max_{\mathbf{x}} |F(\mathbf{x})| \gg_\alpha M^{\frac{1}{2}}. \tag{169}$$

Theorem 22D now follows from (146) and (169).

7.3 Roth's conjecture on disc-segments

For geometric reasons we prefer to reformulate Theorem 23A in the following way: Let D_n be the disc $\{\mathbf{x} = (x_1, x_2) \in \mathbb{R}^2 : x_1^2 + x_2^2 \leqslant n^2\}$ of radius n. Suppose that $\mathscr{P} = \{\mathbf{p}_1, \mathbf{p}_2, \ldots, \mathbf{p}_N\}$, where N is arbitrary, is a distribution of N points in D_n. For each disc-segment A in D_n, let $Z[\mathscr{P}; A]$ denote the number of points of $\mathscr{P}$ which lie in A, and write

$$\Delta[\mathscr{P}; A] = |Z[\mathscr{P}; A] - \mu(A)|,$$

where $\mu(A)$ denotes the usual area of A. We consider

$$\Delta[\mathscr{P}] = \sup_A \Delta[\mathscr{P}; A],$$

where the supremum is taken over all disc-segments A in D_n, and write

$$\Delta[n] = \inf_{\mathscr{P}} \Delta[\mathscr{P}],$$

where the infimum is taken over all distributions $\mathscr{P}$ of points in D_n.

Theorem 23C. *We have, for any* $n \geqslant 2$,

$$\Delta[n] > c_{36} n^{\frac{1}{2}} (\log n)^{-\frac{7}{2}}.$$

Clearly Theorem 23C implies Theorem 23A.

Let $\mathscr{P} = \{\mathbf{p}_1, \mathbf{p}_2, \ldots, \mathbf{p}_N\}$ be a set of points in D_n. The result is obvious if $N < n^2$ or $N > 4n^2$, so we may assume, without loss of generality, that

$$n^2 \leqslant N \leqslant 4n^2. \tag{170}$$

We may further assume that n is sufficiently large.

For any $H \subset \mathbb{R}^2$, let

$$Z_0(H) = \sum_{\mathbf{p}_j \in H \cap \mathscr{P}} 1,$$

i.e. Z_0 denotes the counting measure generated by $\mathscr{P}$. Also, let μ_0 denote the restriction of the usual area to D_n, i.e.

$$\mu_0(H) = \mu(H \cap D_n)$$

for any Lebesgue measurable set $H \subset \mathbb{R}^2$.

Let

$$m = n(\log n)^3, \tag{171}$$

and let

$$S = \{\mathbf{x} = (x_1, x_2) \in \mathbb{R}^2 : |x_1| \leqslant m \text{ and } |x_2| \leqslant \tfrac{1}{100}\},$$

and let $S(\theta)$ denote S rotated about the origin by an angle θ. Write

$$r = n(\log n)^2,$$

and consider a kernel ('truncating function') of the form

$$E(\mathbf{x}) = \exp(-(x_1/r)^2 - (x_2/r)^2) = \mathrm{e}^{-|\mathbf{x}|^2/r^2}.$$

If χ_θ denotes the characteristic function of the tilted rectangle $S(\theta)$, write

$$F_\theta = E(\chi_\theta * (\mathrm{d}Z_0 - \mathrm{d}\mu_0)), \tag{172}$$

where $*$ denotes the convolution operation. More explicitly,

$$\begin{aligned} F_\theta(\mathbf{x}) &= E(\mathbf{x}) \int_{\mathbb{R}^2} \chi_\theta(\mathbf{x} - \mathbf{y}) (\mathrm{d}Z_0(\mathbf{y}) - \mathrm{d}\mu_0(\mathbf{y})) \\ &= E(\mathbf{x}) \Bigg(\sum_{\mathbf{p}_j \in (S(\theta) + \mathbf{x}) \cap \mathscr{P}} 1 - \mu((S(\theta) + \mathbf{x}) \cap D_n) \Bigg). \end{aligned} \tag{173}$$

Note that if $|\mathbf{x}| < m - n$, then $(S(\theta) + \mathbf{x}) \cap D_n$ is the difference of two disc-segments.

In particular, Theorem 23C follows immediately from the following two lemmas.

Lemma 7.15. *We have, for* $N \leqslant 4n^2$ *and* $\theta \in [0, \pi)$,

$$\int_{\mathbf{R}^2} |F_\theta(\mathbf{x})|^2\,d\mathbf{x} \leqslant c_{37} n^2 (\log n)^6 (\Delta[\mathscr{P}])^2 + O(1).$$

Lemma 7.16. *We have, for* $n^2 \leqslant N \leqslant 4n^2$,

$$\int_0^\pi \int_{\mathbf{R}^2} |F_\theta(\mathbf{x})|^2\,d\mathbf{x}\,d\theta \geqslant c_{38} n^3 (\log n)^{-1}.$$

Proof of Lemma 7.15. The proof is based on the fact that $E(\mathbf{x})$ is 'small' when $|\mathbf{x}|$ is large. By (173),

$$\begin{aligned}\int_{\mathbf{R}^2} |F_\theta(\mathbf{x})|^2\,d\mathbf{x} &= \int_{\mathbf{R}^2} E^2(\mathbf{x}) \Bigg(\sum_{\mathbf{p}_j \in (S(\theta)+\mathbf{x}) \cap \mathscr{P}} 1 - \mu((S(\theta)+\mathbf{x}) \cap D_n) \Bigg)^2 d\mathbf{x} \\ &= \Bigg(\int_{|\mathbf{x}| \leqslant \frac{1}{2}m} + \int_{|\mathbf{x}| > \frac{1}{2}m} \Bigg) E^2(\mathbf{x}) \Bigg(\sum_{\mathbf{p}_j \in (S(\theta)+\mathbf{x}) \cap \mathscr{P}} 1 - \mu((S(\theta)+\mathbf{x}) \cap D_n) \Bigg)^2 d\mathbf{x} \\ &\leqslant \frac{m^2 \pi}{4} \sup_{|\mathbf{x}| \leqslant \frac{1}{2}m} \Bigg| \sum_{\mathbf{p}_j \in (S(\theta)+\mathbf{x}) \cap \mathscr{P}} 1 - \mu((S(\theta)+\mathbf{x}) \cap D_n) \Bigg|^2 \\ &\quad + \frac{1}{n^4} \sup_{|\mathbf{x}| > \frac{1}{2}m} \Bigg| \sum_{\mathbf{p}_j \in (S(\theta)+\mathbf{x}) \cap \mathscr{P}} 1 - \mu((S(\theta)+\mathbf{x}) \cap D_n) \Bigg|^2 \end{aligned} \tag{174}$$

since

$$\int_{|\mathbf{x}| > \frac{1}{2}m} E^2(\mathbf{x})\,d\mathbf{x} < \frac{1}{n^4}$$

for all sufficiently large n. Since $N \leqslant 4n^2$, we have that

$$\sup_{\mathbf{x} \in \mathbf{R}^2} \Bigg| \sum_{\mathbf{p}_j \in (S(\theta)+\mathbf{x}) \cap \mathscr{P}} 1 - \mu((S(\theta)+\mathbf{x}) \cap D_n) \Bigg|^2 \leqslant 16 n^4. \tag{175}$$

It follows from (174) and (175) that

$$\int_{\mathbf{R}^2} |F_\theta(\mathbf{x})|^2\,d\mathbf{x} \ll m^2 \sup_{|\mathbf{x}| \leqslant \frac{1}{2}m} \Bigg| \sum_{\mathbf{p}_j \in (S(\theta)+\mathbf{x}) \cap \mathscr{P}} 1 - \mu((S(\theta)+\mathbf{x}) \cap D_n) \Bigg|^2 + O(1). \tag{176}$$

If $\frac{1}{2}m \leqslant m - n$, i.e. if n is sufficiently large, then for each $|\mathbf{x}| \leqslant \frac{1}{2}m$ and

$\theta \in [0, \pi)$,

$$\left| \sum_{\mathbf{p}_j \in (S(\theta)+\mathbf{x}) \cap \mathscr{P}} 1 - \mu((S(\theta)+\mathbf{x}) \cap D_n) \right| \leqslant 2\Delta[\mathscr{P}]. \tag{177}$$

Lemma 7.15 follows from (171), (176) and (177). ■

To prove Theorem 23C, it remains to prove Lemma 7.16. By the Parseval–Plancherel identity (14), it suffices to prove that if $n^2 \leqslant N \leqslant 4n^2$, then

$$\int_0^\pi \int_{\mathbf{R}^2} |\hat{F}_\theta(\mathbf{t})|^2 \, \mathrm{d}\mathbf{t} \, \mathrm{d}\theta \geqslant c_{38} n^3 (\log n)^{-1}. \tag{178}$$

For the sake of simplicity, write

$$\phi(\mathbf{t}) = \widehat{(\mathrm{d}Z_0 - \mathrm{d}\mu_0)}(\mathbf{t}).$$

Then by (172), (12) and (13), we have

$$\hat{F}_\theta = \hat{E} * (\hat{\chi}_\theta \phi),$$

so that

$$\hat{F}_\theta(\mathbf{t}) = \int_{\mathbf{R}^2} \hat{E}(\mathbf{t}-\mathbf{u}) \hat{\chi}_\theta(\mathbf{u}) \phi(\mathbf{u}) \, \mathrm{d}\mathbf{u} = \int_{\mathbf{R}^2} \hat{E}(\mathbf{u}-\mathbf{t}) \hat{\chi}_\theta(\mathbf{u}) \phi(\mathbf{u}) \, \mathrm{d}\mathbf{u}$$

$$= \int_{\mathbf{R}^2} \hat{E}(\mathbf{u}) \hat{\chi}_\theta(\mathbf{t}+\mathbf{u}) \phi(\mathbf{t}+\mathbf{u}) \, \mathrm{d}\mathbf{u}. \tag{179}$$

Firstly, we need the analogue of Lemma 7.13 (see §7.2). Write, for $k, l \in \mathbb{N}^*$ and $k + l \geqslant 1$,

$$\phi_{k,l}(\mathbf{t}) = \frac{\partial^k \partial^l}{\partial t_1^k \partial t_2^l} \phi(\mathbf{t}),$$

where $\mathbf{t} = (t_1, t_2)$. Let c_0 be any positive number, and let

$$T(c_0) = \{\mathbf{t} \in [-100, 100]^2 : |\phi_{k,l}(\mathbf{t})| \leqslant c_0 2^{k+l} n^{k+l} |\phi(\mathbf{t})| \text{ for all } k, l \in \mathbb{N}^*, k + l \geqslant 1\}.$$

Lemma 7.17.

(i) *If* $\mathbf{t} \in T(c_0)$, *then* $|\phi(\mathbf{t}+\mathbf{u}) - \phi(\mathbf{t})| \leqslant c_0 |\phi(\mathbf{t})| (\mathrm{e}^{4n|\mathbf{u}|} - 1)$.

(ii) *If* c_0 *is a sufficiently large constant, independent of n, then*

$$\int_{T(c_0)} |\phi(\mathbf{t})|^2 \, \mathrm{d}\mathbf{t} \gg n^2.$$

Proof. We can proceed along exactly the same lines as in the proof of Lemma 7.13 in §7.2. The only (irrelevant) difference is that here the underlying set is a disc rather than a square. ■

Next we study the function $\hat{\chi}_\theta(\mathbf{t})$ for $\mathbf{t}\in[-100, 100]^2$. Let $\alpha = \alpha(\mathbf{t})$ be defined by

$$t_1 \cos\alpha + t_2 \sin\alpha = 0, \tag{180}$$

and write

$$\Delta\theta = \theta - \alpha(\mathbf{t}).$$

For simplicity of notation, let

$$B(z) = \frac{\sin z}{z}.$$

Note that $|B(z)| \leqslant 1$ for all real z. Write

$$\beta(\mathbf{t}, \theta) = B(\tfrac{1}{100}(t_2 \cos\theta - t_1 \sin\theta)).$$

Lemma 7.18. *If* $\mathbf{t}\in[-100, 100]^2$, *then for all* θ *and* $\mathbf{u}$,

$$\hat{\chi}_\theta(\mathbf{t}+\mathbf{u}) = \hat{\chi}_\theta(\mathbf{u})\beta(\mathbf{t}, \theta) + O(m|\mathbf{u}| + m^2|\Delta\theta|). \tag{181}$$

Proof. By elementary calculation, for all $\mathbf{t}\in\mathbb{R}^2$,

$$\hat{\chi}_\theta(\mathbf{t}) = \frac{m}{50\pi} B(m(t_1\cos\theta + t_2\sin\theta))B(\tfrac{1}{100}(t_2\cos\theta - t_1\sin\theta)).$$

Let

$$\begin{cases} T = t_1\cos\theta + t_2\sin\theta \quad \text{and} \quad T^* = t_2\cos\theta - t_1\sin\theta, \\ U = u_1\cos\theta + u_2\sin\theta \quad \text{and} \quad U^* = u_2\cos\theta - u_1\sin\theta. \end{cases}$$

Then

$$\hat{\chi}_\theta(\mathbf{t}+\mathbf{u}) = \frac{m}{50\pi} B(m(T+U))B(\tfrac{1}{100}(T^*+U^*)), \tag{182}$$

and

$$\hat{\chi}_\theta(\mathbf{u}) = \frac{m}{50\pi} B(mU)B(\tfrac{1}{100}U^*). \tag{183}$$

It is not difficult to show that there exists an absolute constant c_{39} such that

$$|B(x) - B(y)| = \left|\frac{\sin x}{x} - \frac{\sin y}{y}\right| \leqslant c_{39}|x - y|. \tag{184}$$

For any angle γ, let $\mathbf{e}(\gamma) = (\cos\gamma, \sin\gamma)$. We have, in view of (180), that

$$\begin{aligned} |t_1\cos\theta + t_2\sin\theta| &= |\mathbf{t}\cdot\mathbf{e}(\theta)| = |\mathbf{t}\cdot(\mathbf{e}(\theta) - \mathbf{e}(\alpha))| \\ &\leqq |\mathbf{t}||\mathbf{e}(\theta) - \mathbf{e}(\alpha)| \leqslant |\mathbf{t}||\theta - \alpha| = |\mathbf{t}||\Delta\theta|. \end{aligned} \tag{185}$$

Combining (184) and (185), we have

$$|B(m(T+U))-B(mU)|\leqslant c_{39}m|T|=c_{39}m|t_1\cos\theta+t_2\sin\theta|$$
$$\leqslant c_{39}m|\mathbf{t}|\,|\Delta\theta|\leqslant c_{39}m(100\sqrt{2})|\Delta\theta|\ll m|\Delta\theta|.$$

We now return to (182). It follows that

$$\hat{\chi}_\theta(\mathbf{t}+\mathbf{u})=\frac{m}{50\pi}(B(mU)+O(m|\Delta\theta|))B(\tfrac{1}{100}(T^*+U^*)). \tag{186}$$

Combining (186) and (183), we have

$$\hat{\chi}_\theta(\mathbf{t}+\mathbf{u})=\frac{m}{50\pi}B(mU)B(\tfrac{1}{100}(T^*+U^*))+O(m^2|\Delta\theta|)$$
$$=\frac{m}{50\pi}B(mU)B(\tfrac{1}{100}U^*)B(\tfrac{1}{100}T^*)+O(m^2|\Delta\theta|)$$
$$+\frac{m}{50\pi}B(mU)(B(\tfrac{1}{100}(T^*+U^*))-B(\tfrac{1}{100}U^*)B(\tfrac{1}{100}T^*))$$
$$=\hat{\chi}_\theta(\mathbf{u})\beta(\mathbf{t},\theta)+O(m^2|\Delta\theta|)$$
$$+\frac{m}{50\pi}B(mU)(B(\tfrac{1}{100}(T^*+U^*))-B(\tfrac{1}{100}U^*)B(\tfrac{1}{100}T^*)).$$

To prove (181), it remains to check that

$$|B(\tfrac{1}{100}(T^*+U^*))-B(\tfrac{1}{100}U^*)B(\tfrac{1}{100}T^*)|=O(|U^*|). \tag{187}$$

For each $y\in\mathbb{R}$, consider the (continuous) function

$$f_y(x)=B(x+y)-B(x)B(y).$$

Then

$$f_y(0)=B(y)-B(0)B(y)=0.$$

Hence, in view of (184),

$$|f_y(x)|=|f_y(x)-f_y(0)|\leqslant|B(x+y)-B(y)|+|B(y)|\,|B(x)-B(0)|\leqslant 2c_{39}|x|.$$

Choosing $y=\frac{1}{100}T^*$, the left-hand side of (187) is

$$|f_y(\tfrac{1}{100}U^*)|\leqslant 2c_{39}\frac{|U^*|}{100}.$$ ■

Lemma 7.19. *Suppose that* $\mathbf{t}\in[-100,100]^2$. *Then*

$$|\hat{F}_\theta(\mathbf{t})|\geqslant c_{40}r|\phi(\mathbf{t})|-\int_{\mathbb{R}^2}|\hat{E}(\mathbf{u})\hat{\chi}_\theta(\mathbf{u})(\phi(\mathbf{t}+\mathbf{u})-\phi(\mathbf{t}))|\,d\mathbf{u}$$
$$-\int_{\mathbb{R}^2}|\hat{E}(\mathbf{u})O(m|\mathbf{u}|+m^2|\Delta\theta|)\phi(\mathbf{t}+\mathbf{u})|\,d\mathbf{u}.$$

Proof. Recall (179). Using (181), we have

$$\begin{aligned}|\hat{F}_\theta(t)| &= \left|\int_{\mathbb{R}^2} \hat{E}(\mathbf{u})\hat{\chi}_\theta(\mathbf{t}+\mathbf{u})\phi(\mathbf{t}+\mathbf{u})\,d\mathbf{u}\right| \\ &\geqslant \left|\int_{\mathbb{R}^2} \hat{E}(\mathbf{u})\beta(\mathbf{t},\theta)\hat{\chi}_\theta(\mathbf{u})\phi(\mathbf{t}+\mathbf{u})\,d\mathbf{u}\right| \\ &\quad - \int_{\mathbb{R}^2} |\hat{E}(\mathbf{u})O(m|\mathbf{u}| + m^2|\Delta\theta|)\phi(\mathbf{t}+\mathbf{u})|\,d\mathbf{u} \\ &\geqslant \left|\beta(\mathbf{t},\theta)\phi(\mathbf{t})\int_{\mathbb{R}^2} \hat{E}(\mathbf{u})\hat{\chi}_\theta(\mathbf{u})\,d\mathbf{u}\right| \\ &\quad - \int_{\mathbb{R}^2} |\hat{E}(\mathbf{u})\beta(\mathbf{t},\theta)\hat{\chi}_\theta(\mathbf{u})(\phi(\mathbf{t}+\mathbf{u}) - \phi(\mathbf{t}))|\,d\mathbf{u} \\ &\quad - \int_{\mathbb{R}^2} |\hat{E}(\mathbf{u})O(m|\mathbf{u}| + m^2|\Delta\theta|)\phi(\mathbf{t}+\mathbf{u})|\,d\mathbf{u}.\end{aligned}$$

The result follows from the fact that $\beta(\mathbf{t},\theta) \geqslant c_{41} > 0$ for all $\mathbf{t}\in[-100, 100]^2$ and θ, and on noting that, in view of (15), we have

$$\int_{\mathbb{R}^2} \hat{E}(\mathbf{u})\hat{\chi}_\theta(\mathbf{u})\,d\mathbf{u} = \int_{\mathbb{R}^2} E(\mathbf{x})\chi_\theta(\mathbf{x})\,d\mathbf{x} = \int_{S(\theta)} E(\mathbf{x})\,d\mathbf{x} = \int_S e^{-|\mathbf{x}|^2/r^2}\,d\mathbf{x} \gg r. \quad \blacksquare$$

Finally, we need the following lemma. We recall that

$$m = n(\log n)^3 \quad \text{and} \quad r = n(\log n)^2.$$

Lemma 7.20. *Suppose that* $\mathbf{t}\in T(c_0)$ *and* $|\theta - \alpha(\mathbf{t})| = |\Delta\theta| \leqslant n/mr$. *Then*

$$|\hat{F}_\theta(\mathbf{t})| \gg |\phi(\mathbf{t})|n(\log n)^2.$$

Proof. By Lemma 7.19, it suffices to prove that for $\mathbf{t}\in T(c_0)$,

$$\int_{\mathbb{R}^2} |\hat{E}(\mathbf{u})\hat{\chi}_\theta(\mathbf{u})(\phi(\mathbf{t}+\mathbf{u}) - \phi(\mathbf{t}))|\,d\mathbf{u} \ll |\phi(\mathbf{t})|n\log n, \tag{188}$$

$$\int_{\mathbb{R}^2} |\hat{E}(\mathbf{u})m|\mathbf{u}|\phi(\mathbf{t}+\mathbf{u})|\,d\mathbf{u} \ll |\phi(\mathbf{t})|n\log n, \tag{189}$$

and

$$\int_{\mathbb{R}^2} |\hat{E}(\mathbf{u})\phi(\mathbf{t}+\mathbf{u})|\,d\mathbf{u} \ll |\phi(\mathbf{t})|. \tag{190}$$

Now standard calculation gives

$$\hat{E}(\mathbf{u}) = \tfrac{1}{2}r^2\exp(-r^2|\mathbf{u}|^2/4). \tag{191}$$

By (191), (183) and Lemma 7.17(i), the left-hand side of (188) is

$$\ll c_0 mr^2|\phi(\mathbf{t})|\int_{\mathbb{R}^2}\exp(-r^2|\mathbf{u}|^2/4)(e^{4n|\mathbf{u}|}-1)\,d\mathbf{u}$$

$$= c_0 mr^2|\phi(\mathbf{t})|\int_0^{2\pi}\int_0^{\infty}\exp(-r^2u^2/4)(e^{4nu}-1)u\,du\,d\theta$$

$$\ll c_0 mnr^{-1}|\phi(\mathbf{t})| \ll |\phi(\mathbf{t})|n\log n.$$

By (191) and Lemma 7.17(i), the left-hand side of (189) is

$$\ll c_0 mr^2|\phi(\mathbf{t})|\int_{\mathbb{R}^2}\exp(-r^2|\mathbf{u}|^2/4)e^{4n|\mathbf{u}|}|\mathbf{u}|\,d\mathbf{u}$$

$$= c_0 mr^2|\phi(\mathbf{t})|\int_0^{2\pi}\int_0^{\infty}\exp(-r^2u^2/4)e^{4nu}u^2\,du\,d\theta$$

$$\ll c_0 mr^{-1}|\phi(\mathbf{t})| \ll |\phi(\mathbf{t})|n\log n.$$

By (191) and Lemma 7.17(i), the left-hand side of (190) is

$$\ll c_0 r^2|\phi(\mathbf{t})|\int_{\mathbb{R}^2}\exp(-r^2|\mathbf{u}|^2/4)e^{4n|\mathbf{u}|}\,d\mathbf{u}$$

$$= c_0 r^2|\phi(\mathbf{t})|\int_0^{2\pi}\int_0^{\infty}\exp(-r^2u^2/4)e^{4nu}u\,du\,d\theta \ll c_0|\phi(\mathbf{t})| \ll |\phi(\mathbf{t})|.$$

This proves Lemma 7.20. ■

We can now finish the proof of Lemma 7.16. Let c_0 be sufficiently large in view of Lemma 7.17(ii). By Lemmas 7.20 and 7.17(ii),

$$\int_0^{\pi}\int_{\mathbb{R}^2}|\hat{F}_\theta(\mathbf{t})|^2\,d\mathbf{t}\,d\theta \geqslant \int_{T(c_0)}\int_{|\Delta\theta|\leqslant n/mr}|\hat{F}_\theta(\mathbf{t})|^2\,d\theta\,d\mathbf{t}$$

$$\gg (n/mr)n^2(\log n)^4\int_{T(c_0)}|\phi(\mathbf{t})|^2\,d\mathbf{t} \gg n^3(\log n)^{-1}.$$

This proves (178), and Lemma 7.16 follows.

The proof of Theorem 23C is now complete.

7.4 Spherical caps and a geometric application

In this section, we prove Theorem 24B.

Let $\mathscr{P}=\{\mathbf{p}_1,\mathbf{p}_2,\ldots,\mathbf{p}_N\}$ be an N-element distribution on the surface S^K of the unit sphere in $\mathbb{R}^{K+1}$. We introduce two measures in $\mathbb{R}^{K+1}$. For any $H\subset\mathbb{R}^{K+1}$, let

$$Z_0(H)=\sum_{\mathbf{p}_j\in H}1,$$

i.e. $Z_0(H)$ denotes the number of points $\mathbf{p}_1, \mathbf{p}_2, \ldots, \mathbf{p}_N$ that lie in H. For any Lebesgue measurable set $H \subset \mathbb{R}^{K+1}$, let

$$\sigma_0(H) = N\sigma^*(H \cap S^K) = N\frac{\sigma(H \cap S^K)}{\sigma(S^K)}.$$

Let $B(r) = B^{K+1}(r)$ denote the $(K+1)$-dimensional ball

$$B(r) = \{\mathbf{x} \in \mathbb{R}^{K+1} : |\mathbf{x}| \leqslant r\}$$

of radius r. Consider the function

$$F_r = \chi_r * (\mathrm{d}Z_0 - \mathrm{d}\sigma_0), \tag{192}$$

where χ_r denotes the characteristic function of the ball $B(r)$ and $*$ denotes the convolution operation. More explicitly,

$$F_r(\mathbf{x}) = \int_{\mathbb{R}^{K+1}} \chi_r(\mathbf{x}-\mathbf{y})(\mathrm{d}Z_0(\mathbf{y}) - \mathrm{d}\sigma_0(\mathbf{y})) = \sum_{\mathbf{p}_j \in B(r)+\mathbf{x}} 1 - N\sigma^*((B(r)+\mathbf{x}) \cap S^K), \tag{193}$$

where $B + \mathbf{x}$ denotes the translated image of B by the vector $\mathbf{x}$.

We claim that (see (6))

$$\int_1^2 \int_{\mathbb{R}^{K+1}} |F_r(\mathbf{x})|^2 \,\mathrm{d}\mathbf{x}\,\mathrm{d}r \ll \int_{-1}^1 \left(\frac{1}{\sigma(S^K)} \int_{S^K} |Z[\mathscr{P}, \mathbf{v}, t] - N\sigma^*(t)|^2 \,\mathrm{d}\sigma(\mathbf{v})\right)\mathrm{d}t. \tag{194}$$

Here $Z[\mathscr{P}, \mathbf{v}, t]$ denotes the number of points of $\mathscr{P}$ in the spherical cap $C(\mathbf{v}, t) = \{\mathbf{y} \in S^K : \mathbf{v}\cdot\mathbf{y} \leqslant t\}$, and $\sigma^*(t) = \sigma^*(C(\mathbf{v}, t))$. Also, the implicit constant in $\ll$, as in all subsequent argument, depends only on the dimension K.

To prove (194), first observe that the intersection $(B(r)+\mathbf{x}) \cap S^K$ forms a spherical cap $C(\mathbf{v}, t)$. We determine the parameters $\mathbf{v}$ and t. Let $\mathbf{y} \in S^K$ be a point such that $|\mathbf{x}-\mathbf{y}| = r$. Let $x = |\mathbf{x}|$, and let

$$t(r, x) = -\frac{\mathbf{x}\cdot\mathbf{y}}{|\mathbf{x}|}.$$

Then we have

$$(B(r)+\mathbf{x}) \cap S^K = C(\mathbf{v}, t),$$

where

$$\mathbf{v} = -\mathbf{x}/|\mathbf{x}| \quad \text{and} \quad t = t(r, x).$$

It therefore suffices to verify that

$$\inf_{\substack{1<r<2 \\ r-1<x<r+1}} \left|\frac{\partial t(r, x)}{\partial x}\right| \geqslant c_{42} > 0.$$

Simple geometric consideration gives

$$1 - t^2 = r^2 - (x + t)^2,$$

and so

$$t(r, x) = \frac{r^2 - 1 - x^2}{2x}.$$

Simple partial differentiation gives the result and (194) follows.

Consequently, in order to complete the proof of Theorem 24B, it suffices to establish the inequality

$$\int_1^2 \int_{\mathbb{R}^{K+1}} |F_r(\mathbf{x})|^2 \, d\mathbf{x} \, dr \gg N^{1-1/K}. \tag{195}$$

We proceed along similar lines to the proof of Theorem 17E in §6.1. First of all, we consider 'trivial errors'. Let $S(r) = S^K(r)$ denote the surface of the ball $B(r)$, i.e.

$$S(r) = \{\mathbf{x} \in \mathbb{R}^{K+1} : |\mathbf{x}| = r\}.$$

We choose ρ so that $\sigma^*(S(\rho)) = (2N)^{-1}$, i.e. $\rho = c_{43}(K)N^{-1/K}$. Since $N\sigma^*(S(\rho)) = \frac{1}{2}$, we have (the reader may wish to draw a picture for $K = 1$)

$$\left| \sum_{\mathbf{p}_j \in B(r) + \mathbf{x}} 1 - N\sigma^*((B(r) + \mathbf{x}) \cap S^K) \right| \geqslant N\sigma^*((B(r) + \mathbf{x}) \cap S^K) \gg 1 \tag{196}$$

for all r and $\mathbf{x}$ satisfying $\frac{1}{2}\rho \leqslant r \leqslant \rho$ and $1 - \rho/4 \leqslant |\mathbf{x}| \leqslant 1 + \rho/4$. By (193) and (196), we have

$$\frac{2}{\rho} \int_{\frac{1}{2}\rho}^{\rho} \int_{\mathbb{R}^{K+1}} |F_r(\mathbf{x})|^2 \, d\mathbf{x} \, dr \gg \int_{1-\rho/4 \leqslant |\mathbf{x}| \leqslant 1+\rho/4} d\mathbf{x} \gg \rho \gg N^{-1/K}. \tag{197}$$

Next we 'blow up' this trivial error by showing that

$$\frac{\int_1^2 |\hat{\chi}_r(\mathbf{t})|^2 \, dr}{\frac{2}{\rho} \int_{\frac{1}{2}\rho}^{\rho} |\hat{\chi}_r(\mathbf{t})|^2 \, dr} \gg N \quad \text{uniformly for all } \mathbf{t} \in \mathbb{R}^{K+1}. \tag{198}$$

We recall (53) and (54) in Chapter 6 (dimension $K + 1$):

$$\hat{\chi}_r(\mathbf{t}) = c_{44}(K)(r/t)^{\frac{1}{2}(K+1)} J_{\frac{1}{2}(K+1)}(rt), \tag{199}$$

where $t = |\mathbf{t}|$ and

$$J_\nu(x) = (2/\pi x)^{\frac{1}{2}}(\cos(x - (2\nu + 1)\pi/4) + c_{45}(\nu)x^{-1}).$$

Hence

$$\hat{\chi}_r(\mathbf{t}) = c_{46}(K) r^{\frac{1}{2}K} t^{-\frac{1}{2}K - 1}(\cos(rt - (K + 2)\pi/4) + \mathrm{O}(r^{-1}t^{-1})), \tag{200}$$

and so

$$\frac{1}{y}\int_y^{2y} |\hat{\chi}_r(\mathbf{t})|^2 \, dr \gg \ll \frac{y^K}{t^{K+2}} \tag{201}$$

provided that $yt > c_{47}(K)$. On the other hand, by (52) in Chapter 6, we see that with $v = \frac{1}{2}(K+1)$,

$$\frac{1}{z}\int_z^{2z} (J_v(x))^2 \, dx \gg \ll z^{2v} \tag{202}$$

if $0 \leqslant z \leqslant c_{47}(K)$. In view of (199) and (202), we have

$$\frac{1}{y}\int_y^{2y} |\hat{\chi}_r(\mathbf{t})|^2 \, dr \gg \ll y^{2K+2} \tag{203}$$

provided that $0 \leqslant yt \leqslant c_{47}(K)$. (198) follows immediately from (201) and (203).

We can now prove (195). By the Parseval–Plancherel identity (14),

$$\int_y^{2y}\int_{\mathbf{R}^{K+1}} |F_r(\mathbf{x})|^2 \, d\mathbf{x} \, dr = \int_y^{2y}\int_{\mathbf{R}^{K+1}} |\hat{F}_r(\mathbf{t})|^2 \, d\mathbf{t} \, dr.$$

In view of (192) and (12), we have $\hat{F}_r = \hat{\chi}_r \phi$, where $\phi = \widehat{dZ_0 - d\sigma_0}$. Therefore

$$\frac{1}{y}\int_y^{2y}\int_{\mathbf{R}^{K+1}} |F_r(\mathbf{x})|^2 \, d\mathbf{x} \, dr = \int_{\mathbf{R}^{K+1}} \left(\frac{1}{y}\int_y^{2y} |\hat{\chi}_r(\mathbf{t})|^2 \, dr\right) |\phi(\mathbf{t})|^2 \, d\mathbf{t}. \tag{204}$$

Combining (197), (198) and (204), we have

$$\begin{aligned}
\int_1^2\int_{\mathbf{R}^{K+1}} |F_r(\mathbf{x})|^2 \, d\mathbf{x} \, dr &= \int_{\mathbf{R}^{K+1}} \left(\int_1^2 |\hat{\chi}_r(\mathbf{t})|^2 \, dr\right) |\phi(\mathbf{t})|^2 \, d\mathbf{t} \\
&\gg N \int_{\mathbf{R}^{K+1}} \left(\frac{2}{\rho}\int_{\frac{1}{2}\rho}^{\rho} |\hat{\chi}_r(\mathbf{t})|^2 \, dr\right) |\phi(\mathbf{t})|^2 \, d\mathbf{t} \\
&= N\left(\frac{2}{\rho}\int_{\frac{1}{2}\rho}^{\rho}\int_{\mathbf{R}^{K+1}} |F_r(\mathbf{x})|^2 \, d\mathbf{x} \, dr\right) \gg N^{1-1/K}.
\end{aligned}$$

This proves (195), and the proof of Theorem 24B is now complete.

8

More upper bounds

8.1 A probabilistic method

In this section we prove Theorems 18A and 18B (see Chapter 6) and Theorems 23B, 24A and 24D (see Chapter 7).

The proof of Theorem 18A is based on the following lemma. Let T denote the set of proper orthogonal transformations in $\mathbb{R}^K$.

Lemma 8.1. *Let $A \subset \mathbb{R}^K$ be a compact convex body and let $M \geqslant 2$ be an integer. There exists an $N = M^K$-element set $S(A, M) = \{\mathbf{z}_1, \ldots, \mathbf{z}_N\} \subset [0, M)^K$ such that every cube $[l_1, l_1+1) \times \cdots \times [l_K, l_K+1)$, where $\mathbf{l} = (l_1, \ldots, l_K) \in [0, M)^K \cap \mathbb{Z}^K$, contains exactly one element of $S(A, M)$ and that for every triple $(\lambda, \mathbf{v}, \tau)$ with $0 < \lambda \leqslant 1$, $\mathbf{v} \in \mathbb{R}^K$ and $\tau \in T$, we have*

$$\left| \sum_{\mathbf{z}_i \in A(\lambda, \mathbf{v}, \tau)} 1 - \mu(A(\lambda, \mathbf{v}, \tau) \cap [0, M)^K) \right|$$
$$< c_1(K)(\max\{\lambda, 1/r(A)\})^{\frac{1}{2}(K-1)}(\sigma(\partial A))^{\frac{1}{2}} \max\{(\log M)^{\frac{1}{2}}, (\log(2 + d(A)))^{\frac{1}{2}}\}. \quad (1)$$

Proof of Theorem 18A. Let $M = [d(A)] + 1$, where $d(A)$ denotes the diameter of A. Extending $S(A, M)$ periodically modulo $[0, M)^K$ over the whole space $\mathbb{R}^K$, we obtain the desired set $\mathscr{Q}_0 = \mathscr{Q}_0(A)$. Indeed, the result easily follows from the facts that

$$d(A) \leqslant (r(A))^{K-2} d(A) \ll \sigma(\partial A)$$

and that every set $A(\lambda, \mathbf{v}, \tau)$, where $0 < \lambda \leqslant 1$, $\mathbf{v} \in \mathbb{R}^K$ and $\tau \in T$, is the disjoint union of not more than 2^K sets of the type

$$A(\lambda, \mathbf{v}, \tau) \cap ([0, M)^K + M\mathbf{l}) \quad (\mathbf{l} \in \mathbb{Z}^K).$$ ■

We now prove Lemma 8.1. Let

$$\mathscr{G} = \{(\lambda, \mathbf{v}, \tau) : 0 \leqslant \lambda \leqslant 2, \mathbf{v} \in \mathbb{R}^K, \tau \in T\}.$$

One can easily find a subset $\mathscr{G}^*$ of $\mathscr{G}$ such that

$$\#\mathscr{G}^* \leqslant \max\{M^{c_2(K)}, (2 + d(A))^{c_2(K)}\} \quad (2)$$

and that for every $A_0 = A(\lambda_0, \mathbf{v}_0, \tau_0)$, where $0 < \lambda_0 \leqslant 1$, $\mathbf{v}_0 \in \mathbb{R}^K$ and $\tau_0 \in T$, there exist $A_1 = A(\lambda_1, \mathbf{v}_1, \tau_1)$ and $A_2 = A(\lambda_2, \mathbf{v}_2, \tau_2)$ such that

(i) both triples $(\lambda_1, \mathbf{v}_1, \tau_1)$ and $(\lambda_2, \mathbf{v}_2, \tau_2)$ belong to $\mathscr{G}^*$;
(ii) $A_1 \cap [0, M)^K \subset A_0 \cap [0, M)^K \subset A_2 \cap [0, M)^K$; and
(iii) $\mu((A_2 \backslash A_1) \cap [0, M)^K) \leq 1$.

For every lattice point $\mathbf{l} = (l_1, \ldots, l_K) \in \mathbb{Z}^K$, let

$$Q(\mathbf{l}) = [l_1, l_1 + 1) \times \cdots \times [l_K, l_K + 1).$$

Clearly

$$[0, M)^K = \bigcup_{\mathbf{l} \in [0,M)^K \cap \mathbb{Z}^K} Q(\mathbf{l}).$$

Let us now associate with each $\mathbf{l} \in [0, M)^K \cap \mathbb{Z}^K$ a uniformly distributed random point $\eta_{\mathbf{l}}$ in $Q(\mathbf{l})$. In other words, the random variable $\eta_{\mathbf{l}}$ satisfies

$$\text{Prob}(\eta_{\mathbf{l}} \in H) = \mu(H \cap Q(\mathbf{l}))$$

for every measurable $H \subset \mathbb{R}^K$. Furthermore, assume that the random variables $\eta_{\mathbf{l}}$ $(\mathbf{l} \in [0, M)^K \cap \mathbb{Z}^K)$ are independent of each other.

Our aim is to show that the random M^K-element set

$$\{\eta_{\mathbf{l}} : \mathbf{l} \in [0, M)^K \cap \mathbb{Z}^K\}$$

satisfies (1) simultaneously for all $A(\lambda, \mathbf{v}, \tau)$ with $(\lambda, \mathbf{v}, \tau) \in \mathscr{G}^*$. By using properties (i), (ii) and (iii) of $\mathscr{G}^*$, this yields the lemma.

Let $A_1 = A(\lambda_1, \mathbf{v}_1, \tau_1)$, where $(\lambda_1, \mathbf{v}_1, \tau_1) \in \mathscr{G}^*$ is arbitrary but fixed. Let $\tilde{A}_1$ be the union of those cubes $Q(\mathbf{l})$ $(\mathbf{l} \in \mathbb{Z}^K)$ which are contained in $A_1 \cap [0, M)^K$. Obviously $\tilde{A}_1$ contains exactly as many random points as its volume. It therefore suffices to study the discrepancy of $(A_1 \cap [0, M)^K) \backslash \tilde{A}_1$. Let $\mathscr{L}(A_1)$ denote the set of lattice points $\mathbf{l} \in [0, M)^K \cap \mathbb{Z}^K$ such that both $A_1 \cap Q(\mathbf{l})$ and $(\mathbb{R}^K \backslash A_1) \cap Q(\mathbf{l})$ are non-empty. Then clearly

$$(A_1 \cap [0, M)^K) \backslash \tilde{A}_1 = \bigcup_{\mathbf{l} \in \mathscr{L}(A_1)} (A_1 \cap Q(\mathbf{l})).$$

Simple geometric consideration yields

$$\#\mathscr{L}(A_1) \ll \max\{\sigma(\partial(\lambda_1 A)), \sigma(\partial((1/r(A))A))\} = (\max\{\lambda_1, 1/r(A)\})^{K-1} \sigma(\partial A). \tag{3}$$

Let us define the random variables $\xi_{\mathbf{l}}$, where $\mathbf{l} \in \mathscr{L}(A_1)$, as follows: Let

$$\xi_{\mathbf{l}} = \begin{cases} 1 & (\eta_{\mathbf{l}} \in A_1 \cap Q(\mathbf{l})), \\ 0 & (\text{otherwise}). \end{cases}$$

Then

$$\sum_{\eta_{\mathbf{l}} \in A_1} 1 - \mu(A_1 \cap [0, M)^K) = \sum_{\mathbf{l} \in \mathscr{L}(A_1)} \xi_{\mathbf{l}} - \sum_{\mathbf{l} \in \mathscr{L}(A_1)} \mu(A_1 \cap Q(\mathbf{l})) = \sum_{\mathbf{l} \in \mathscr{L}(A_1)} (\xi_{\mathbf{l}} - \mathbb{E}\xi_{\mathbf{l}}), \tag{4}$$

where $\mathbb{E}\xi$ denotes, as usual, the expected value of ξ. Since the random variables $\xi_{\mathbf{l}}$ ($\mathbf{l}\in\mathscr{L}(A_1)$) are independent of each other, in order to estimate the sum $\sum(\xi_{\mathbf{l}}-\mathbb{E}\xi_{\mathbf{l}})$, we can apply the classical large-deviation type inequality due to Bernstein and Chernoff.

Lemma 8.2. *Let $\xi_1,\ldots,\xi_m$ be independent random variables with $|\xi_i|\leqslant 1$ for $i=1,\ldots,m$. Let $\beta=\sum_{i=1}^{m}\mathbb{E}(\xi_i-\mathbb{E}\xi_i)^2$. Then*

$$\operatorname{Prob}\left(\left|\sum_{i=1}^{m}(\xi_i-\mathbb{E}\xi_i)\right|\geqslant\gamma\right)\leqslant\begin{cases}2e^{-\gamma/4} & (\gamma\geqslant\beta),\\ 2e^{-\gamma^2/4\beta} & (\gamma\leqslant\beta).\end{cases}$$

For the sake of completeness, we include the proof of Lemma 8.2 at the end of this section.

Now set

$$\beta_1=\sum_{\mathbf{l}\in\mathscr{L}(A_1)}\mathbb{E}(\xi_{\mathbf{l}}-\mathbb{E}\xi_{\mathbf{l}})^2.$$

In view of (3), we have

$$\beta_1\leqslant\#\mathscr{L}(A_1)\ll(\max\{\lambda_1,1/r(A)\})^{K-1}\sigma(\partial A).$$

Let γ_1 be a sufficiently large absolute constant (depending only on K) multiple of

$$(\max\{\lambda_1,1/r(A)\})^{\frac{1}{2}(K-1)}(\sigma(\partial A))^{\frac{1}{2}}\max\{(\log M)^{\frac{1}{2}},(\log(2+d(A)))^{\frac{1}{2}}\}.$$

By elementary calculation, we obtain

$$\tfrac{1}{2}\min\{M^{-c_2(K)},(2+d(A))^{-c_2(K)}\}\geqslant\begin{cases}2e^{-\gamma_1/4} & (\gamma_1\geqslant\beta_1),\\ 2e^{-\gamma_1^2/4\beta_1} & (\gamma_1\leqslant\beta_1).\end{cases}\tag{5}$$

It follows from Lemma 8.2, (2) and (5) that

$$\operatorname{Prob}\left(\left|\sum_{\mathbf{l}\in\mathscr{L}(A_1)}(\xi_{\mathbf{l}}-\mathbb{E}\xi_{\mathbf{l}})\right|\geqslant\gamma_1\right)\leqslant\tfrac{1}{2}(\#\mathscr{G}^*)^{-1}.\tag{6}$$

Combining (4) and (6), we conclude that

$$\operatorname{Prob}\left(\left|\sum_{\eta_1\in A(\lambda_1,\mathbf{v}_1,\tau_1)}1-\mu(A(\lambda_1,\mathbf{v}_1,\tau_1)\cap[0,M)^K)\right|\geqslant\gamma_1\right.$$
$$\left.\text{for some }(\lambda_1,\mathbf{v}_1,\tau_1)\in\mathscr{G}^*\right)\leqslant\tfrac{1}{2},$$

and this completes the proof of Lemma 8.1.

The proof of Theorem 18B is based on iterated application of the following lemma.

Lemma 8.3. *Let $A \subset \mathbb{R}^K$ be a compact convex body and let $M \geqslant 1$ be an integer. Let S be a $(2M)^K$-element set in $[0, 2M)^K$ such that any unit cube $Q(\mathbf{l}) = [l_1, l_1 + 1) \times \cdots \times [l_K, l_K + 1)$, where $\mathbf{l} \in [0, 2M)^K \cap \mathbb{Z}^K$, contains exactly one point of S. Then one can find a partition*

$$S = \bigcup_{v=1}^{2^K} S_v$$

such that

(i) *$\#S_v = M^K$ for every v satisfying $1 \leqslant v \leqslant 2^K$;*
(ii) *for every $\mathbf{m} \in [0, M)^K \cap \mathbb{Z}^K$ and v satisfying $1 \leqslant v \leqslant 2^K$, the cube $2Q(\mathbf{m}) = [2m_1, 2m_1 + 2) \times \cdots \times [2m_K, 2m_K + 2)$ contains exactly one point of S_v; and*
(iii) *for every index v satisfying $1 \leqslant v \leqslant 2^K$ and triple $(\lambda, \mathbf{v}, \tau)$ with $0 < \lambda \leqslant 1$, $\mathbf{v} \in \mathbb{R}^K$ and $\tau \in T$, we have*

$$\begin{aligned} &|\#(S_v \cap A(\lambda, \mathbf{v}, \tau)) - 2^{-K}\#(S \cap A(\lambda, \mathbf{v}, \tau))| \\ &\quad < c_3(K)(\max\{\lambda, 1/r(A)\})^{\frac{1}{2}(K-1)}(\sigma(\partial A))^{\frac{1}{2}} \\ &\quad \max\{(\log(2M))^{\frac{1}{2}}, (\log(2 + d(A)))^{\frac{1}{2}}\}. \end{aligned} \tag{7}$$

The idea of the proof of Lemma 8.3 is the same as that in the proof of Lemma 8.1. Write

$$S = \{z(\mathbf{l}) : z(\mathbf{l}) \in Q(\mathbf{l}), \mathbf{l} \in [0, 2M)^K \cap \mathbb{Z}^K\}.$$

Denote by $P = (P(1), P(2), \ldots, P(2^K))$ a permutation of the integers from 1 to 2^K. For every $\mathbf{m} \in [0, M)^K \cap \mathbb{Z}^K$, let

$$\zeta_{\mathbf{m}} = (\zeta_{\mathbf{m}}(1), \zeta_{\mathbf{m}}(2), \ldots, \zeta_{\mathbf{m}}(2^K))$$

be a random permutation of the integers from 1 to 2^K, i.e.

$$\text{Prob}(\zeta_{\mathbf{m}} = P) = 1/(2^K)!$$

for every permutation P. Assume further that the random variables $\zeta_{\mathbf{m}}$ ($\mathbf{m} \in [0, M)^K \cap \mathbb{Z}^K$) are independent of each other.

The set $\{\zeta_{\mathbf{m}} : \mathbf{m} \in [0, M)^K \cap \mathbb{Z}^K\}$ of random permutations defines a random partition

$$S = \bigcup_{v=1}^{2^K} S_v$$

as follows: We have

$$2Q(\mathbf{m}) = \prod_{i=1}^{K} [2m_i, 2m_i + 2) = \bigcup_{\varepsilon} Q(2\mathbf{m} + \varepsilon),$$

where $\boldsymbol{\varepsilon}=(\varepsilon_1,\varepsilon_2,\ldots,\varepsilon_K)$ and for $i=1,\ldots,K$, $\varepsilon_i=0$ or 1. Let

$$N(\boldsymbol{\varepsilon})=\sum_{i=1}^{K}\varepsilon_i 2^{i-1}.$$

Clearly $1\leqslant N(\boldsymbol{\varepsilon})\leqslant 2^K$. Now let

$$z(2\mathbf{m}+\boldsymbol{\varepsilon})\in S_\nu \Leftrightarrow \zeta_{\mathbf{m}}(N(\boldsymbol{\varepsilon}))=\nu. \tag{8}$$

We shall prove that the probability of the event that the random partition defined in (8) satisfies (7) is at least $\frac{1}{2}$. From this the lemma immediately follows.

Again let

$$\mathscr{G}=\{(\lambda,\mathbf{v},\tau):0\leqslant\lambda\leqslant 2, \mathbf{v}\in\mathbb{R}^K, \tau\in T\}.$$

One can easily find a subset $\mathscr{G}^*$ of $\mathscr{G}$ such that

$$\#\mathscr{G}^*\leqslant\max\{(2M)^{c_2(K)},(2+d(A))^{c_2(K)}\} \tag{9}$$

and that for any $A_0=A(\lambda_0,\mathbf{v}_0,\tau_0)$ with $0<\lambda_0\leqslant 1$, $\mathbf{v}_0\in\mathbb{R}^K$ and $\tau_0\in T$, there exist $A_1=A(\lambda_1,\mathbf{v}_1,\tau_1)$ and $A_2=A(\lambda_2,\mathbf{v}_2,\tau_2)$ such that

(i) both triples $(\lambda_1,\mathbf{v}_1,\tau_1)$ and $(\lambda_2,\mathbf{v}_2,\tau_2)$ belong to $\mathscr{G}^*$;
(ii) $A_1\cap[0,2M)^K\subset A_0\cap[0,2M)^K\subset A_2\cap[0,2M)^K$; and
(iii) $\mu((A_2\backslash A_1)\cap[0,2M)^K)<1$.

Let ν be an integer satisfying $1\leqslant\nu\leqslant 2^K$ and let $A_1=A(\lambda_1,\mathbf{v}_1,\tau_1)$, where $(\lambda_1,\mathbf{v}_1,\tau_1)\in\mathscr{G}^*$ is arbitrary but fixed. Let $\tilde{A}_1$ be the union of those cubes $2Q(\mathbf{m})$ ($\mathbf{m}\in\mathbb{Z}^K$) which are contained in $A_1\cap[0,2M)^K$. Obviously

$$\#(S_\nu\cap\tilde{A}_1)=2^{-K}\#(S\cap\tilde{A}_1)=2^{-K}\mu(\tilde{A}_1). \tag{10}$$

It therefore suffices to study $(A_1\cap[0,2M)^K)\backslash\tilde{A}_1$. Let $\mathscr{L}(A_1)$ denote the set of lattice points $\mathbf{m}\in[0,M)^K\cap\mathbb{Z}^K$ such that both sets $A_1\cap 2Q(\mathbf{m})$ and $(\mathbb{R}^K\backslash A_1)\cap 2Q(\mathbf{m})$ are non-empty. Then clearly

$$(A_1\cap[0,2M)^K)\backslash\tilde{A}_1=\bigcup_{\mathbf{m}\in\mathscr{L}(A_1)}(A_1\cap 2Q(\mathbf{m})).$$

Simple geometric consideration yields

$$\#\mathscr{L}(\mathrm{A}_1)\ll\max\{\sigma(\partial(\lambda_1 A)),\sigma(\partial((1/r(A))A))\}=(\max\{\lambda_1,1/r(A)\})^{K-1}\sigma(\partial A). \tag{11}$$

For every $\mathbf{m}\in\mathscr{L}(A_1)$, let

$$a_1(\mathbf{m})=\#(S\cap A_1\cap 2Q(\mathbf{m}))=\sum_{\boldsymbol{\varepsilon}:z(2\mathbf{m}+\boldsymbol{\varepsilon})\in A_1}1, \tag{12}$$

where $\boldsymbol{\varepsilon}=(\varepsilon_1,\ldots,\varepsilon_K)$ and for $1\leqslant i\leqslant K$, $\varepsilon_i=0$ or 1. Moreover, introduce the random variables

$$\xi_{\mathbf{m},v} = \#(S_v \cap A_1 \cap 2Q(\mathbf{m})). \tag{13}$$

Clearly $\xi_{\mathbf{m},v}$ has only the values 0 or 1.

By (8) and (12), we have

$$\mathbb{E}\xi_{\mathbf{m},v} = \mathrm{Prob}\,(\xi_{\mathbf{m},v} = 1) = \frac{a_1(\mathbf{m})(2^K - 1)!}{(2^K)!} = \frac{a_1(\mathbf{m})}{2^K}. \tag{14}$$

It follows from (10) and (12)–(14) that

$$\begin{aligned} &\#(S_v \cap A_1) - 2^{-K}\#(S \cap A_1) \\ &= \sum_{\mathbf{m}\in\mathscr{L}(A_1)} (\#(S_v \cap A_1 \cap 2Q(\mathbf{m})) - 2^{-K}\#(S \cap A_1 \cap 2Q(\mathbf{m}))) \\ &= \sum_{\mathbf{m}\in\mathscr{L}(A_1)} (\xi_{\mathbf{m},v} - \mathbb{E}\xi_{\mathbf{m},v}). \end{aligned} \tag{15}$$

Since the random variables $\xi_{\mathbf{m},v}$ ($\mathbf{m}\in\mathscr{L}(A_1)$) are independent, we can apply Lemma 8.2. Write

$$\beta_1 = \beta_1(A_1, v) = \sum_{\mathbf{m}\in\mathscr{L}(A_1)} \mathbb{E}(\xi_{\mathbf{m},v} - \mathbb{E}\xi_{\mathbf{m},v})^2.$$

By (11), we have

$$\beta_1 \leqslant \#\mathscr{L}(A_1) \ll (\max\{\lambda_1, 1/r(A)\})^{K-1}\sigma(\partial A). \tag{16}$$

Let γ_1 be a sufficiently large absolute constant (dependent only on K) multiple of

$$(\max\{\lambda_1, 1/r(A)\})^{\frac{1}{2}(K-1)}(\sigma(\partial A))^{\frac{1}{2}} \max\{(\log(2M))^{\frac{1}{2}}, (\log(2 + d(A)))^{\frac{1}{2}}\}. \tag{17}$$

Then from (9), (16) and (17), we obtain, via elementary calculation, that

$$\begin{aligned} 2^{-(K+1)}(\#\mathscr{G}^*)^{-1} &\geqslant 2^{-(K+1)} \min\{(2M)^{-c_2(K)}, (2 + d(A))^{-c_2(K)}\} \\ &\geqslant \begin{cases} 2e^{-\gamma_1/4} & (\gamma_1 \geqslant \beta_1), \\ 2e^{-\gamma_1^2/4\beta_1} & (\gamma_1 \leqslant \beta_1). \end{cases} \end{aligned} \tag{18}$$

Hence, by Lemma 8.2 and (18), we have

$$\mathrm{Prob}\left(\left|\sum_{\mathbf{m}\in\mathscr{L}(A_1)} (\xi_{\mathbf{m},v} - \mathbb{E}\xi_{\mathbf{m},v})\right| \geqslant \gamma_1\right) \leqslant 2^{-(K+1)}(\#\mathscr{G}^*)^{-1}. \tag{19}$$

It now follows from (15) and (19) that

$$\begin{aligned} &\mathrm{Prob}\,(|\#(S_v \cap A(\lambda_1, \mathbf{v}_1, \tau_1)) - 2^{-K}\#(S \cap A(\lambda_1, \mathbf{v}_1, \tau_1))| \geqslant \gamma_1 \text{ for some} \\ &\quad (\lambda_1, \mathbf{v}_1, \tau_1) \in \mathscr{G}^* \text{ and } v \text{ satisfying } 1 \leqslant v \leqslant 2^K) \leqslant \tfrac{1}{2}, \end{aligned}$$

and this completes the proof of Lemma 8.3.

To deduce Theorem 18B, we now renormalize Lemmas 8.1 and 8.3 as follows.

Lemma 8.1′. *Let $A \subset \mathbb{R}^K$ be a compact convex body and let $n \geqslant 1$ be an integer. There exists a 2^{nK}-element set $S = S(A, n)$ in the unit cube U_0^K such that every cube of the type $[l_1 2^{-n}, (l_1+1)2^{-n}) \times \cdots \times [l_K 2^{-n}, (l_K+1)2^{-n})$, where $\mathbf{l} = (l_1, \ldots, l_K) \in [0, 2^n)^K \cap \mathbb{Z}^K$, contains exactly one point of S and that for every triple $(\lambda, \mathbf{v}, \tau)$ with $0 < \lambda \leqslant 1$, $\mathbf{v} \in \mathbb{R}^K$ and $\tau \in T$, we have*

$$|\#(S \cap A(\lambda, \mathbf{v}, \tau)) - 2^{nK}\mu(A(\lambda, \mathbf{v}, \tau) \cap U_0^K)| < c_4(A)(\max\{2^n\lambda, 1\})^{\frac{1}{2}(K-1)} n^{\frac{1}{2}}.$$

Lemma 8.3′. Let *$A \subset \mathbb{R}^K$ be a compact convex body. Let $n \geqslant 1$ be an integer, and let S be a 2^{nK}-element set in the unit cube U_0^K such that every cube of the type $[l_1 2^{-n}, (l_1+1)2^{-n}) \times \cdots \times [l_K 2^{-n}, (l_K+1)2^{-n})$, where $\mathbf{l} = (l_1, \ldots, l_K) \in [0, 2^n)^K \cap \mathbb{Z}^K$, contains exactly one point of S. Then one can find a partition*

$$S = \bigcup_{\nu=1}^{2^K} S_\nu$$

such that

(i) *$\# S_\nu = 2^{(n-1)K}$ for every ν satisfying $1 \leqslant \nu \leqslant 2^K$;*
(ii) *for every $\mathbf{m} = (m_1, \ldots, m_K) \in [0, 2^{n-1})^K \cap \mathbb{Z}^K$ and ν satisfying $1 \leqslant \nu \leqslant 2^K$, the cube $[m_1 2^{-(n-1)}, (m_1+1)2^{-(n-1)}) \times \cdots \times [m_K 2^{-(n-1)}, (m_K+1) 2^{-(n-1)})$ contains exactly one point of S_ν; and*
(iii) *for every index ν satisfying $1 \leqslant \nu \leqslant 2^K$ and triple $(\lambda, \mathbf{v}, \tau)$ with $0 < \lambda \leqslant 1$, $\mathbf{v} \in \mathbb{R}^K$ and $\tau \in T$, we have*

$$|\#(S_\nu \cap A(\lambda, \mathbf{v}, \tau)) - 2^{-K}\#(S \cap A(\lambda, \mathbf{v}, \tau))| < c_5(A)(\max\{2^n\lambda, 1\})^{\frac{1}{2}(K-1)} n^{\frac{1}{2}}.$$

Next we need

Lemma 8.4. *Let $A \subset \mathbb{R}^K$ be a compact convex body and let $n \geqslant 1$ be an integer. There exists a 2^{nK}-element sequence $\mathbf{z}_1, \mathbf{z}_2, \ldots, \mathbf{z}_N$, where $N = 2^{nK}$, in the unit cube U_0^K such that for every triple $(\lambda, \mathbf{v}, \tau)$ with $0 < \lambda \leqslant 1$, $\mathbf{v} \in \mathbb{R}^K$ and $\tau \in T$ and for every integer j satisfying $1 \leqslant j \leqslant N = 2^{nK}$, we have*

$$\begin{aligned} &|\#(\{\mathbf{z}_1, \mathbf{z}_2, \ldots, \mathbf{z}_j\} \cap A(\lambda, \mathbf{v}, \tau)) - j\mu(A(\lambda, \mathbf{v}, \tau) \cap U_0^K)| \\ &\quad < c_6(A)(\max\{2^n\lambda, 1\})^{\frac{1}{2}(K-1)} n^{\frac{1}{2}} \log(1 + 1/\lambda). \end{aligned} \tag{20}$$

Proof. Let $S = S(A, n)$ denote the $N = 2^{nK}$-element subset of U_0^K satisfying Lemma 8.1′. Applying Lemma 8.3′ to $S = S(A, n)$, we obtain a partition

$$S = \bigcup_{\nu=1}^{2^K} S_\nu$$

satisfying properties (i), (ii) and (iii). We now apply Lemma 8.3′ to every S_{ν_1} satisfying $1 \leqslant \nu_1 \leqslant 2^K$ and obtain a partition

$$S_{\nu_1} = \bigcup_{\nu=1}^{2^K} S_{\nu_1,\nu}.$$

We can again apply Lemma 8.3′ to every S_{ν_1,ν_2} satisfying $1 \leqslant \nu_1, \nu_2 \leqslant 2^K$ and obtain a partition

$$S_{\nu_1,\nu_2} = \bigcup_{\nu=1}^{2^K} S_{\nu_1,\nu_2,\nu},$$

and so on. In the last step, we obtain one-element sets

$$S_{\nu}\,(\nu = (\nu_1, \nu_2, \ldots, \nu_n), 1 \leqslant \nu_i \leqslant 2^K, 1 \leqslant i \leqslant n).$$

Every integer j satisfying $1 \leqslant j \leqslant N = 2^{nK}$ can be uniquely represented in the form

$$j = 1 + \sum_{i=1}^{n} (\theta_i - 1)2^{(n-i)K} \quad (1 \leqslant \theta_i \leqslant 2^K, 1 \leqslant i \leqslant n). \tag{21}$$

Then write $\boldsymbol{\theta} = (\theta_1, \ldots, \theta_n) \in \mathbb{Z}^n$, and let $\mathbf{z}_j$ be the only element of $S_{\boldsymbol{\theta}}$. We now show that this sequence $\mathbf{z}_1, \mathbf{z}_2, \ldots, \mathbf{z}_N$ satisfies inequality (20). For every m satisfying $1 \leqslant m \leqslant n$ and every $\boldsymbol{\nu} = (\nu_1, \ldots, \nu_n)$ satisfying $1 \leqslant \nu_i \leqslant 2^K$ and $1 \leqslant i \leqslant n$, write

$$\boldsymbol{\nu}[m] = (\nu_1, \ldots, \nu_m) \in \mathbb{Z}^m.$$

Clearly (convention $S_{\boldsymbol{\nu}[0]} = S$)

$$\begin{aligned}
&\#(S_{\boldsymbol{\nu}[m]} \cap A(\lambda, \mathbf{v}, \tau)) - 2^{(n-m)K}\mu(A(\lambda, \mathbf{v}, \tau) \cap U_0^K) \\
&\quad = 2^{-mK}(\#(S \cap A(\lambda, \mathbf{v}, \tau)) - 2^{nK}\mu(A(\lambda, \mathbf{v}, \tau) \cap U_0^K)) \\
&\qquad + \sum_{i=1}^{m} 2^{-(m-i)K}(\#(S_{\boldsymbol{\nu}[i]} \cap A(\lambda, \mathbf{v}, \tau)) - 2^{-K}\#(S_{\boldsymbol{\nu}[i-1]} \cap A(\lambda, \mathbf{v}, \tau))).
\end{aligned}$$

Hence, by Lemmas 8.1′ and 8.3′,

$$\begin{aligned}
&|\#(S_{\boldsymbol{\nu}[m]} \cap A(\lambda, \mathbf{v}, \tau)) - 2^{(n-m)K}\mu(A(\lambda, \mathbf{v}, \tau) \cap U_0^K)| \\
&\quad < 2^{-mK}c_4(A)(\max\{2^n\lambda, 1\})^{\frac{1}{2}(K-1)}n^{\frac{1}{2}} \\
&\qquad + \sum_{i=1}^{m} 2^{-(m-i)K}c_5(A)(\max\{2^{n-i+1}\lambda, 1\})^{\frac{1}{2}(K-1)}n^{\frac{1}{2}} \\
&\quad < c_4(A)2^{-mK}(\max\{2^n\lambda, 1\})^{\frac{1}{2}(K-1)}n^{\frac{1}{2}} + c_7(A)(\max\{2^{n-m+1}\lambda, 1\})^{\frac{1}{2}(K-1)}n^{\frac{1}{2}}.
\end{aligned} \tag{22}$$

On the other hand, it follows from (21) that

$$\{\mathbf{z}_1,\ldots,\mathbf{z}_j\}=\left(\bigcup_{1\leqslant v<\theta_1} S_v\right)\cup\left(\bigcup_{1\leqslant v<\theta_2} S_{\theta_1,v}\right)\cup\cdots$$
$$\cup\left(\bigcup_{1\leqslant v<\theta_{n-1}} S_{\theta_1,\ldots,\theta_{n-2},v}\right)\cup\left(\bigcup_{1\leqslant v<\theta_n} S_{\theta_1,\ldots,\theta_{n-1},v}\right). \tag{23}$$

It follows from (22) and (23) that

$$|\#(\{\mathbf{z}_1,\ldots,\mathbf{z}_j\}\cap A(\lambda,\mathbf{v},\tau))-j\mu(A(\lambda,\mathbf{v},\tau)\cap U_0^K)|$$
$$<\sum_{i=1}^{n}\theta_i(c_4(A)2^{-iK}(\max\{2^n\lambda,1\})^{\frac{1}{2}(K-1)}n^{\frac{1}{2}}$$
$$+c_7(A)(\max\{2^{n-i+1}\lambda,1\})^{\frac{1}{2}(K-1)}n^{\frac{1}{2}})$$
$$<c_8(A)(\max\{2^n\lambda,1\})^{\frac{1}{2}(K-1)}n^{\frac{1}{2}}\log(1+1/\lambda),$$

and Lemma 8.4 follows. ■

Proof of Theorem 18B. Every integer $t\geqslant 1$ can be uniquely written in the form

$$t=\left(\sum_{n=1}^{q-1}2^{nK}\right)+r \quad (q\geqslant 1,1\leqslant r<2^{qK}).$$

For every $n\geqslant 1$, let $\mathbf{z}_1^{(n)},\mathbf{z}_2^{(n)},\ldots,\mathbf{z}_N^{(n)}$ $(N=2^{nK})$ be a sequence satisfying Lemma 8.4. Then let

$$\mathbf{s}_t=\mathbf{z}_r^{(q)}.$$

It follows immediately from (20) and trivial calculation that this sequence $\mathbf{s}_1,\mathbf{s}_2,\mathbf{s}_3,\ldots$ satisfies the uniformity properties in Theorem 18B. ■

Proof of Theorem 24A. We have to find an N-element subset $\mathscr{P}$ of S^K such that

$$\int_{-1}^{1}\left(\frac{1}{\sigma(S^K)}\int_{S^K}|Z[\mathscr{P},\mathbf{x},t]-N\sigma^*(t)|^2\,\mathrm{d}\sigma(\mathbf{x})\right)\mathrm{d}t\ll N^{1-1/K}, \tag{24}$$

where $Z[\mathscr{P},\mathbf{x},t]$ denotes the number of points of $\mathscr{P}$ in the spherical cap $C(\mathbf{x},t)=\{\mathbf{y}\in S^K:\mathbf{x}\cdot\mathbf{y}\leqslant t\}$ and $\sigma^*(t)=\sigma(C(\mathbf{x},t))/\sigma(S^K)$. One can easily find a partition

$$S^K=\bigcup_{l=1}^{N}Q_l$$

such that for $1 \leqslant l \leqslant N$,

$$\sigma(Q_l) = N^{-1}\sigma(S^K)$$

and

$$d(Q_l) \ll N^{-1/K},$$

where $d(Q_l)$ is the diameter of Q_l. Let us now associate with each Q_l, where $1 \leqslant l \leqslant N$, a uniformly distributed random point η_l in Q_l. In other words, the random variable η_l satisfies

$$\mathrm{Prob}\,(\eta_l \in H) = \frac{\sigma(H \cap Q_l)}{\sigma(Q_l)}$$

for every measurable $H \subset S^K$. Furthermore, assume that these random variables $\eta_l\,(1 \leqslant l \leqslant N)$ are independent of each other. Our aim is to study the left-hand side of (24) when $\mathscr{P}$ is replaced by the random point set $\{\eta_l : 1 \leqslant l \leqslant N\}$. Let $C = C(\mathbf{x}, t)$ be an arbitrary but fixed spherical cap of S^K. Let $\tilde{C}$ be the union of those Q_l which are contained in the spherical cap C. Obviously $\tilde{C}$ contains exactly as many random points η_l as it is expected. It therefore suffices to study the discrepancy of $C \backslash \tilde{C}$. Let $\mathscr{L}(C)$ denote the set of indices l such that both sets $C \cap Q_l$ and $(S^K \backslash C) \cap Q_l$ are non-empty. Then clearly

$$C \backslash \tilde{C} = \bigcup_{l \in \mathscr{L}(C)} (C \cap Q_l).$$

Using the properties of the partition of S^K, we obtain that

$$\#\mathscr{L}(C) \ll N^{1-1/K}. \tag{25}$$

Let us now define the random variables $\xi_l\ (l \in \mathscr{L}(C))$ as follows: Let

$$\xi_l = \begin{cases} 1 & (\eta_l \in C \cap Q_l), \\ 0 & (\eta_l \notin C \cap Q_l). \end{cases}$$

Then

$$\sum_{\eta_l \in C} 1 - N\frac{\sigma(C)}{\sigma(S^K)} = \sum_{l \in \mathscr{L}(C)} \xi_l - \sum_{l \in \mathscr{L}(C)} \frac{\sigma(C \cap Q_l)}{\sigma(S^K)} = \sum_{l \in \mathscr{L}(C)} (\xi_l - \mathbb{E}\xi_l), \tag{26}$$

where $\mathbb{E}\xi$ denotes the expected value of ξ. Since the random variables ξ_l are independent of each other, we clearly have

$$\begin{aligned} \mathbb{E}\left(\sum_{l \in \mathscr{L}(C)} (\xi_l - \mathbb{E}\xi_l)\right)^2 &= \sum_{l_1 \in \mathscr{L}(C)} \sum_{l_2 \in \mathscr{L}(C)} \mathbb{E}(\xi_{l_1} - \mathbb{E}\xi_{l_1})(\xi_{l_2} - \mathbb{E}\xi_{l_2}) \\ &= \sum_{l_1 \in \mathscr{L}(C)} \sum_{l_2 \in \mathscr{L}(C)} \mathbb{E}(\xi_{l_1} - \mathbb{E}\xi_{l_1})\mathbb{E}(\xi_{l_2} - \mathbb{E}\xi_{l_2}) \\ &= \sum_{l \in \mathscr{L}(C)} \mathbb{E}(\xi_l - \mathbb{E}\xi_l)^2. \end{aligned} \tag{27}$$

It follows from (25)–(27) that

$$\mathbb{E}\left(\sum_{\eta_l\in C} 1 - N\frac{\sigma(C)}{\sigma(S^K)}\right)^2 = \sum_{l\in\mathscr{L}(C)} \mathbb{E}(\xi_l - \mathbb{E}\xi_l)^2 \leqslant \#\mathscr{L}(C) \ll N^{1-1/K}.$$

Consequently,

$$\mathbb{E}\left(\int_{-1}^{1}\left(\frac{1}{\sigma(S^K)}\int_{S^K}\left|\sum_{\eta_l\in C(\mathbf{x},t)} 1 - N\frac{\sigma(C(\mathbf{x},t))}{\sigma(S^K)}\right|^2 \mathrm{d}\sigma(\mathbf{x})\right)\mathrm{d}t\right) \ll N^{1-1/K}. \quad (28)$$

The existence of an N-element set $\mathscr{P} \subset S^K$ satisfying (24) follows immediately from (28). ■

Proof of Theorem 24D. We can repeat the proof of Theorem 24A up to identity (26) without any modification. Then we apply Lemma 8.2 to the right-hand side of (26), and finish the proof along the same lines as in the proof of Theorem 18A. ■

Proof of Theorem 23B. We can easily find a partition

$$D = \bigcup_{l=1}^{N} Q_l$$

of the circular disc D of unit area such that for $1 \leqslant l \leqslant N$,

$$\mu(Q_l) = N^{-1}$$

and

$$d(Q_l) \ll N^{-\frac{1}{2}}.$$

We then associate with each Q_l, where $1 \leqslant l \leqslant N$, a uniformly distributed random point η_l in Q_l. In other words, the random variable η_l satisfies

$$\mathrm{Prob}\,(\eta_l \in H) = \frac{\mu(H \cap Q_l)}{\mu(Q_l)} = N\mu(H \cap Q_l)$$

for every measurable $H \subset D$. Furthermore, assume that these random variables η_l $(1 \leqslant l \leqslant N)$ are independent of each other. We can now proceed along the same lines as in the proof of Theorem 18A. ■

As we promised, we conclude this section by proving Lemma 8.2. We may assume, without loss of generality, that

$$\mathbb{E}\xi_i = 0$$

for every $i = 1, \ldots, m$. Set

$$X = \sum_{i=1}^{m} \xi_i.$$

Clearly

$$\text{Prob}(X \geqslant \gamma) = \text{Prob}(e^{yX} \geqslant e^{y\gamma}) \leqslant e^{-y\gamma}\mathbb{E}(e^{yX}), \tag{29}$$

where the parameter y, satisfying $0 < y \leqslant 1$, will be fixed later. Since X is the sum of independent random variables, we have

$$\mathbb{E}(e^{yX}) = \prod_{i=1}^{m} \mathbb{E}(e^{y\xi_i}).$$

We shall give an upper bound on $\mathbb{E}(e^{y\xi_i})$. Using the formula of the exponential series, we get, after some easy calculation,

$$\mathbb{E}(e^{y\xi_i}) = \sum_{n=0}^{\infty} \frac{y^n \mathbb{E}\xi_i^n}{n!} \leqslant 1 + \frac{y^2\mathbb{E}\xi_i^2}{2} + \frac{y^3\mathbb{E}\xi_i^2}{6-2y}. \tag{30}$$

If we substitute (30) into (29), we obtain

$$\text{Prob}(X \geqslant \gamma) \leqslant \exp\left(\frac{y^2\beta}{2}\left(1 + \frac{y}{3-y}\right) - y\gamma\right). \tag{31}$$

We distinguish two cases. If $\gamma \geqslant \beta$, then let $y = 1$. In view of (31), we see that

$$\text{Prob}(X \geqslant \gamma) \leqslant e^{-\gamma/4}.$$

If $\gamma \leqslant \beta$, then let $y = \gamma/\beta$. Again by (31),

$$\text{Prob}(X \geqslant \gamma) \leqslant e^{-\gamma^2/4\beta}.$$

Repeating the same calculation for $\text{Prob}(X \leqslant -\gamma)$, we obtain the desired upper bounds. Lemma 8.2 follows.

8.2 A hypergraph 2-colouring approach

In this section, we study the general problem of 2-colouring a set as uniformly as possible with respect to a given family of subsets. What we want to achieve is that the colouring be nearly balanced in each of the subsets considered.

Let $X = \{x_1, x_2, \ldots, x_p\}$ be an arbitrary finite set and $\mathscr{Y} = \{Y_1, Y_2, \ldots, Y_q\}$ an arbitrary family of subsets of X. The pair $(X, \mathscr{Y})$ is often known as a hypergraph in the literature on combinatorics. We would like to find a 2-colouring $f: X \to \{-1, +1\}$ of the underlying set X such that

$$\max_{Y \in \mathscr{Y}} \left| \sum_{x \in Y} f(x) \right|$$

is as small as possible. In other words, we investigate the function

$$\delta(\mathscr{Y}, X) = \min_f \max_{Y \in \mathscr{Y}} \left| \sum_{x \in Y} f(x) \right|,$$

where the minimum is taken over all 2-colourings $f:X\to\{-1,+1\}$. Observe that $\delta(\mathscr{Y},X')=\delta(\mathscr{Y},X'')$ for any two underlying sets X' and X'' with

$$X'\cap X''\supset\bigcup_{Y\in\mathscr{Y}}Y.$$

We may therefore omit reference to X and write $\delta(\mathscr{Y})$ for $\delta(\mathscr{Y},X)$, and call this the 2-colouring discrepancy of the family $\mathscr{Y}$.

Let $\deg(\mathscr{Y})$ be the 'maximum degree' of $\mathscr{Y}$, i.e.

$$\deg(\mathscr{Y})=\max_{x\in X}\#\{Y\in\mathscr{Y}:x\in Y\}.$$

The following lemma gives an upper bound on $\delta(\mathscr{Y})$ which depends only on $\deg(\mathscr{Y})$, i.e. on the 'local size' of $\mathscr{Y}$.

Lemma 8.5 (Beck and Fiala (1981)). *For any finite family* $\mathscr{Y}$,

$$\delta(\mathscr{Y})<2\deg(\mathscr{Y}).$$

To illustrate the applicability of Lemma 8.5, we derive the upper bound in Theorem 10A (see §2.4), which is clearly equivalent to the following result.

Theorem 10C. *Let* $A=(a_{ij})$, *where* $a_{ij}=0$ *or* 1, *be a matrix of size* $N\times N$. *Then there exist 'signs'* $\varepsilon_{ij}=\pm 1$ *such that*

$$\left|\sum_{i=1}^{s}\sum_{j=1}^{t}\varepsilon_{ij}a_{ij}\right|\ll(\log N)^4$$

for all $s,t\in\{1,\dots,N\}$.

Proof. We may assume that $N=2^l$, where l is an integer. For $0\leqslant p,q\leqslant l$, we partition A into 2^{p+q} submatrices, splitting the horizontal side of the matrix into 2^p equal pieces and the vertical side of the matrix into 2^q equal pieces. There are $(l+1)^2\gg\ll(\log N)^2$ such partitions. Let us call a submatrix of A special if it occurs in one of these partitions, and let $\mathscr{Y}$ be the collection of all these special submatrices. Then by Lemma 8.5, there exists an assignment of ± 1's so that the absolute value of the sum of the signed entries in each of the special submatrices is less than $2\deg(\mathscr{Y})\leqslant 2(l+1)^2$. Note, however, that any submatrix of A containing the lower left corner of A is the union of at most l^2 disjoint special submatrices. The deduction of Theorem 10C from Lemma 8.5 is now complete. ■

A slight modification of the technique in the proof of Lemma 8.5 will give the following 'integer-making' lemma.

Lemma 8.6. *Let $X=\{x_1,x_2,\ldots,x_p\}$ be a finite set. For $i=1,2,\ldots,$ let $\mathscr{Y}^{(i)}=\{Y_1^{(i)},Y_2^{(i)},\ldots\}$ be a partition of X, i.e.*

$$X=\bigcup_{j\geqslant 1} Y_j^{(i)}$$

is a union of mutually disjoint sets $Y_j^{(i)}$. Let us associate with every x_k $(1\leqslant k\leqslant p)$ a real $\alpha_k\in[0,1]$. Then there are integers $a_k=0$ or 1 $(1\leqslant k\leqslant p)$ such that

$$\left|\sum_{x_k\in Y_j^{(i)}}(a_k-\alpha_k)\right|<c_9(\varepsilon)i^{1+\varepsilon}$$

for all $Y_j^{(i)}$ satisfying $i\geqslant 1$ and $j\geqslant 1$. Here $\varepsilon>0$ is arbitrarily small but fixed, and the positive constant $c_9(\varepsilon)$ depends only on ε.

Remark. When $\alpha_1=\cdots=\alpha_p=\frac{1}{2}$, the set of values $a_1,\ldots,a_p=0$ or 1 corresponds to a 2-colouring $f:X\to\{-1,+1\}$ given by $f(x_k)=2a_k-1$.

We suspect that Lemma 8.5 can be improved to

$$\delta(\mathscr{Y})<d^{\frac{1}{2}+\varepsilon}$$

whenever $d=\deg(\mathscr{Y})>d_0(\varepsilon)$. The following lemma justifies this conjecture when both $\#\mathscr{Y}$ and $\#X$ are 'subexponential' functions of $d=\deg(\mathscr{Y})$ (see inequality (33) below). For later use, we state it in the following very general form.

Lemma 8.7 (Beck). *Let X be a finite set, and let $\mathscr{H}=\{H_1,H_2,\ldots\}$ and $\mathscr{Y}=\{Y_1,Y_2,\ldots\}$ be families of subsets of X. Suppose that*

(*i*) $\max_{H\in\mathscr{H}}\#H\leqslant m$.

Suppose further that there is a third family $\mathscr{Z}=\{Z_1,Z_2,\ldots\}$ of subsets of X such that

(*ii*) $\deg(\mathscr{Z})\leqslant d$; *and*
(*iii*) *every $Y\in\mathscr{Y}$ can be represented as the union of at most k disjoint elements of $\mathscr{Z}$.*

Then

$$\delta(\mathscr{H}\cup\mathscr{Y})\leqslant c_{10}(kd\log(d+2)\log(\#\mathscr{Y}+2)+m\log(\#\mathscr{H}+2))^{\frac{1}{2}}\log(\#X+2). \tag{32}$$

Remark. In the particular case $\mathscr{H}=\varnothing$ and $\mathscr{Y}=\mathscr{Z}$, Lemma 8.7 immediately yields the relation (note that $k=1$ and $\deg(\mathscr{Z})\leqslant\#\mathscr{Z}$)

$$\delta(\mathscr{Y})\ll(\deg(\mathscr{Y}))^{\frac{1}{2}}\log(\#\mathscr{Y}+2)\log(\#X+2). \tag{33}$$

The proof of Lemma 8.7 works especially for 2-colourings only. In order to get an 'integer-making' result, we need the following general argument. We consider two extensions of the concept of 2-colouring discrepancy.

Let $X=\{x_1,\dots,x_p\}$ be a finite set and $\mathscr{G}=\{G_1,\dots,G_n\}$ a family of subsets of X. For any $A\subset X$, the restriction $\mathscr{G}|_A$ is the family $\{G_1\cap A,\dots, G_n\cap A\}$. The hereditary discrepancy of $\mathscr{G}$ is defned by

$$\operatorname{herdis}(\mathscr{G})=\max_{A\subset X}\delta(\mathscr{G}|_A),$$

where $\delta(\mathscr{H})$ denotes the 2-colouring discrepancy of $\mathscr{H}$.

The second extension is exactly the 'integer-making' concept introduced in Lemma 8.6. We define the linear discrepancy of $\mathscr{G}$ by

$$\operatorname{lindis}(\mathscr{G})=\max_{\alpha_1,\dots,\alpha_p}\ \min_{a_1,\dots,a_p}\ \max_{G\in\mathscr{G}}\left|\sum_{x_i\in G}(a_i-\alpha_i)\right|,$$

where $\alpha_1,\dots,\alpha_p\in[0,1]$ and $a_1,\dots,a_p\in\{0,1\}$.

Obviously

$$\delta(\mathscr{G})\leqslant 2\operatorname{lindis}(\mathscr{G}) \tag{34}$$

(the factor 2 on the right-hand side is technical, due to the translation from $[-1,1]$ to $[0,1]$).

We may think of $a_1,\dots,a_p=0$ or 1 as a choice of the set $B=\{x_i\in X: a_i=1\}\subset X$. When, for example, $\alpha_1=\cdots=\alpha_p=\alpha\in(0,1)$, we wish to find $B\subset X$ so that $\#(B\cap G)$ is roughly $\alpha\#G$ for all $G\in\mathscr{G}$. The value of lindis $(\mathscr{G})$ gives us a notion of how well this may be done.

Our two definitions are related by the following result.

Lemma 8.8. *For any finite family* $\mathscr{G}$,

$$\operatorname{lindis}(\mathscr{G})\leqslant\operatorname{herdis}(\mathscr{G}).$$

Remark. A particular case of Lemma 8.8 was proved and applied in Beck and Spencer (1984). This nice general formulation of Lemma 8.8 is due to Spencer.

Let $X,\mathscr{H},\mathscr{Y}$ and $\mathscr{Z}$ be as in Lemma 8.7. For any $A\subset X$, the restrictions $\mathscr{H}|_A,\mathscr{Y}|_A$ and $\mathscr{Z}|_A$ satisfy properties (i), (ii) and (iii) with the same m,d and k respectively. Hence, by Lemma 8.7,

$$\operatorname{herdis}(\mathscr{H}\cup\mathscr{Y})\leqslant\text{right-hand side of (32)}.$$

Applying Lemma 8.8, we have

$$\operatorname{lindis}(\mathscr{H}\cup\mathscr{Y})\leqslant\text{right-hand side of (32)}.$$

We therefore obtain the following particular case ($\alpha = \alpha_1 = \cdots = \alpha_p$) of Lemma 8.7.

Lemma 8.7′. *Let X, $\mathscr{H}$, $\mathscr{Y}$ and $\mathscr{X}$ be as in Lemma 8.7. Let $0 < \alpha < 1$. There is a function $g: X \to \{-\alpha, 1-\alpha\}$ such that for any $G = H \in \mathscr{H}$ and any $G = Y \in \mathscr{Y}$,*

$$\left|\sum_{x \in G} g(x)\right| \leqslant \textit{right-hand side of } (32).$$

The rest of this section is devoted to the proof of Lemmas 8.5–8.8.

Proof of Lemma 8.5. We shall actually prove the more general result (see (34))

$$\text{lindis}(\mathscr{Y}) < \deg(\mathscr{Y}). \tag{35}$$

Let $X = \{x_1, \ldots, x_p\}$, and let $\alpha_1, \ldots, \alpha_p$ be real numbers in the unit interval $U = [0, 1]$. The construction of the desired integers $a_k = 0$ or 1 ($1 \leqslant k \leqslant p$) will be based on the well-known result in linear algebra that if a system of homogeneous linear equations has more variables than equations, then there exists a non-trivial solution. We may assume that $0 < \alpha_k < 1$ ($1 \leqslant k \leqslant p$). Let $\boldsymbol{\alpha} = (\alpha_1, \alpha_2, \ldots, \alpha_p)$. Obviously $\boldsymbol{\alpha} \in U^p$. We shall define a sequence

$$\boldsymbol{\alpha}_0, \boldsymbol{\alpha}_1, \boldsymbol{\alpha}_2, \ldots, \boldsymbol{\alpha}_\nu = (\alpha_{1,\nu}, \alpha_{2,\nu}, \ldots, \alpha_{p,\nu}), \ldots$$

of vectors in U^p satisfying the following properties: Let

$$X_\nu = \{x_k \in X : \alpha_{k,\nu} \neq 0 \text{ and } \alpha_{k,\nu} \neq 1\}.$$

Then we require

$$X_{\nu+1} \subsetneqq X_\nu, \tag{36a}$$

$$\alpha_{k,\nu} = 0 \text{ or } 1 \Rightarrow \alpha_{k,\nu} = \alpha_{k,\nu+1}, \tag{36b}$$

and

$$\sum_{x_k \in Y} \alpha_{k,\nu} = \sum_{x_k \in Y} \alpha_{k,\nu+1} \tag{36c}$$

for all $Y \in \mathscr{Y}$ with $\#(Y \cap X_\nu) > \deg(\mathscr{Y})$. The construction of the sequence $\boldsymbol{\alpha}_\nu$ of vectors is by induction. Let $\boldsymbol{\alpha}_0 = \boldsymbol{\alpha}$. Suppose now that $\boldsymbol{\alpha}_\nu$ is defined and X_ν (see (36a)) is non-empty. Let

$$\mathscr{Y}_\nu = \{Y \in \mathscr{Y} : \#(Y \cap X_\nu) > \deg(\mathscr{Y})\}.$$

We claim that

$$\#\mathscr{Y}_\nu < \#X_\nu. \tag{37}$$

Indeed

$$(\#\mathscr{Y}_\nu)\deg(\mathscr{Y}) < \sum_{Y\in\mathscr{Y}_\nu} \#(Y\cap X_\nu) \leqslant (\#X_\nu)\deg(\mathscr{Y}_\nu) \leqslant (\#X_\nu)\deg(\mathscr{Y}).$$

Let us now associate a real variable y_k with each x_k $(1\leqslant k\leqslant p)$, and consider the system of linear equations

$$\sum_{x_k\in Y\cap X_\nu} y_k = 0$$

for all $Y\in\mathscr{Y}_\nu$, and where $y_k=0$ for all $x_k\in X\backslash X_\nu$. By (37), the number of equations is less than the number of variables. Therefore, a non-trivial solution $\mathbf{y}=(y_1,\ldots,y_p)$ exists. Now let t_0 be the largest positive value for which the inequalities

$$0\leqslant \alpha_{k,\nu}+t_0 y_k\leqslant 1 \quad (x_k\in X_\nu)$$

hold, and set

$$\alpha_{k,\nu+1}=\alpha_{k,\nu}+t_0 y_k \quad (1\leqslant k\leqslant p).$$

By the maximality of t_0, (36a) holds. Also, if $\alpha_{k,\nu}=0$ or 1, then $x_k\in X\backslash X_\nu$, so that $y_k=0$ and so (36b) holds. (36c) also follows easily. It now follows from (36a) that the sequence $\boldsymbol{\alpha}_0,\boldsymbol{\alpha}_1,\boldsymbol{\alpha}_2,\ldots$ certainly remains constant after a finite number of steps (s steps, say), Then $X_s=\varnothing$ and the vector $\boldsymbol{\alpha}_s$ has coordinates 0 and 1 only. Choosing $a_k=\alpha_{k,s}$ $(1\leqslant k\leqslant p)$, it follows from (36b) and (36c) that for all $Y\in\mathscr{Y}$,

$$\left|\sum_{x_k\in Y}(a_k-\alpha_k)\right| < \deg(\mathscr{Y}).$$

This proves (35), and Lemma 8.5 follows. ■

Proof of Lemma 8.6. Let $\mathscr{Y}=\bigcup_{i\geqslant 1}\mathscr{Y}^{(i)}$. We proceed along exactly the same lines as in the proof of Lemma 8.5. The only difference is that instead of (36c), we require

$$\sum_{x_k\in Y_j^{(i)}}\alpha_{k,\nu}=\sum_{x_k\in Y_j^{(i)}}\alpha_{k,\nu+1} \tag{36c*}$$

for all $Y_j^{(i)}$ with $\#(Y_j^{(i)}\cap X_\nu)\geqslant c_9(\varepsilon)i^{1+\varepsilon}$. The corresponding definition of $\mathscr{Y}_\nu$ is

$$\mathscr{Y}_\nu^*=\{Y_j^{(i)}: i\geqslant 1, j\geqslant 1 \text{ and } \#(Y_j^{(i)}\cap X_\nu)\geqslant c_9(\varepsilon)i^{1+\varepsilon}\}$$

accordingly. It therefore remains to check the analogue of (37), i.e.

$$\#\mathscr{Y}_\nu^* < \#X_\nu.$$

This is clear, since $Y_j^{(i)} \cap Y_k^{(i)} = \varnothing$ whenever $j \neq k$ and so we have

$$\#\mathcal{Y}_v^* = \sum_{i \geq 1} \#\{j : \#(Y_j^{(i)} \cap X_v) \geq c_9(\varepsilon) i^{1+\varepsilon}\} < \sum_{i=1}^{\infty} \frac{\#X_v}{c_9(\varepsilon) i^{1+\varepsilon}} = \#X_v$$

with

$$c_9(\varepsilon) = \sum_{i=1}^{\infty} \frac{1}{i^{1+\varepsilon}} < \infty. \qquad \blacksquare$$

Before we prove Lemma 8.7, we need the following very simple estimate.

Lemma 8.9. *Let $\xi_1, \xi_2, \ldots, \xi_n$ be independent random variables with common distribution function*

$$\text{Prob}(\xi_i = -1) = \text{Prob}(\xi_i = 1) = \tfrac{1}{2} \quad (1 \leq i \leq n).$$

Then for any $\gamma > 0$,

$$\text{Prob}\left(\left|\sum_{i=1}^{n} \xi_i\right| \geq \gamma\right) = 2^{-n} \sum_{|i - \frac{1}{2}n| \geq \frac{1}{2}\gamma} \binom{n}{i} \leq 2\mathrm{e}^{-\gamma^2/2n}.$$

Proof. We borrow the idea of the proof of Lemma 8.2. Let

$$S = \sum_{i=1}^{n} \xi_i.$$

Clearly

$$\text{Prob}(S \geq \gamma) = \text{Prob}(\mathrm{e}^{yS} \geq \mathrm{e}^{y\gamma}) \leq \mathrm{e}^{-y\gamma}\mathbb{E}(\mathrm{e}^{yS}),$$

where the parameter $y > 0$ will be fixed later and where $\mathbb{E}$ denotes the expected value. Since S is the sum of independent random variables, we have

$$\mathbb{E}(\mathrm{e}^{yS}) = \prod_{i=1}^{n} \mathbb{E}(\mathrm{e}^{y\xi_i}) = (\tfrac{1}{2}(\mathrm{e}^{y} + \mathrm{e}^{-y}))^n.$$

Using the elementary inequality

$$\tfrac{1}{2}(\mathrm{e}^{y} + \mathrm{e}^{-y}) \leq \mathrm{e}^{\frac{1}{2}y^2},$$

we conclude that

$$\text{Prob}(S \geq \gamma) \leq \mathrm{e}^{\frac{1}{2}ny^2 - y\gamma}.$$

Choosing $y = \gamma/n$, we obtain

$$\text{Prob}(S \geq \gamma) \leq \mathrm{e}^{-\gamma^2/2n}.$$

Since S has a symmetric distribution function, Lemma 8.9 follows. $\blacksquare$

The proof of Lemma 8.7 is based on iterated application of the following lemma.

Lemma 8.10. *Let X, $\mathscr{H}$, $\mathscr{Y}$ and $\mathscr{Z}$ be as in Lemma* 8.7. *Then there is a function $g:X \to \{-1,0,1\}$ such that*

$$\#\{x\in X: g(x)=0\} \leqslant \tfrac{9}{10}\#X, \tag{38}$$

and for any $H\in\mathscr{H}$ and $Y\in\mathscr{Y}$,

$$\left|\sum_{x\in H} g(x)\right| \ll (m\log(\#\mathscr{H}+2))^{\frac{1}{2}} \tag{39}$$

and

$$\left|\sum_{x\in Y} g(x)\right| \ll (kd\log(d+2)\log(\#\mathscr{Y}+2))^{\frac{1}{2}}. \tag{40}$$

Proof. By hypothesis (iii) in Lemma 8.7, any $Y_i\in\mathscr{Y}$ can be represented as a disjoint union

$$Y_i = Z_{i_1}\cup Z_{i_2}\cup\cdots\cup Z_{i_l}, \tag{41}$$

where $l = l(Y_i)\leqslant k$. Let $\mathscr{Z}=\mathscr{Z}^*\cup\mathscr{Z}^{**}$, where

$$\mathscr{Z}^* = \{Z\in\mathscr{Z}: \#Z < 100\, d\log(d+2)\}$$

and

$$\mathscr{Z}^{**} = \{Z\in\mathscr{Z}: \#Z \geqslant 100\, d\log(d+2)\}.$$

According to this splitting of $\mathscr{Z}$, let $Y_i = Y_i^*\cup Y_i^{**}$, where (see (41))

$$Y_i^* = \bigcup_{\substack{1\leqslant j\leqslant l\\ Z_{i_j}\in\mathscr{Z}^*}} Z_{i_j} \quad\text{and}\quad Y_i^{**} = \bigcup_{\substack{1\leqslant j\leqslant l\\ Z_{i_j}\in\mathscr{Z}^{**}}} Z_{i_j}.$$

Our aim is to find many 'partial 2-colourings' $g:X\to\{-1,0,1\}$ such that all $H\in\mathscr{H}$ and $Y_i^*(Y_i\in\mathscr{Y})$ have 'small error', and all $Y_i^{**}(Y_i\in\mathscr{Y})$ have 'zero error'. Let $\#X = N$. We may assume that N is sufficiently large. Denote by $\mathscr{F}$ the set of the 2^N 2-colourings $f:X\to\{-1,+1\}$. By Lemma 8.9, we have that for any fixed $A\subset X$ and $\gamma>0$,

$$\#\left\{f\in\mathscr{F}: \left|\sum_{x\in A} f(x)\right| \geqslant \gamma\right\} = 2^{N-\#A}\sum_{|i-\frac{1}{2}\#A|\geqslant\frac{1}{2}\gamma}\binom{\#A}{i} \leqslant 2^N 2\mathrm{e}^{-\gamma^2/2\#A}.$$

It follows from this that for a sufficiently large absolute constant c_{11}, the cardinality of the set

$$\mathscr{F}'_1 = \left\{f\in\mathscr{F}: \left|\sum_{x\in H} f(x)\right| \leqslant c_{11}(m\log(\#\mathscr{H}+2))^{\frac{1}{2}} \text{ for all } H\in\mathscr{H}\right\}$$

is greater than

$$2^N(1-2(\#\mathscr{H})(\#\mathscr{H}+2)^{-\frac{1}{2}c_{11}^2}) \geqslant \tfrac{3}{4}\cdot 2^N.$$

Similarly, since $\#Y_i^* < k(100\, d\log(d+2))$, we obtain that for a sufficiently large absolute constant c_{11}, the cardinality of the set

$$\mathscr{F}_1'' = \left\{ f \in \mathscr{F} : \left| \sum_{x \in Y_i^*} f(x) \right| \leqslant c_{11}(kd \log(d+2) \log(\#\mathscr{Y}+2))^{\frac{1}{2}} \text{ for all } Y_i \in \mathscr{Y} \right\}$$

is greater than

$$2^N(1 - 2(\#\mathscr{Y})(\#\mathscr{Y}+2)^{-c_{11}^2/200}) \geqslant \tfrac{3}{4} \cdot 2^N.$$

Hence the set $\mathscr{F}_1 = \mathscr{F}_1' \cap \mathscr{F}_1''$ has cardinality $\geqslant 2^{N-1}$. Let $\#\mathscr{Z}^{**} = M$ (note that $\mathscr{Z}$ is finite), and write

$$\mathscr{Z}^{**} = \{Z_1, Z_2, \ldots, Z_M\}.$$

Let us now associate with every 2-colouring $f \in \mathscr{F}$ the M-dimensional integer vector

$$\mathbf{v}(f) = (v_1(f), v_2(f), \ldots, v_M(f)),$$

where, for $i = 1, \ldots, M$,

$$v_i(f) = \sum_{x \in Z_i} f(x).$$

We are going to estimate the number of different vectors of the type $\mathbf{v}(f)$, where $f \in \mathscr{F}$. Since $|v_i(f)| \leqslant \#Z_i$ and $v_i(f) \equiv \#Z_i \pmod 2$, we have the trivial upper bound

$$\#\{\mathbf{v}(f) : f \in \mathscr{F}\} \leqslant \prod_{i=1}^{M} (\#Z_i + 1). \tag{42}$$

We shall use the elementary inequality that for any positive real numbers $b_1, b_2, \ldots, b_M$,

$$\prod_{i=1}^{M} b_i \leqslant \exp\left(\sum_{i=1}^{M} \frac{b_i}{\mathrm{e}} \right). \tag{43}$$

In view of (43), we have that

$$\prod_{i=1}^{M} (\#Z_i + 1) \leqslant \prod_{i=1}^{M} (2\#Z_i) = t^M \prod_{i=1}^{M} \left(\frac{2\#Z_i}{t} \right) \leqslant t^M \exp\left(2 \sum_{i=1}^{M} \frac{\#Z_i}{\mathrm{e}t} \right), \tag{44}$$

where the real parameter $t > 1$ will be fixed later. Clearly

$$\sum_{i=1}^{M} \#Z_i \leqslant \sum_{Z \in \mathscr{Z}} \#Z \leqslant \deg(\mathscr{Z})(\#X) \leqslant dN, \tag{45}$$

and so we have

$$M \leqslant \frac{1}{100\, d \log(d+2)} \sum_{i=1}^{M} \#Z_i \leqslant \frac{N}{100 \log(d+2)}. \tag{46}$$

Let $t = 10d$. Combining (42) and (44)–(46), we have

$$\#\{\mathbf{v}(f): f \in \mathscr{F}\} \leqslant \prod_{i=1}^{M} (\#Z_i + 1) < (10d)^{N/100 \log(d+2)} e^{2N/10e} < 2^{N/5}.$$

It follows by the pigeonhole principle that there is a subset $\mathscr{F}_2 \subset \mathscr{F}_1 = \mathscr{F}'_1 \cap \mathscr{F}''_1$ such that

$$\#\mathscr{F}_2 \geqslant \frac{\#\mathscr{F}_1}{\#\{\mathbf{v}(f): f \in \mathscr{F}\}} \geqslant 2^{4N/5-1}$$

and $\mathbf{v}(f_1) = \mathbf{v}(f_2)$ for all $f_1, f_2 \in \mathscr{F}_2$. Fixing an element $f_1 \in \mathscr{F}_2$, set

$$\mathscr{G}_2 = \{\tfrac{1}{2}(f - f_1): f \in \mathscr{F}_2\}.$$

Clearly

$$\sum_{x \in Y_i^{**}} g(x) = 0$$

for all $g \in \mathscr{G}_2$ and $Y_i \in \mathscr{Y}$. Summarizing, $\mathscr{G}_2$ is a set of partial 2-colourings $g: X \to \{-1, 0, 1\}$ such that

$$\#\mathscr{G}_2 \geqslant 2^{4N/5-1},$$

and that

$$\left| \sum_{x \in H} g(x) \right| < c_{11} (m \log(\#\mathscr{H} + 2))^{\frac{1}{2}}$$

for all $H \in \mathscr{H}$, and

$$\left| \sum_{x \in Y_i} g(x) \right| = \left| \sum_{x \in Y_i^{*}} g(x) \right| < c_{11} (kd \log(d+2) \log(\#\mathscr{Y} + 2))^{\frac{1}{2}}$$

for all $Y_i \in \mathscr{Y}$. Finally, let $\mathscr{G}_3$ denote the set of functions $g: X \to \{-1, 0, 1\}$ such that

$$\#\{x \in X: g(x) = 0\} \geqslant 9N/10.$$

Clearly

$$\#\mathscr{G}_3 = \sum_{i=0}^{[N/10]} \binom{N}{i} 2^i,$$

and so by Lemma 8.9,

$$\#\mathscr{G}_3 < 2^{N/10} \left(\sum_{i=0}^{[N/10]} \binom{N}{i} \right) \leqslant 2^{N/10} 2^{N+1} e^{-8N/25} < 2^{4N/5-1} \leqslant \#\mathscr{G}_2.$$

Hence the set-theoretic difference $\mathscr{G}_2 \backslash \mathscr{G}_3$ is non-empty, and any element $g \in \mathscr{G}_2 \backslash \mathscr{G}_3$ satisfies properties (38)–(40). Lemma 8.10 follows. ■

Proof of Lemma 8.7. Let $g_1: X \to \{-1, 0, 1\}$ be a partial 2-colouring of X satisfying (38)–(40). Let

$$X_1 = \{x \in X: g_1(x) = 0\}.$$

Let $\mathcal{H}_1, \mathcal{Y}_1$ and $\mathcal{Z}_1$ be the restriction of $\mathcal{H}, \mathcal{Y}$ and $\mathcal{Z}$ to X_1 respectively, i.e. $\mathcal{H}_1 = \{H \cap X_1 : H \in \mathcal{H}\}, \mathcal{Y}_1 = \{Y \cap X_1 : Y \in \mathcal{Y}\}$ and $\mathcal{Z}_1 = \{Z \cap X_1 : Z \in \mathcal{Z}\}$. Obviously $\mathcal{H}_1, \mathcal{Y}_1$ and $\mathcal{Z}_1$ satisfy properties (i), (ii) and (iii) in Lemma 8.7 with the same m, d and k respectively. It follows from Lemma 8.10 that there is a partial 2-colouring $g_2 : X_1 \to \{-1, 0, 1\}$ satisfying the analogues of (38)–(40). Repeating this argument, at the i-th step, we obtain a function $g_i : X_{i-1} \to \{-1, 0, 1\}$ such that for any $H \in \mathcal{H}_{i-1}$ and $Y \in \mathcal{Y}_{i-1}$,

$$\left| \sum_{x \in H} g_i(x) \right| \ll (m \log(\#\mathcal{H}_{i-1} + 2))^{\frac{1}{2}} \tag{47}$$

and

$$\left| \sum_{x \in Y} g_i(x) \right| \ll (kd \log(d+2) \log(\#\mathcal{Y}_{i-1} + 2))^{\frac{1}{2}}, \tag{48}$$

and such that the set

$$X_i = \{x \in X_{i-1} : g_i(x) = 0\}$$

satisfies

$$\#X_i \leqslant \tfrac{9}{10} \#X_{i-1}.$$

This procedure stops within

$$\left[\frac{\log N}{\log(10/9)} \right] + 1 \tag{49}$$

steps, since

$$\#X_i \leqslant (\tfrac{9}{10})^i \#X = (\tfrac{9}{10})^i N.$$

Suppose that $x \in X$. Then there is precisely one value i^* for which $g_{i^*}(x) \neq 0$, noting that $x \notin X_i$ for any $i \geqslant i^*$ and consequently $g_i(x)$ is not defined for any $i > i^*$. Let

$$f_0(x) = \sum_{i=1}^{i^*} g_i(x).$$

Clearly f_0 is a 2-colouring of X, i.e. f_0 has values ± 1 only. Lemma 8.7 now follows from (47)–(49). ∎

Remark. Note that the definition of f_0 in the proof of Lemma 8.7 can be taken to be

$$f_0 = \sum_{i \geqslant 1} g_i,$$

provided we have the convention that $g_i(x) = 0$ for all $x \in X \setminus X_{i-1}$ and that $X = X_0$.

Proof of Lemma 8.8. Fix values $\alpha_1, \ldots, \alpha_p \in [0, 1]$. Assume that all the α_i's have finite binary expansion, i.e. there is a natural number n so that $2^n \alpha_i \in \mathbb{Z}$ for all $i = 1, \ldots, p$. Let n be minimal with this property. Let $Y \subset X$ be the set of points $x_i \in X$ such that α_i has 1 as its n-th binary digit. As $\delta(\mathscr{G}_{|Y}) \leqslant$ herdis $(\mathscr{G})$, there exist $\varepsilon_i = \pm 1$ for all $x_i \in Y$ such that

$$\left|\sum_{x_i \in G \cap Y} \varepsilon_i\right| \leqslant \operatorname{herdis}(\mathscr{G})$$

for all $G \in \mathscr{G}$. Define approximations $\alpha_1^{(1)}, \alpha_2^{(1)}, \ldots, \alpha_p^{(1)}$ by

$$\alpha_i^{(1)} = \begin{cases} \alpha_i + \varepsilon_i 2^{-n} & (x_i \in Y), \\ \alpha_i & (x_i \in X \setminus Y). \end{cases}$$

For any $G \in \mathscr{G}$,

$$\left|\sum_{x_i \in G} (\alpha_i^{(1)} - \alpha_i)\right| = \left|\sum_{x_i \in G \cap Y} 2^{-n} \varepsilon_i\right| \leqslant 2^{-n} \operatorname{herdis}(\mathscr{G}).$$

The values $\alpha_i^{(1)}$ have binary expansions of length at most $(n-1)$. We repeat this procedure (note that Y will be a different set), giving $\alpha_i^{(2)}$ with

$$\left|\sum_{x_i \in G} (\alpha_i^{(2)} - \alpha_i^{(1)})\right| \leqslant 2^{-(n-1)} \operatorname{herdis}(\mathscr{G})$$

for all $G \in \mathscr{G}$. We apply this procedure n times, finally reaching $\alpha_i^{(n)}$ with binary expansions of length zero, i.e. $\alpha_i^{(n)} = 0$ or 1. These $\alpha_i^{(n)}$ are our desired integers a_i $(1 \leqslant i \leqslant p)$, since for all $G \in \mathscr{G}$,

$$\left|\sum_{x_i \in G} (\alpha_i^{(n)} - \alpha_i)\right| \leqslant \sum_{j=0}^{n-1} \left|\sum_{x_i \in G} (\alpha_i^{(j+1)} - \alpha_i^{(j)})\right|$$

$$\leqslant \sum_{j=0}^{n-1} 2^{-(n-j)} \operatorname{herdis}(\mathscr{G}) \leqslant \operatorname{herdis}(\mathscr{G}).$$

Finally, as this argument holds whenever $\alpha_1, \alpha_2, \ldots, \alpha_p$ have finite binary expansions, a compactness argument implies its truth for arbitrary $\alpha_1, \alpha_2, \ldots, \alpha_p \in [0, 1]$, and the proof of Lemma 8.8 is complete. ■

8.3 Applications of the combinatorial lemmas

In this section, we prove Theorems 11A (see §3.2), 20A, 20B and 20C (see Chapter 7).

To prove Theorem 11A, we apply a sequence of reduction steps. The first such step is summarized by the following lemma.

Lemma 8.11. *Let $A \subset \mathbb{R}^K$ be a Lebesgue measurable set with $\mu(A)=1$. For every $\eta>0$, there exists an integer $M=M(A,\eta)$ such that for some M-element subset $\mathcal{Q}$ of A,*

$$\left|\frac{1}{M}\sum_{\mathbf{q}\in\mathcal{Q}\cap B^{\infty}(\mathbf{x})} 1-\mu(A\cap B^{\infty}(\mathbf{x}))\right|<\eta \tag{50}$$

for all $\mathbf{x}\in\mathbb{R}^K$.

Proof. We use first the following elementary fact from Lebesgue's measure theory: There is a set $G\subset\mathbb{R}^K$ such that G is the union of a finite number of K-dimensional balls and $\mu(A\Delta G)<\eta/6$, where $A\Delta G$ denotes the symmetric difference of A and G. For any $m\in\mathbb{N}$ and $\mathbf{v}\in\mathbb{R}^K$, let

$$m^{-1}\mathbb{Z}^K+\mathbf{v}=\{m^{-1}\mathbf{l}+\mathbf{v}:\mathbf{l}\in\mathbb{Z}^K\}.$$

Since G is a 'nice' set, it is not difficult to see that there is an integer $m_0(G,\eta)$ such that for any $m\geqslant m_0(G,\eta)$ and any $\mathbf{v}\in\mathbb{R}^K$, the lattice $\Lambda=m^{-1}\mathbb{Z}^K+\mathbf{v}$ satisfies the inequality

$$\left|\frac{1}{m^K}\sum_{\mathbf{y}\in\Lambda\cap G\cap B^{\infty}(\mathbf{x})} 1-\mu(G\cap B^{\infty}(\mathbf{x}))\right|<\frac{\eta}{6} \tag{51}$$

for all $\mathbf{x}\in\mathbb{R}^K$. Let $D=A\Delta G$. Clearly

$$\int_{m^{-1}U_0^K}\#((m^{-1}\mathbb{Z}^K+\mathbf{v})\cap D)\,d\mathbf{v}=\mu(D)<\frac{\eta}{6}.$$

Hence there exists a vector $\mathbf{v}_0=\mathbf{v}_0(D,m)\in m^{-1}U_0^K$ such that

$$m^{-K}\#((m^{-1}\mathbb{Z}^K+\mathbf{v}_0)\cap D)<\eta/6. \tag{52}$$

Let $m_0=m_0(G,\eta)$, $\mathbf{v}_0=\mathbf{v}_0(D,m_0)$ and $\Lambda_0=m_0^{-1}\mathbb{Z}^K+\mathbf{v}_0$. Then it follows from (51) and (52) that for all $\mathbf{x}\in\mathbb{R}^K$.

$$\left|\frac{1}{m^K}\sum_{\mathbf{y}\in\Lambda_0\cap A\cap B^{\infty}(\mathbf{x})} 1-\mu(A\cap B^{\infty}(\mathbf{x}))\right|\leqslant\left|\frac{1}{m^K}\sum_{\mathbf{y}\in\Lambda_0\cap G\cap B^{\infty}(\mathbf{x})} 1-\mu(G\cap B^{\infty}(\mathbf{x}))\right|$$
$$+\frac{1}{m^K}\sum_{\mathbf{y}\in\Lambda_0\cap D} 1+\mu(D)<\frac{\eta}{6}+\frac{\eta}{6}+\frac{\eta}{6}=\frac{\eta}{2}. \tag{53}$$

On letting $x_1,x_2,\ldots,x_K\to+\infty$ in (53), where $\mathbf{x}=(x_1,\ldots,x_K)$, we have

$$\left|\frac{1}{m^K}\sum_{\mathbf{y}\in\Lambda_0\cap A} 1-1\right|\leqslant\frac{\eta}{2}. \tag{54}$$

Now let $\mathcal{Q}=\Lambda_0\cap A$ and $M=\#(\Lambda_0\cap A)$. Then we have, by (53) and (54), that for all $\mathbf{x}\in\mathbb{R}^K$,

$$\left|\frac{1}{M}\sum_{\mathbf{y}\in\mathcal{Q}\cap B^\infty(\mathbf{x})} 1 - \mu(A\cap B^\infty(\mathbf{x}))\right|$$

$$\leqslant \left|\frac{1}{m^K}\sum_{\mathbf{y}\in\mathcal{Q}\cap B^\infty(\mathbf{x})} 1 - \mu(A\cap B^\infty(\mathbf{x}))\right| + \left|\frac{1}{M} - \frac{1}{m^K}\right| \sum_{\mathbf{y}\in\mathcal{Q}\cap B^\infty(\mathbf{x})} 1$$

$$< \frac{\eta}{2} + \left|1 - \frac{M}{m^K}\right| \leqslant \frac{\eta}{2} + \frac{\eta}{2} = \eta$$

as required. ■

Choosing $\eta = 1/N$ in Lemma 8.11, we obtain an M-element subset $\mathcal{Q} = \{\mathbf{q}_1, \mathbf{q}_2, \ldots, \mathbf{q}_M\}$ of A such that

$$\left|\frac{N}{M}\sum_{\mathbf{q}_i\in B^\infty(\mathbf{x})} 1 - N\mu(A\cap B^\infty(\mathbf{x}))\right| < 1$$

for all $\mathbf{x}\in\mathbb{R}^K$. Note also that $M = M(A, N)$ can be arbitrarily large compared to N. We may therefore assume that $M \geqslant N$.

It is now clear that Theorem 11A is essentially equivalent to

Theorem 11B. *Let $\mathcal{Q} = \{\mathbf{q}_1, \mathbf{q}_2, \ldots, \mathbf{q}_M\}$ be a set of M points in $\mathbb{R}^K$ and let N be a natural number satisfying $2 \leqslant N \leqslant M$. Then there exists an N-element subset $\mathcal{P} = \{\mathbf{p}_1, \ldots, \mathbf{p}_N\}$ of $\mathcal{Q}$ such that for all $\mathbf{x}\in\mathbb{R}^K$,*

$$\left|\sum_{\mathbf{p}_i\in B^\infty(\mathbf{x})} 1 - \frac{N}{M}\sum_{\mathbf{q}_i\in B^\infty(\mathbf{x})} 1\right| \ll_{K,\varepsilon} (\log N)^{K+\frac{3}{2}+\varepsilon}.$$

Theorem 11B can be reformulated as follows.

Theorem 11C. *Let $(a_\mathbf{n})$ $(\mathbf{n}\in[1, M]^K\cap\mathbb{Z}^K)$ be a K-dimensional $0-1$ matrix (i.e. $a_\mathbf{n} = 0$ or 1) of size $M \times \cdots \times M$. Assume that*

$$\#\{\mathbf{n}\in[1, M]^K\cap\mathbb{Z}^K : a_\mathbf{n} = 1\} = M.$$

Suppose that $2 \leqslant N \leqslant M$. Then there exists a choice of $\varepsilon_\mathbf{n} = 0$ or 1 $(\mathbf{n}\in[1, M]^K\cap\mathbb{Z}^K)$ such that

$$\#\{\mathbf{n}\in[1, M]^K\cap\mathbb{Z}^K : \varepsilon_\mathbf{n} = 1\} = N$$

and

$$\left|\sum_{\mathbf{n}\leqslant\mathbf{m}} \varepsilon_\mathbf{n} a_\mathbf{n} - \frac{N}{M}\sum_{\mathbf{n}\leqslant\mathbf{m}} a_\mathbf{n}\right| \ll_{K,\varepsilon} (\log N)^{K+\frac{3}{2}+\varepsilon}$$

for all $\mathbf{m} = (m_1, \ldots, m_K)$ satisfying $1 \leqslant m_i \leqslant M$ $(1 \leqslant i \leqslant K)$. Here $\mathbf{n} \leqslant \mathbf{m}$ if and only if $n_i \leqslant m_i$ for all i satisfying $1 \leqslant i \leqslant K$.

We shall derive Theorem 11C from the following result.

Lemma 8.12. *Let $F=(b_{\mathbf{n}})$ $(\mathbf{n}\in[1,M]^K\cap\mathbb{Z}^K)$ be a K-dimensional $0-1$ matrix of size $M\times\cdots\times M$. Let $\#\{\mathbf{n}\in[1,M]^K\cap\mathbb{Z}^K:b_{\mathbf{n}}=1\}=L$. Suppose that $0<\alpha<1$. Then there is a function $h(\mathbf{n})=-\alpha$ or $(1-\alpha)$ $(\mathbf{n}\in[1,M]^K\cap\mathbb{Z}^K)$ such that*

$$\left|\sum_{\mathbf{n}\leqslant\mathbf{m}} h(\mathbf{n})b_{\mathbf{n}}\right| \ll_{K,\varepsilon} (\log L)^{K+\frac{3}{2}+\varepsilon}$$

for all $\mathbf{m}=(m_1,\ldots,m_K)$ satisfying $1\leqslant m_i\leqslant M\,(1\leqslant i\leqslant K)$.

Proof of Theorem 11C. Write

$$\frac{N}{M}=\alpha_1\alpha_2\cdots\alpha_t,$$

where $\alpha_1=\cdots=\alpha_{t-1}=\frac{1}{2}$ and $\frac{1}{2}\leqslant\alpha_t<1$. Applying Lemma 8.12 to the matrix E with $\alpha=\alpha_1$, we obtain a function $h_1(\mathbf{n})=-\alpha_1$ or $(1-\alpha_1)$ such that

$$\left|\sum_{\mathbf{n}\leqslant\mathbf{m}} h_1(\mathbf{n})a_{\mathbf{n}}\right| \ll_{K,\varepsilon} (\log M)^{K+\frac{3}{2}+\varepsilon}$$

for all $\mathbf{m}=(m_1,\ldots,m_K)$ satisfying $1\leqslant m_i\leqslant M$ $(1\leqslant i\leqslant K)$. Let $E^{(1)}=(a_{\mathbf{n}}^{(1)})$ $(\mathbf{n}\in[1,M]^K\cap\mathbb{Z}^K)$, where

$$a_{\mathbf{n}}^{(1)}=\begin{cases}a_{\mathbf{n}} & (h_1(\mathbf{n})=1-\alpha_1),\\ 0 & (\text{otherwise}),\end{cases}$$

and let

$$\#\{\mathbf{n}\in[1,M]^K\cap\mathbb{Z}^K:a_{\mathbf{n}}^{(1)}=1\}=M_1.$$

Applying Lemma 8.12 now to the matrix $E^{(1)}$ with $\alpha=\alpha_2$, we obtain a function $h_2(\mathbf{n})=-\alpha_2$ or $(1-\alpha_2)$ such that

$$\left|\sum_{\mathbf{n}\leqslant\mathbf{m}} h_2(\mathbf{n})a_{\mathbf{n}}^{(1)}\right| \ll_{K,\varepsilon} (\log M_1)^{K+\frac{3}{2}+\varepsilon}$$

for all $\mathbf{m}=(m_1,\ldots,m_K)$ satisfying $1\leqslant m_i\leqslant M\,(1\leqslant i\leqslant K)$. Let $E^{(2)}=(a_{\mathbf{n}}^{(2)})$ $(\mathbf{n}\in[1,M]^K\cap\mathbb{Z}^K)$, where

$$a_{\mathbf{n}}^{(2)}=\begin{cases}a_{\mathbf{n}}^{(1)} & (h_2(\mathbf{n})=1-\alpha_2),\\ 0 & (\text{otherwise}),\end{cases}$$

and let

$$\#\{\mathbf{n}\in[1,M]^K\cap\mathbb{Z}^K:a_{\mathbf{n}}^{(2)}=1\}=M_2.$$

Repeating this argument, we obtain a sequence of functions h_j, matrices $E^{(j)}$ and integers M_j $(1\leqslant j\leqslant t)$ such that $h_j(\mathbf{n})=-\alpha_j$ or $(1-\alpha_j)$, $E^{(j)}=(a_{\mathbf{n}}^{(j)})$ $(\mathbf{n}\in[1,M]^K\cap\mathbb{Z}^K)$, where

$$a_{\mathbf{n}}^{(j)}=\begin{cases}a_{\mathbf{n}}^{(j-1)} & (h_j(\mathbf{n})=1-\alpha_j),\\ 0 & (\text{otherwise}),\end{cases}$$

and where

$$\#\{\mathbf{n}\in[1,M]^K\cap\mathbb{Z}^K : a_{\mathbf{n}}^{(j)}=1\}=M_j,$$

and

$$\left|\sum_{\mathbf{n}\leqslant\mathbf{m}} h_j(\mathbf{n})a_{\mathbf{n}}^{(j-1)}\right| \ll_{K,\varepsilon} (\log M_{j-1})^{K+\frac{3}{2}+\varepsilon} \tag{55}$$

for all $\mathbf{m}=(m_1,\ldots,m_K)$ satisfying $1\leqslant m_i\leqslant M\,(1\leqslant i\leqslant K)$. Clearly

$$\sum_{\mathbf{n}\leqslant\mathbf{m}} h_j(\mathbf{n})a_{\mathbf{n}}^{(j-1)} = \sum_{\mathbf{n}\leqslant\mathbf{m}} (a_{\mathbf{n}}^{(j)}-\alpha_j a_{\mathbf{n}}^{(j-1)}).$$

Also we have (recall that $\alpha_1\alpha_2\cdots\alpha_t=N/M$)

$$\begin{aligned}\left|\sum_{\mathbf{n}\leqslant\mathbf{m}}\left(a_{\mathbf{n}}^{(t)}-\frac{N}{M}a_{\mathbf{n}}\right)\right| \leqslant{}& \left|\sum_{\mathbf{n}\leqslant\mathbf{m}}(a_{\mathbf{n}}^{(t)}-\alpha_t a_{\mathbf{n}}^{(t-1)})\right| \\ &+\alpha_t\left|\sum_{\mathbf{n}\leqslant\mathbf{m}}(a_{\mathbf{n}}^{(t-1)}-\alpha_{t-1}a_{\mathbf{n}}^{(t-2)})\right| \\ &+\alpha_t\alpha_{t-1}\left|\sum_{\mathbf{n}\leqslant\mathbf{m}} a_{\mathbf{n}}^{(t-2)}-\alpha_{t-2}a_{\mathbf{n}}^{(t-3)}\right|+\cdots \\ &+\alpha_t\alpha_{t-1}\cdots\alpha_2\left|\sum_{\mathbf{n}\leqslant\mathbf{m}}(a_{\mathbf{n}}^{(1)}-\alpha_1 a_{\mathbf{n}})\right|.\end{aligned} \tag{56}$$

Combining (55) and (56), we see that

$$\left|\sum_{\mathbf{n}\leqslant\mathbf{m}}\left(a_{\mathbf{n}}^{(t)}-\frac{N}{M}a_{\mathbf{n}}\right)\right| \ll_{K,\varepsilon} \sum_{j=1}^{t}\left(\prod_{\alpha=j+1}^{t}\alpha_v\right)(\log M_{j-1})^{K+\frac{3}{2}+\varepsilon} \tag{57}$$

for all $\mathbf{m}$. Choosing $\mathbf{m}=(M,\ldots,M)$ in (55), we obtain

$$|M_j-\alpha_j M_{j-1}| \ll_{K,\varepsilon} (\log M_{j-1})^{K+\frac{3}{2}+\varepsilon},$$

and so obviously

$$M_j \geqslant (\alpha_j-\tfrac{1}{10})M_{j-1} \tag{58}$$

if $M_j>c_{12}(K,\varepsilon)$. It follows from (57) and (58) that

$$\left|\sum_{\mathbf{n}\leqslant\mathbf{m}}\left(a_{\mathbf{n}}^{(t)}-\frac{N}{M}a_{\mathbf{n}}\right)\right| \ll_{K,\varepsilon} (\log M_t)^{K+\frac{3}{2}+\varepsilon} \tag{59}$$

for all $\mathbf{m}$. Choosing $\mathbf{m}=(M,\ldots,M)$ in (59), we have

$$|M_t-N| \ll_{K,\varepsilon} (\log M_t)^{K+\frac{3}{2}+\varepsilon}. \tag{60}$$

Let $J^{(1)}$ be a $(\min\{N,M_t\})$-element subset of

$$\{\mathbf{n}\in[1,M]^K\cap\mathbb{Z}^K : a_{\mathbf{n}}^{(t)}=1\},$$

and let $J^{(2)}$ be an $(N-\min\{N,M_t\})$-element subset of

$$\{\mathbf{n}\in[1,M]^K\cap\mathbb{Z}^K : a_{\mathbf{n}}=1 \text{ and } a_{\mathbf{n}}^{(t)}=0\}.$$

Finally, let $J=J^{(1)}\cup J^{(2)}$, and write

$$\varepsilon_{\mathbf{n}}=\begin{cases}1 & (\mathbf{n}\in J),\\ 0 & (\text{otherwise}).\end{cases}$$

Theorem 11C follows immediately from (59) and (60). ■

Proof of Lemma 8.12. Applying a trivial reduction of the size of the matrix F (for example, removing 0-entries), one can easily achieve $M\leqslant L$. On the other hand, for technical reasons, it is desirable that M is a power of 2. We may therefore assume that $M=2^q$, where q is an integer, and that $M<2L$. For $\mathbf{r}=(r_1,r_2,\ldots,r_K)\in\mathbb{Z}^K$ satisfying $0\leqslant r_i\leqslant q$ $(1\leqslant i\leqslant K)$, we partition F into $2^{r_1+r_2+\cdots+r_K}$ submatrices, splitting the i-th side of the matrix into 2^{r_i} equal pieces $(1\leqslant i\leqslant K)$. There are $(q+1)^K$ such partitions. Let us call a submatrix special if it occurs in one of these partitions. Observe that any submatrix of F of the type $(b_{\mathbf{n}})$ $(\mathbf{n}\leqslant\mathbf{m})$, where $\mathbf{m}\in\mathbb{Z}^K$, is the union of at most q^K disjoint special submatrices. We now apply Lemma 8.7′, where $X=\{\mathbf{n}\in[1,M]^K\cap\mathbb{Z}^K : b_{\mathbf{n}}=1\}$, $\mathscr{Y}$ is the family of all submatrices of the type $(b_{\mathbf{n}})$ $(\mathbf{n}\leqslant\mathbf{m})$, $\mathscr{Z}$ is the family of all special submatrices, $\mathscr{H}$ is the empty set, $d=(q+1)^K$ and $k=q^K$. Then there is a function $h(\mathbf{n})=-\alpha$ or $(1-\alpha)$ $(\mathbf{n}\in[1,M]^K\cap\mathbb{Z}^K)$ such that

$$\left|\sum_{\mathbf{n}\leqslant\mathbf{m}} h(\mathbf{n})b_{\mathbf{n}}\right| \ll_{K,\varepsilon} (\log L)^{K+\frac{3}{2}+\varepsilon}$$

for all $\mathbf{m}$, noting that $M<2L$. ■

We now investigate Theorem 20A. Following the idea of the proof of Theorem 10C, we first attempt to approximate an arbitrary polygon $A\in POL(S_1,\ldots,S_l)$ by triangles and parallelograms of certain special types.

Note that $K=2$ for the remainder of this chapter.

Consider a cartesian coordinate system on $\mathbb{R}^2$ with axes X_1 and X_2. Instead of studying the family $POL(S_1,\ldots,S_l)$, we shall study the slightly bigger family $POL(X_1,X_2,S_1,\ldots,S_l)$. In other words, we include the horizontal and vertical directions.

Definition. Let $\mathbf{n}=(n_1,n_2)\in\mathbb{Z}^2$ and $\mathbf{m}=(m_1,m_2)\in\mathbb{Z}^2$. By a special rectangle of order $\mathbf{n}$ and position $\mathbf{m}$, we mean the rectangle

$$\{\mathbf{x}=(x_1,x_2)\in\mathbb{R}^2 : m_1 2^{n_1}\leqslant x_1<(m_1+1)2^{n_1},\, m_2 2^{n_2}\leqslant x_2<(m_2+1)2^{n_2}\}. \quad (61)$$

Furthermore, we let $SR(\mathbf{n})$ denote the family of all special rectangles of order $\mathbf{n}$.

Definition. Let $i \in \{1, \ldots, l\}$. By a triangle of type i, we simply mean a triangle with sides parallel to X_1, X_2 and S_i.

For any $i \in \{1, \ldots, l\}$, let Δ_i be a triangle of type i. Let $t_i^{(1)}$ and $t_i^{(2)}$ denote the lengths of the sides of Δ_i parallel to X_1 and X_2 respectively, and write

$$\lambda_i = \frac{t_i^{(1)}}{t_i^{(2)}}.$$

Note that for a fixed value of i, the value of λ_i is independent of the choice of the triangle Δ_i.

For any $i \in \{1, \ldots, l\}$ and $n \in \mathbb{Z}$, let $\Lambda(i, n)$ denote the rectangular lattice of points

$$\mathbf{v}(i, n, \mathbf{m}) = (m_1 2^n \lambda_i^{\frac{1}{2}}, m_2 2^n \lambda_i^{-\frac{1}{2}}) \in \mathbb{R}^2,$$

where $\mathbf{m} = (m_1, m_2) \in \mathbb{Z}^2$. In other words, $\Lambda(i, n)$ is the lattice generated by $(2^n \lambda_i^{\frac{1}{2}}, 0)$ and $(0, 2^n \lambda_i^{-\frac{1}{2}})$.

For convenience of notation, let $\mathbf{e}_1 = (1, 0)$ and $\mathbf{e}_2 = (0, 1)$.

Definition. Let $i \in \{1, \ldots, l\}$ and $n \in \mathbb{Z}$. By a special triangle of type i and order n, we mean a triangle with vertices

$$\mathbf{v}(i, n, \mathbf{m}), \mathbf{v}(i, n, \mathbf{m} + \mathbf{e}_1), \mathbf{v}(i, n, \mathbf{m} + \mathbf{e}_2),$$

or a triangle with vertices

$$\mathbf{v}(i, n, \mathbf{m}), \mathbf{v}(i, n, \mathbf{m} - \mathbf{e}_1), \mathbf{v}(i, n, \mathbf{m} - \mathbf{e}_2),$$

where $\mathbf{m} \in \mathbb{Z}^2$. Furthermore, we let $ST(i, n)$ denote the family of all special triangles of type i and order n.

Definition. Let $i \in \{1, \ldots, l\}$ and $j \in \{1, 2\}$. By a parallelogram of type (i, j), we mean a parallelogram with sides parallel to S_i and X_j.

Consider the unit square U^2. For $i \in \{1, \ldots, l\}$, let L_i^* denote the linear transformation of determinant 1 represented in matrix notation by

$$L_i^* \begin{pmatrix} x_1 \\ x_2 \end{pmatrix} = \begin{pmatrix} \lambda_i^{\frac{1}{2}} & -\lambda_i^{\frac{1}{2}} \\ 0 & \lambda_i^{-\frac{1}{2}} \end{pmatrix} \begin{pmatrix} x_1 \\ x_2 \end{pmatrix}.$$

Let $P_i^* = \{L_i^*(\mathbf{x}) : \mathbf{x} \in U^2\}$. It is not difficult to see that P_i^* is the parallelogram with vertices

$$\mathbf{v}(i, 0, \mathbf{0}), \mathbf{v}(i, 0, \mathbf{e}_1), \mathbf{v}(i, 0, \mathbf{e}_2), \mathbf{v}(i, 0, \mathbf{e}_2 - \mathbf{e}_1).$$

Definition. Let $i \in \{1, \ldots, l\}$, $\mathbf{n} \in \mathbb{Z}^2$ and $\mathbf{m} \in \mathbb{Z}^2$. By the special parallelogram of type $(i, 1)$, order $\mathbf{n}$ and position $\mathbf{m}$, we simply mean the image of the special rectangle (61) under L_i^*.

Similarly, for $i \in \{1, \ldots, l\}$, let L_i^{**} denote the linear transformation of determinant 1 represented in matrix notation by

$$L_i^{**}\begin{pmatrix} x_1 \\ x_2 \end{pmatrix} = \begin{pmatrix} \lambda_i^{\frac{1}{2}} & 0 \\ -\lambda_i^{-\frac{1}{2}} & \lambda_i^{-\frac{1}{2}} \end{pmatrix}\begin{pmatrix} x_1 \\ x_2 \end{pmatrix}.$$

Let $P_i^{**} = \{L_i^{**}(\mathbf{x}) : \mathbf{x} \in U^2\}$. Then P_i^{**} is the parellelogram with vertices

$$\mathbf{v}(i, 0, \mathbf{0}), \mathbf{v}(i, 0, \mathbf{e}_1), \mathbf{v}(i, 0, \mathbf{e}_2), \mathbf{v}(i, 0, \mathbf{e}_1 - \mathbf{e}_2).$$

Definition. Let $i \in \{1, \ldots, l\}$, $\mathbf{n} \in \mathbb{Z}^2$ and $\mathbf{m} \in \mathbb{Z}^2$. By the special parallelogram of type $(i, 2)$, order $\mathbf{n}$ and position $\mathbf{m}$, we simply mean the image of the special rectangle (61) under L_i^{**}.

Definition. For any $i \in \{1, \ldots, l\}$, $j \in \{1, 2\}$ and $\mathbf{n} \in \mathbb{Z}^2$, we let $SP(i, j, \mathbf{n})$ denote the family of all special parallelograms of type (i, j) and order $\mathbf{n}$.

For convenience of notation, we sometimes write

$$SP(0, 0, \mathbf{n}) = SR(\mathbf{n}).$$

With the above definitions, we can now state the following geometric lemma on approximating the characteristic function of an arbitrary polygon by those of special triangles and special parallelograms.

Lemma 8.13. *Let $A \in POL(X_1, X_2, S_1, \ldots, S_l)$ be arbitrary. Then there exist special triangles $T'_1, \ldots, T'_m$ and $T''_1, \ldots, T''_M$ of types $\in \{1, \ldots, l\}$, special parallelograms $P'_1, \ldots, P'_n$ and $P''_1, \ldots, P''_N$ of types $\in \{(0,0)\} \cup \{(i,j) : 1 \leqslant i \leqslant l$ and $j = 1, 2\}$ and signs $\varepsilon'_1 = \pm 1, \ldots, \varepsilon'_m = \pm 1$, $\varepsilon''_1 = \pm 1, \ldots, \varepsilon''_M = \pm 1$, $\delta'_1 = \pm 1, \ldots, \delta'_n = \pm 1$, $\delta''_1 = \pm 1, \ldots, \delta''_N = \pm 1$ such that*

$$\sum_{\nu=1}^{m} \varepsilon'_\nu \chi_{T'_\nu} + \sum_{\beta=1}^{n} \delta'_\beta \chi_{P'_\beta} \leqslant \chi_A \leqslant \sum_{\nu=1}^{M} \varepsilon''_\nu \chi_{T''_\nu} + \sum_{\beta=1}^{N} \delta''_\beta \chi_{P''_\beta} \tag{62}$$

and

$$\sum_{\nu=1}^{M} \varepsilon''_\nu \mu(T''_\nu) + \sum_{\beta=1}^{N} \delta''_\beta \mu(P''_\beta) - \sum_{\nu=1}^{m} \varepsilon'_\nu \mu(T'_\nu) - \sum_{\beta=1}^{n} \delta'_\beta \mu(P'_\beta) \ll l \log(d(A) + 2).$$

Furthermore,

$$\max_{\nu, \beta} \{d(T'_\nu), d(P'_\beta), d(T''_\nu), d(P''_\beta)\} \ll d(A)$$

and the numbers m, M, n and N satisfy

$$\max\{m, M\} \ll l \log(d(A) + 2)$$

and

$$\max\{n, N\} \ll l(\log(d(A)+2))^3.$$

Remark. We shall actually prove Lemma 8.13 with $P''_1, \ldots, P''_4$ being special rectangles of the same order, $\delta''_1 = \cdots = \delta''_4 = +1$, $\delta''_5 = \cdots = \delta''_N = \varepsilon''_1 = \cdots = \varepsilon''_M = -1$ and $P''_5, \ldots, P''_N, T''_1, \ldots, T''_M$ pairwise disjoint.

We shall prove Lemma 8.13 in a series of lemmas, the first of which reduces the problem to investigating rectangles and triangles.

Lemma 8.14. *Every $A \in POL(X_1, X_2, S_1, \ldots, S_l)$ can be written in the form*

$$A = (P_1 \cup P_2 \cup P_3 \cup P_4) \setminus \left(\left(\bigcup_{\beta=1}^{q_1} R_\beta \right) \cup \left(\bigcup_{\nu=1}^{q_2} T_\nu \right) \right),$$

where

(i) *$P_1, \ldots, P_4$ are special rectangles of the same order and $d(P_\alpha) < 3d(A)$ for every $\alpha = 1, \ldots, 4$;*
(ii) *for every $\beta = 1, \ldots, q_1$, R_β is an aligned rectangle, i.e. a rectangle with sides parallel to the coordinate axes, and $d(R_\beta) < 5d(A)$;*
(iii) *for every $\nu = 1, \ldots, q_2$, T_ν is a triangle of type $\in \{1, \ldots, l\}$ and $d(T_\nu) \leqslant d(A)$;*
(iv) *$q_1 \leqslant 4l + 8$ and $q_2 \leqslant 4l + 6$; and*
(v) *R_β $(\beta = 1, \ldots, q_1)$ and T_ν $(\nu = 1, \ldots, q_2)$ are pairwise disjoint (in the sense that the intersection has measure zero).*

Proof. For $j = 1, 2$, let $A^{(j)}$ denote the projection of A onto the X_j-axis, and let $L^{(j)}$ denote the length of the interval $A^{(j)}$. Let $n_j \in \mathbb{Z}$ be chosen such that $2^{n_j - 1} < L^{(j)} \leqslant 2^{n_j}$. Clearly $A^{(j)}$ is contained in the union of at most two intervals of the type $[m_j 2^{n_j}, (m_j + 1)2^{n_j})$. Let $\mathbf{n} = (n_1, n_2)$. Then A is contained in the union of at most four special rectangles of order $\mathbf{n}$. Denote these rectangles by P_α $(\alpha = 1, \ldots, 4)$ with the convention that they may not be distinct, and note that

$$d(P_\alpha) = (2^{2n_1} + 2^{2n_2})^{\frac{1}{2}} < (4d^2 + 4d^2)^{\frac{1}{2}} < 3d,$$

where $d = d(A)$. Let $P = P_1 \cup \cdots \cup P_4$, and for $j = 1, 2$, let $P^{(j)}$ denote the projection of P onto the X_j-axis. Let q be the number of sides of A. Then $q \leqslant 2l + 4$, since A is convex. For every vertex $\mathbf{v}$ of A, we draw a straight line parallel to the X_1-axis and passing through $\mathbf{v}$. These lines give a decomposition of A into at most two triangles and $(q-3) \leqslant (2l+1)$ trapeziums. Let B denote one of these triangles or trapeziums, and for $j = 1, 2$, let $B^{(j)}$ denote the projection of B onto the X_j-axis. Clearly

$$B^{(1)} \times B^{(2)} = B \cup T' \cup T'',$$

where T' and T'' are disjoint triangles of types $\in\{1,\ldots,l\}$ and with diameters not exceeding $d(A)$. Furthermore,

$$P^{(1)}\times B^{(2)}=(B^{(1)}\times B^{(2)})\cup R'\cup R'',$$

where R' and R'' are disjoint aligned rectangles of diameter not exceeding $((4d)^2+d^2)^{\frac{1}{2}}$. Clearly $A\subset P^{(1)}\times A^{(2)}$, and $(P^{(1)}\times A^{(2)})\backslash A$ is a (pairwise disjoint) union of at most $(4l+6)$ triangles of types $\in\{1,\ldots,l\}$ and $(4l+6)$ aligned rectangles. The result follows from the observation that $P\backslash(P^{(1)}\times A^{(2)})$ is a union of at most two disjoint rectangles of diameter not exceeding $((4d)^2+(2d)^2)^{\frac{1}{2}}$. ■

Our next step is clearly to approximate these rectangles and triangles. We begin with rectangles.

Lemma 8.15. *Let R be an aligned rectangle.*

(*i*) *There exist mutually disjoint special rectangles $R'_1,\ldots,R'_s$, where $s\ll(\log(\mu(R)+2))^2$, such that*

$$\bigcup_{\beta=1}^{s}R'_\beta\subset R$$

and

$$\mu\left(R\backslash\left(\bigcup_{\beta=1}^{s}R'_\beta\right)\right)\leqslant 1.$$

(*ii*) *There exist mutually disjoint special rectangles $R''_1,\ldots,R''_4$, satisfying $\mu(R''_\beta)<4\mu(R)$ for $1\leqslant\beta\leqslant 4$, and mutually disjoint special rectangles $R''_5,\ldots,R''_t$, where $t\ll(\log(\mu(R)+2))^2$, such that*

$$R\subset(R''_1\cup R''_2\cup R''_3\cup R''_4)\backslash\left(\bigcup_{\beta=5}^{t}R''_\beta\right)$$

and

$$\mu\left(\left((R''_1\cup R''_2\cup R''_3\cup R''_4)\backslash\left(\bigcup_{\beta=5}^{t}R''_\beta\right)\right)\backslash R\right)\leqslant 1.$$

To prove Lemma 8.15, we need a simple 1-dimensional result. By a special interval, we mean an interval of the type $[m2^n,(m+1)2^n)$, where $m,n\in\mathbb{Z}$. Note that special rectangles are then simply the cartesian products of two special intervals.

Lemma 8.16. *Let $[a,b)\subset\mathbb{R}$ be an interval. For every natural number D, there exist special intervals $I_1,\ldots,I_D$ such that*

$$\bigcup_{\alpha=1}^{D}I_\alpha\subset[a,b)$$

and

$$\mu_0\left([a,b)\backslash\left(\bigcup_{\alpha=1}^{D} I_\alpha\right)\right) \leqslant \left(\frac{7}{8}\right)^D (b-a),$$

where μ_0 denotes the usual measure on $\mathbb{R}$.

Proof. Let I_1 denote a longest special interval in $[a,b)$. We define I_α for $\alpha \geqslant 2$ inductively such that (i) I_α is a longest special interval contained in $[a,b)\backslash(I_1 \cup \cdots \cup I_{\alpha-1})$; (ii) $I_1 \cup \cdots \cup I_\alpha$ is an interval; and (iii) if $[a,b)\backslash(I_1 \cup \cdots \cup I_{\alpha-1})$ is a union of two disjoint intervals, then I_α belongs to the longer of the two (any one if of equal length). Note, first of all, that $\mu_0(I_1) \geqslant (b-a)/4$. Indeed, let $n \in \mathbb{Z}$ satisfy $2^{n+1} \leqslant b-a < 2^{n+2}$. Then $2^n > (b-a)/4$ and there exists $m \in \mathbb{Z}$ such that $[m2^n, (m+1)2^n) \subset [a,b)$. A similar argument will give the inequality $\mu_0(I_\alpha) \geqslant \mu_0([a,b)\backslash(I_1 \cup \cdots \cup I_{\alpha-1}))/8$. The result follows easily. ∎

Proof of Lemma 8.15. Let $R = [a_1, b_1) \times [a_2, b_2)$. For $j = 1, 2$, we apply Lemma 8.16 to the interval $[a_j, b_j)$, and obtain special intervals $I_1^{(j)}, \ldots, I_{D_j}^{(j)}$ such that

$$\bigcup_{\alpha_j=1}^{D_j} I_{\alpha_j}^{(j)} \subset [a_j, b_j)$$

and

$$\mu_0([a_j, b_j)\backslash(I_1^{(j)} \cup \cdots \cup I_{D_j}^{(j)})) \leqslant \frac{b_j - a_j}{2\mu(R)},$$

and such that $D_j \ll \log(\mu(R)+2)$. Clearly the family of special rectangles

$$I_{\alpha_1}^{(1)} \times I_{\alpha_2}^{(2)} \quad (1 \leqslant \alpha_1 \leqslant D_1 \text{ and } 1 \leqslant \alpha_2 \leqslant D_2)$$

satisfies the requirements of (i). To prove (ii), note, first of all, that if n_j $(j = 1, 2)$ is the integer satisfying $2^{n_j - 1} < b_j - a_j \leqslant 2^{n_j}$, then for some $m_j \in \mathbb{Z}$,

$$[a_j, b_j) \subset [m_j 2^{n_j}, (m_j + 2)2^{n_j}).$$

Hence

$$\begin{aligned} R \subset \bigcup_{\varepsilon_1=0}^{1} \bigcup_{\varepsilon_2=0}^{1} & ([(m_1+\varepsilon_1)2^{n_1}, (m_1+\varepsilon_1+1)2^{n_1}) \\ & \times [(m_2+\varepsilon_2)2^{n_2}, (m_2+\varepsilon_2+1)2^{n_2})) = R_1'' \cup R_2'' \cup R_3'' \cup R_4'', \end{aligned}$$

say. Clearly $\mu(R_\beta'') < 4\mu(R)$ for $1 \leqslant \beta \leqslant 4$. The set $(R_1'' \cup R_2'' \cup R_3'' \cup R_4'')\backslash R$ is the disjoint union of at most four aligned rectangles. Applying (i) to each of these gives the desired result. ∎

Next, we deal with triangles of types $\in \{1, \ldots, l\}$.

Definition. Let $i \in \{1, \ldots, l\}$. By a nice triangle of type i, we mean a triangle which is the intersection of a special triangle T^* of type i and a half-plane with the boundary parallel to one of the sides of T^*.

Let $i \in \{1, \ldots, l\}$. Let $T_0 \subset T$ be the largest inscribed special triangle of type i. If we extend the edges of T_0 to the boundary of T, we see that T is the disjoint (in the sense of measure) union of T_0 and at most three trapeziums and three parallelograms. Each of these trapeziums is clearly the disjoint union of a nice triangle of type i and a parallelogram. Note that all the parallelograms are of types $(0, 0)$, $(i, 1)$ or $(i, 2)$. To summarize,

> T is the disjoint union of one special triangle of type i and at most three nice triangles of type i and six parallelograms of types $(0, 0)$, $(i, 1)$ or $(i, 2)$. (63)

We therefore have to deal with parallelograms of various types as well as nice triangles of type i.

Since special parallelograms of type (i, j) and order $\mathbf{n}$ are obtained from special rectangles of the same order by a linear transformation of determinant 1, we obtain the following analogue of Lemma 8.15.

Lemma 8.15′. *Let $i \in \{1, \ldots, l\}$ and $j \in \{1, 2\}$ be fixed, and let P be a parallelogram of type (i, j).*

(i) *There exist mutually disjoint special parallelograms $P'_1, \ldots, P'_s$ of type (i, j), where $s \ll (\log(\mu(P) + 2))^2$, such that*

$$\bigcup_{\beta=1}^{s} P'_\beta \subset P$$

and

$$\mu\left(P \setminus \left(\bigcup_{\beta=1}^{s} P'_\beta\right)\right) \leqslant 1.$$

(ii) *There exist mutually disjoint special parallelograms $P''_1, \ldots, P''_4$ of type (i, j), satisfying $\mu(P''_\beta) < 4\mu(P)$ for $1 \leqslant \beta \leqslant 4$, and mutually disjoint special parallelograms $P''_5, \ldots, P''_t$, where $t \ll (\log(\mu(P) + 2))^2$, such that*

$$P \subset (P''_1 \cup P''_2 \cup P''_3 \cup P''_4) \setminus \left(\bigcup_{\beta=5}^{t} P''_\beta\right)$$

and

$$\mu\left(\left((P''_1 \cup P''_2 \cup P''_3 \cup P''_4) \setminus \left(\bigcup_{\beta=5}^{t} P''_\beta\right)\right) \setminus P\right) \leqslant 1.$$

Our next step is to deal with nice triangles.

Lemma 8.17. *Let $i\in\{1,\ldots,l\}$ and let T be a nice triangle of type i.*

(i) *There exist mutually disjoint special triangles $T'_1,\ldots,T'_s$ of type i and parallelograms $P'_1,\ldots,P'_s$ of types $\in\{(i,1),\ (i,2),\ (0,0)\}$, where $s\ll\log(\mu(T)+2)$, such that*

$$\left(\bigcup_{\nu=1}^{s} T'_\nu\right)\cup\left(\bigcup_{\nu=1}^{s} P'_\nu\right)\subset T$$

and

$$\mu\left(T\backslash\left(\left(\bigcup_{\nu=1}^{s} T'_\nu\right)\cup\left(\bigcup_{\nu=1}^{s} P'_\nu\right)\right)\right)\leqslant 1.$$

(ii) *There exist a special triangle T''_0 of type i, satisfying $d(T''_0)<2d(T)$, and mutually disjoint special triangles $T''_1,\ldots,T''_t$ of type i and parallelograms $P''_1,\ldots,P''_q$ of types $\in\{(i,1),\ (i,2),\ (0,0)\}$, where* max *$\{t,q\}\ll\log(\mu(T)+2)$, such that*

$$T\subset T''_0\backslash\left(\left(\bigcup_{\nu=1}^{t} T''_\nu\right)\cup\left(\bigcup_{\nu=1}^{q} P''_\nu\right)\right)$$

and

$$\mu\left(\left(T''_0\backslash\left(\left(\bigcup_{\nu=1}^{t} T''_\nu\right)\cup\left(\bigcup_{\nu=1}^{q} P''_\nu\right)\right)\right)\backslash T\right)\leqslant 1.$$

The proof of Lemma 8.17 depends on the following lemma.

Lemma 8.18. *Let $i\in\{1,\ldots,l\}$ and let T be a nice triangle of type i. For every natural number D, there exist mutually disjoint special triangles $T_1,\ldots,T_D$ of type i and parallelograms $P_1,\ldots,P_D$ of types $\in\{(i,1),\ (i,2),\ (0,0)\}$ such that*

$$\left(\bigcup_{\nu=1}^{D} T_\nu\right)\cup\left(\bigcup_{\nu=1}^{D} P_\nu\right)\subset T \tag{64}$$

and

$$\mu\left(T\backslash\left(\left(\bigcup_{\nu=1}^{D} T_\nu\right)\cup\left(\bigcup_{\nu=1}^{D} P_\nu\right)\right)\right)\leqslant 4^{-D}\mu(T). \tag{65}$$

Proof. Since T is a nice triangle of type i, T is the intersection of a special triangle T^* of type i and a half-plane H with the boundary parallel to one of the sides of T. Let $\mathbf{b}'$ and $\mathbf{b}''$ denote the vertices of T on the boundary of H, and let $\mathbf{c}$ denote the third vertex of T. Let $T_1\subset T$ be the largest inscribed special triangle of type i. Observe that $\mathbf{c}$ is a vertex of T_1 and $\mu(T_1)\geqslant\mu(T)/4$. Let $\mathbf{b}'_1$ and $\mathbf{b}''_1$ denote the other two vertices of T_1. Then the trapezium with vertices $\mathbf{b}',\mathbf{b}'',\mathbf{b}'_1,\mathbf{b}''_1$ is clearly the disjoint union of a nice triangle T'_1 and a parallelogram P_1 of type $\in\{(i,1),(i,2),(0,0)\}$. Note that $\mu(T'_1)\leqslant\mu(T)/4$. We can now repeat the argument to T'_1 and obtain a special triangle T_2, parallelogram P_2 and nice triangle T'_2, mutually disjoint and such that $T'_1 = T_2 \cup P_2 \cup T'_2$ and $\mu(T'_2) \leqslant \mu(T'_1)/4$. After D steps, we obtain the desired result. ■

Proof of Lemma 8.17. (i) follows immediately from Lemma 8.18. To prove (ii), let T''_0 denote the smallest special triangle of type i such that T is the intersection of T''_0 with some half-plane H. Then clearly $d(T''_0) < 2d(T)$. Moreover, the set $T''_0 \setminus T$ is the disjoint union of a nice triangle of type i and a parallelogram of type $\in \{(i,1),(i,2),(0,0)\}$. The result follows on applying (i) to this latter nice triangle. ■

Lemma 8.13 now follows on combining (63) and Lemmas 8.14, 8.15, 8.15′ and 8.17.

While Lemma 8.13 gives an approximation to the characteristic function of an arbitrary polygon by those of special triangles and special parallelograms, the next lemma enables us to use the information to investigate the discrepancy function of the arbitrary polygon in question.

We shall state the lemma in a general form. Recall that given any discrete subset $\mathscr{P} \subset \mathbb{R}^2$ and any compact subset $B \subset \mathbb{R}^2$,

$$\mathscr{D}[\mathscr{P}; B] = \sum_{\mathbf{p} \in B \cap \mathscr{P}} 1 - \mu(B).$$

Lemma 8.19. *Suppose that $A, B'_1, \ldots, B'_q, B''_1, \ldots, B''_r$ are compact subsets of $\mathbb{R}^2$ and $\varepsilon'_1 = \pm 1, \ldots, \varepsilon'_q = \pm 1$, $\varepsilon''_1 = \pm 1, \ldots, \varepsilon''_r = \pm 1$ are 'signs' such that*

$$\sum_{\alpha=1}^{q} \varepsilon'_\alpha \chi_{B'_\alpha} \leqslant \chi_A \leqslant \sum_{\tau=1}^{r} \varepsilon''_\tau \chi_{B''_\tau}$$

and

$$\sum_{\tau=1}^{r} \varepsilon''_\tau \mu(B''_\tau) - \sum_{\alpha=1}^{q} \varepsilon'_\alpha \mu(B'_\alpha) \leqslant D_1.$$

Suppose further that $\mathscr{P}$ is a discrete subset of $\mathbb{R}^2$ such that

$$\max_{\alpha,\tau} \{|\mathscr{D}[\mathscr{P}; B'_\alpha]|, |\mathscr{D}[\mathscr{P}; B''_\tau]|\} \leqslant D_2.$$

Then

$$|\mathscr{D}[\mathscr{P}; A]| \leqslant D_1 + D_2 \max\{q, r\}.$$

Proof. We have

$$\mathscr{D}[\mathscr{P}; A] = \sum_{\mathbf{p} \in A \cap \mathscr{P}} 1 - \mu(A) \leqslant \sum_{\tau=1}^{r} \varepsilon''_\tau \sum_{\mathbf{p} \in B''_\tau \cap \mathscr{P}} 1 - \mu(A)$$

$$= \sum_{\tau=1}^{r} \varepsilon''_\tau \left(\sum_{\mathbf{p} \in B''_\tau \cap \mathscr{P}} 1 - \mu(B''_\tau) \right) + \left(\sum_{\tau=1}^{r} \varepsilon''_\tau \mu(B''_\tau) - \mu(A) \right)$$

$$= \sum_{\tau=1}^{r} \varepsilon_\tau'' \mathscr{D}[\mathscr{P}; B_\tau''] + \left(\sum_{\tau=1}^{r} \varepsilon_\tau'' \mu(B_\tau'') - \mu(A) \right)$$

$$\leqslant \sum_{\tau=1}^{r} |\mathscr{D}[\mathscr{P}; B_\tau'']| + \left(\sum_{\tau=1}^{r} \varepsilon_\tau'' \mu(B_\tau'') - \sum_{\alpha=1}^{q} \varepsilon_\alpha' \mu(B_\alpha') \right)$$

$$\leqslant D_2 r + D_1.$$

A similar argument gives

$$-\mathscr{D}[\mathscr{P}; A] \leqslant D_2 q + D_1.$$

The lemma follows on combining these two inequalities. ∎

For convenience of notation, let

$$SPEC(X_1, X_2, S_1, \ldots, S_l) = \left(\bigcup_{\substack{1 \leqslant i \leqslant l \\ n \in \mathbb{Z}}} ST(i, n) \right)$$

$$\cup \left(\bigcup_{\substack{1 \leqslant i \leqslant l \\ 1 \leqslant j \leqslant 2 \\ \mathbf{n} \in \mathbb{Z}^2}} SP(i, j, \mathbf{n}) \right) \cup \left(\bigcup_{\mathbf{n} \in \mathbb{Z}^2} SR(\mathbf{n}) \right).$$

In other words, $SPEC(X_1, X_2, S_1, \ldots, S_l)$ denotes the big family of all special triangles, special parallelograms and special rectangles defined in this section.

Lemma 8.20. *Let $\mathscr{P} \subset \mathbb{R}^2$ be a finite set, and let $\alpha \in [0, 1]$ be fixed. There is a function $f: \mathscr{P} \to \{-\alpha, 1-\alpha\}$ such that for every polygon $B \in SPEC(X_1, X_2, S_1, \ldots, S_l)$ satisfying $d(B) \geqslant 1$, we have*

$$\left| \sum_{p \in B \cap \mathscr{P}} f(p) \right| \ll_\varepsilon (l(\log(d(B) + 2))^2)^{1+\varepsilon}.$$

Proof. The proof is based on an application of Lemma 8.6 with $X = \mathscr{P}$. We therefore have to introduce a sequence $\mathscr{Y}^{(1)}, \mathscr{Y}^{(2)}, \mathscr{Y}^{(3)}, \ldots$ of partitions of $\mathscr{P}$. Let

$$\begin{aligned} SET(X_1, X_2, S_1, \ldots, S_l) = {} & \{ST(i, n): 1 \leqslant i \leqslant l \text{ and } n \in \mathbb{Z}\} \\ & \cup \{SP(i, j, \mathbf{n}): 1 \leqslant i \leqslant l, 1 \leqslant j \leqslant 2 \text{ and } \mathbf{n} \in \mathbb{Z}^2\} \\ & \cup \{SR(\mathbf{n}): \mathbf{n} \in \mathbb{Z}^2\}. \end{aligned}$$

For every $C \in SET(X_1, X_2, S_1, \ldots, S_l)$, let $d(C)$ denote the common diameter of all the elements of C. We now define a linear ordering on the subset

$$\{C \in SET(X_1, X_2, S_1, \ldots, S_l): d(C) \geqslant 1\}$$

according to the size of $d(C)$ with the convention that in the case of equal diameters, the ordering is defined arbitrarily. Observe that for any positive real number y,

$$\begin{aligned}
&\#\{C \in SET(X_1, X_2, S_1, \ldots, S_l) : 1 \leqslant d(C) \leqslant y\} \\
&\quad = \sum_{i=1}^{l} \#\{n \in \mathbb{Z} : 1 \leqslant d(ST(i,n)) \leqslant y\} \\
&\quad\quad + \sum_{i=1}^{l} \sum_{j=1}^{2} \#\{\mathbf{n} \in \mathbb{Z}^2 : 1 \leqslant d(SP(i,j,\mathbf{n})) \leqslant y\} \\
&\quad\quad + \#\{\mathbf{n} \in \mathbb{Z}^2 : 1 \leqslant d(SR(\mathbf{n})) \leqslant y\} \\
&\quad \ll l \log(y+2) + l(\log(y+2))^2 \ll l(\log(y+2))^2. \qquad (66)
\end{aligned}$$

Let $\mathscr{P}$ be fixed. We now let $\mathscr{Y}^{(1)}, \mathscr{Y}^{(2)}, \mathscr{Y}^{(3)}, \ldots$ be the partitions of $\mathscr{P}$ defined by the families in $\{C \in SET(X_1, X_2, S_1, \ldots, S_l) : 1 \leqslant d(C) \leqslant d(B)\}$ ordered in the way described earlier. The conclusion of Lemma 8.20 now follows from Lemma 8.6 and (66). ■

Let $\kappa = 2^k$, where $k \geqslant 1$ is an integer, and consider the square $[-\kappa, \kappa)^2$. Let

$$\mathscr{P} = \{(a/\kappa, b/\kappa) : a, b \in \mathbb{Z} \text{ and } -\kappa^2 \leqslant a, b < \kappa^2\}.$$

Clearly $\#\mathscr{P} = 4\kappa^4$. Let $\alpha = \kappa^{-2}$. Then $\alpha\#\mathscr{P} = 4\kappa^2$, i.e. the expected number of points of the desired set $\mathscr{Q}$ in $[-\kappa, \kappa)^2$.

By Lemma 8.20, there exists a function $f : \mathscr{P} \to \{-\alpha, 1-\alpha\}$ such that for all polygons $B \in SPEC(X_1, X_2, S_1, \ldots, S_l)$ satisfying $B \subset [-\kappa, \kappa)^2$ and $d(B) \geqslant 1$, we have

$$\left| \sum_{\mathbf{p} \in B \cap \mathscr{P}} f(\mathbf{p}) \right| \ll_{\varepsilon} (l(\log(d(B)+2))^2)^{1+\varepsilon}. \qquad (67)$$

Let $\mathscr{P}_k = \{\mathbf{p} \in \mathscr{P} : f(\mathbf{p}) = 1 - \alpha\}$. Clearly

$$\sum_{\mathbf{p} \in B \cap \mathscr{P}} f(\mathbf{p}) = \sum_{\mathbf{p} \in B \cap \mathscr{P}_k} 1 - \kappa^{-2} \sum_{\mathbf{p} \in B \cap \mathscr{P}} 1. \qquad (68)$$

On the other hand, trivial arguments show that for any convex $B \subset [-\kappa, \kappa)^2$,

$$\left| \sum_{\mathbf{p} \in B \cap \mathscr{P}} 1 - \kappa^2 \mu(B) \right| \ll \kappa \sigma(\partial B) \ll \kappa^2, \qquad (69)$$

where $\sigma(\partial B)$ denotes the length of the perimeter of B. Combining (67)–(69), we have that

$$|\mathscr{D}[\mathscr{P}_k;B]| = \left|\sum_{\mathbf{p}\in B\cap\mathscr{P}_k} 1 - \mu(B)\right| \leqslant \left|\sum_{\mathbf{p}\in B\cap\mathscr{P}_k} 1 - \kappa^{-2}\sum_{\mathbf{p}\in B\cap\mathscr{P}} 1\right|$$
$$+\left|\kappa^{-2}\sum_{\mathbf{p}\in B\cap\mathscr{P}} 1 - \mu(B)\right| \ll_\varepsilon (l(\log(d(B)+2))^2)^{1+\varepsilon}$$

for all $B\in SPEC(X_1,X_2,S_1,\ldots,S_l)$ satisfying $B\subset[-\kappa,\kappa)^2$ and $d(B)\geqslant 1$.

Suppose now that $B\in SPEC(X_1,X_2,S_1,\ldots,S_l)$ satisfies $B\subset[-\kappa,\kappa)^2$ and $d(B)<1$. Then $B\subset B_0$ for some $B_0\in SPEC(X_1,X_2,S_1,\ldots,S_l)$ with $1\leqslant d(B_0)<2$. Then applying (67) and (68) to B_0, we have

$$\sum_{\mathbf{p}\in B_0\cap\mathscr{P}_k} 1 = \kappa^{-2}\sum_{\mathbf{p}\in B_0\cap\mathscr{P}} 1 + \sum_{\mathbf{p}\in B_0\cap\mathscr{P}} f(\mathbf{p}) \leqslant 4 + \sum_{\mathbf{p}\in B_0\cap\mathscr{P}} f(\mathbf{p}) \ll_\varepsilon l^{1+\varepsilon},$$

noting that $\mu(B_0)\leqslant(d(B_0))^2<4$. It follows that

$$\sum_{\mathbf{p}\in B\cap\mathscr{P}_k} 1 \leqslant \sum_{\mathbf{p}\in B_0\cap\mathscr{P}_k} 1 \ll_\varepsilon (l(\log(d(B)+2))^2)^{1+\varepsilon}.$$

Using $\mu(B)<\mu(B_0)<4$, we have

$$|\mathscr{D}[\mathscr{P}_k;B]| \ll_\varepsilon (l(\log(d(B)+2))^2)^{1+\varepsilon}. \tag{70}$$

It follows that (70) holds for all $B\in SPEC(X_1,X_2,S_1,\ldots,S_l)$ satisfying $B\subset[-\kappa,\kappa)^2$. Combining this with Lemmas 8.13 and 8.19, we have that

$$|\mathscr{D}[\mathscr{P}_k;C]| \ll_\varepsilon l^{2+\varepsilon}(\log(d(C)+2))^{5+\varepsilon} \tag{71}$$

for all $C\in POL(X_1,X_2,S_1,\ldots,S_l)$ satisfying $C\subset[-\kappa,\kappa)^2$.

We are now in a position to construct the set $\mathscr{Q}$ in terms of the sets $\mathscr{P}_k$. Observe that

$$\bigcup_{n\in\mathbb{N}}([-2^{2^n},2^{2^n})^2\setminus[-2^{2^{n-1}},2^{2^{n-1}})^2) = \mathbb{R}^2\setminus[-2,2)^2,$$

and that any set in this union is the disjoint union of four aligned rectangles. Let

$$\mathscr{Q} = \mathscr{P}_1 \cup \left(\bigcup_{\substack{k=2^n\\ n\in\mathbb{N}}} (\mathscr{P}_k\cap([-2^k,2^k)^2\setminus[-2^{\frac{1}{2}k},2^{\frac{1}{2}k})^2))\right).$$

Consider any arbitrary $A\in POL(X_1,X_2,S_1,\ldots,S_l)$. The intersection

$$A_k = A\cap([-2^k,2^k)^2\setminus[-2^{\frac{1}{2}k},2^{\frac{1}{2}k})^2)$$

is the disjoint union of at most four sets in $POL(X_1,X_2,S_1,\ldots,S_l)$. It follows from (71) that

$$|\mathscr{D}[\mathscr{Q};A]| = \left|\sum_{\mathbf{q}\in A\cap\mathscr{Q}} 1 - \mu(A)\right|$$

$$= \left|\#(A\cap\mathscr{P}_1) - \mu(A\cap[-2,2)^2) + \sum_{\substack{k=2^n\\ n\in\mathbb{N}}} (\#(A_k\cap\mathscr{P}_k) - \mu(A_k))\right|$$

$$\ll_\varepsilon \sum{}^* l^{2+\varepsilon}(\min\{\log(d(A)+2), k\})^{5+\varepsilon}, \tag{72}$$

where the summation $\sum^*$ is extended over all $k = 2^n$, where $n\in\mathbb{N}$, for which A_k is non-empty. Simple calculation gives

$$\sum{}^*(\min\{\log(d(A)+2), k\})^{5+\varepsilon} \ll (\log(d(A)+2))^{5+\varepsilon}. \tag{73}$$

Theorem 20A now follows from (72) and (73).

Finally, we prove Theorems 20B and 20C. The proofs of both are based on the following lemma. We recall that $r(A)$ denotes the radius of the largest inscribed circle of the convex region A.

Lemma 8.21. *Let $A\subset\mathbb{R}^2$ be a compact convex region with $r(A)\geqslant 1$. Let $M\geqslant 2$ be an integer and let $N = M^2$. Then there exists an N-element set $S = S(A,N) = \{\mathbf{z}_1, \mathbf{z}_2, \ldots, \mathbf{z}_N\}\subset[0,M)^2$ such that for any pair $(\lambda, \mathbf{v})$, where $-1\leqslant\lambda\leqslant 1$ and $\mathbf{v}\in\mathbb{R}^2$, we have*

$$\left|\sum_{\mathbf{z}_i\in A(\lambda,\mathbf{v})} 1 - \mu(A(\lambda,\mathbf{v})\cap[0,M)^2)\right| \ll_\varepsilon \xi(A)(\max\{\log M, \log(d(A)+2)\})^{4+\varepsilon}.$$

We first derive Theorems 20B and 20C. Applying a certain linear transformation of determinant 1, we may assume in Theorem 20B that $r(A)\gg d(A)$. Let $M = [d(A)]+1$, and extend $S(A,M^2)$ periodically modulo $[0,M)^2$ over the whole plane. This gives the desired point distribution in Theorem 20B. The deduction of Theorem 20C is even simpler. Contract the large square $[0,M)^2$ to the unit square. The result follows immediately.

It remains to prove Lemma 8.21.

Let $q = \xi(A)\geqslant 3$. There is an inscribed polygon $A_q\subset A$ of q sides such that $\mu(A\backslash A_q)\leqslant q^2$. For technical reasons, we prefer to work with circumscribed polygons, and shall use the following elementary result: For any compact convex region $B\subset\mathbb{R}^2$, there is a circumscribed polygon $B^{(s)}\supset B$ of s sides, where $s\leqslant c_{13}$ and

$$\mu(B^{(s)}\backslash B)\leqslant\mu(B). \tag{74}$$

The difference $A\backslash A_q$ is the disjoint union of q convex sets ('segments'). Let

$$A\backslash A_q = B_1 \cup B_2 \cup \cdots \cup B_q. \tag{75}$$

Applying (74) to each B_i $(1 \leqslant i \leqslant q)$ in (75), we obtain a circumscribed polygon $A^{(l)} \supset A$ of $l \ll q = \xi(A)$ sides such that

$$\mu(A^{(l)}\backslash A) \leqslant \mu(A\backslash A_q) \leqslant q^2.$$

Let $S_1, S_2, \ldots, S_l$ denote the directions of the sides of $A^{(l)}$. Then for any $A_0 = A(\lambda_0, \mathbf{v}_0)$, where $\lambda_0 \in \mathbb{R}$ and $\mathbf{v}_0 \in \mathbb{R}^2$, one can find a circumscribed polygon $A_0^* \in POL(S_1, \ldots, S_l)$ such that

$$\mu(A_0^*\backslash A_0) \leqslant \lambda_0^2 q^2. \tag{76}$$

Let

$$\mathscr{A} = \{A(\lambda, \mathbf{v}): -2 \leqslant \lambda \leqslant 2, \mathbf{v} \in \mathbb{R}^2\}.$$

Since $r(A) \geqslant 1$, it is easy to find a finite subset $\mathscr{A}_0$ of $\mathscr{A}$ and a positive absolute constant c_{14} such that

$$\#\mathscr{A}_0 \ll \max\{M^{c_{14}}, (d(A)+2)^{c_{14}}\}; \tag{77}$$

and that for any $A(\lambda, \mathbf{v})$, satisfying $-1 \leqslant \lambda \leqslant 1$ and $\mathbf{v} \in \mathbb{R}^2$, there are two elements A' and A'' of $\mathscr{A}_0$ such that

$$A' \cap [0, M)^2 \subset A(\lambda, \mathbf{v}) \cap [0, M)^2 \subset A'' \cap [0, M)^2 \tag{78a}$$

and

$$\mu(A''\backslash A') \leqslant 1. \tag{78b}$$

Let

$$\mathscr{A}_1 = \mathscr{A}_0 \cup \{[0, M)^2\}.$$

We now associate with every $A_1 \in \mathscr{A}_1$ a circumscribed polygon $A_1^* \in POL(X_1, X_2, S_1, \ldots, S_l)$ such that (see (76))

$$\mu(A_1^*\backslash A_1) \leqslant 4q^2 = 4(\xi(A))^2. \tag{79}$$

Let

$$POL(\mathscr{A}_1) = \{A_1^*: A_1 \in \mathscr{A}_1\}.$$

We recall that $SPEC(X_1, X_2, S_1, \ldots, S_l)$ denotes the family of all special triangles and special parallelograms defined earlier in this section, and note that $S_1, \ldots, S_l$ denote the directions of the sides of $A^{(l)}$. For every integer $k \geqslant 1$, let $\mathscr{B}(k)$ denote the family of sets B which can be represented as the union of at most k pairwise disjoint elements of $SPEC(X_1, X_2, S_1, \ldots, S_l)$. Let

$$k = [c_{15} l (\log(d(A)+2))^3],$$

and let $A_1^* \in POL(\mathscr{A}_1)$ be arbitrary. If c_{15} is a sufficiently large constant, then by the remark after Lemma 8.13, there exists a set A_1^{**} such that $A_1^{**} = B''\backslash B'$ for some $B', B'' \in \mathscr{B}(k)$, $A_1^{**} \supset A_1^*$ and

$$\mu(A_1^{**}\backslash A_1^*) \ll l\log(d(A)+2) \ll \xi(A)\log(d(A)+2). \tag{80}$$

Let us associate with every $A_1 \in \mathscr{A}_1$ two elements $B'(A_1)$ and $B''(A_1)$ of $\mathscr{B}(k)$ such that these properties are satisfied with $A^{**} = B''(A_1)\backslash B'(A_1)$. Here, and subsequently, $k = [c_{15}l(\log(d(A)+2))^3]$, where c_{15} is a sufficiently large fixed constant.

Let $L = kM$, and let

$$\mathscr{P} = \{(a/L, b/L): a, b \in \mathbb{Z} \text{ and } 0 \leqslant a, b < ML\} \subset [0, M)^2.$$

Finally, let

$$L^{-2} = \alpha_1\alpha_2\cdots\alpha_t,$$

where $\alpha_1 = \cdots = \alpha_{t-1} = \frac{1}{2}$ and $\frac{1}{2} \leqslant \alpha_t < 1$. We now apply Lemma 8.7′ with $X = \mathscr{P}, \mathscr{Y} = \{B \cap \mathscr{P}: B = B'(A_1)$ or $B = B''(A_1)$ for some $A_1 \in \mathscr{A}_1\}, \mathscr{Z} = \{T \cap \mathscr{P}: T \in SPEC(X_1, X_2, S_1, \ldots, S_l)$ occurs in a disjoint representation of $B'(A_1)$ or $B''(A_1)$ for some $A_1 \in \mathscr{A}_1\}$, $\mathscr{H} = \{(A_1^{**}\backslash A_1) \cap \mathscr{P}: A_1 \in \mathscr{A}_1\}$ and $\alpha = \alpha_1$. We shall first of all establish an upper bound for $\deg(\mathscr{Z})$. Note that if $T \in SPEC(X_1, X_2, S_1, \ldots, S_l)$ satisfies $T \cap \mathscr{P} \neq \emptyset$ and $d(T) < \frac{1}{2}L^{-1}$, then $\#(T \cap \mathscr{P}) = 1$ and there exists $T_0 \in SPEC(X_1, X_2, S_1, \ldots, S_l)$ such that $T_0 \supset T$ and $\frac{1}{2}L^{-1} \leqslant d(T_0) < L^{-1}$. Clearly $\#(T_0 \cap \mathscr{P}) = 1$, and so $T \cap \mathscr{P} = T_0 \cap \mathscr{P}$. It follows that (see also (66))

$$\begin{aligned} \deg(\mathscr{Z}) &\leqslant \#\{C \in SET(X_1, X_2, S_1, \ldots, S_l): \tfrac{1}{2}L^{-1} \leqslant d(C) \leqslant d(A)\} \\ &\ll l(\log L + \log(d(A)+2))^2 \ll l(\log M)^2 + l(\log(d(A)+2))^2. \end{aligned} \tag{81}$$

Furthermore, $l \ll \xi(A)$ and obviously we also have

$$\max\{\#\mathscr{Y}, \#\mathscr{H}\} \leqslant 2\#\mathscr{A}_1 \quad \text{and} \quad \#X = \#\mathscr{P} = (ML)^2 = M^4k^2. \tag{82}$$

We also recall (3) in Chapter 7, i.e.

$$\xi(A) \ll (\mu(A))^{\frac{1}{4}} \leqslant (d(A))^{\frac{1}{2}}. \tag{83}$$

It now follows from (77), (81)–(83) and Lemma 8.7′ that there exists a function $g_1: \mathscr{P} \to \{-\alpha_1, 1-\alpha_1\}$ such that for any $G = Y \in \mathscr{Y}$ and any $G = H \in \mathscr{H}$, we have

$$\left|\sum_{\mathbf{p} \in G} g_1(\mathbf{p})\right| \ll_\varepsilon \xi(A)b^{4+\varepsilon} + m^{\frac{1}{2}}b^{\frac{3}{2}},$$

where

$$b = \max\{\log M, \log(d(A)+2)\}$$

and

$$m = \max_{H \in \mathscr{H}} \#H = \max_{A_1 \in \mathscr{A}_1} \#((A_1^{**}\backslash A_1) \cap \mathscr{P}).$$

Let $\mathscr{P}_1 = \{\mathbf{p}\in\mathscr{P}: g_1(\mathbf{p}) = 1 - \alpha_1\}$. Applying Lemma 8.7′ again with $X = \mathscr{P}_1$, $\mathscr{Y}_1 = \mathscr{Y}_{|\mathscr{P}_1}$, $\mathscr{Z}_1 = \mathscr{Z}_{|\mathscr{P}_1}$, $\mathscr{H}_1 = \mathscr{H}_{|\mathscr{P}_1}$ and $\alpha = \alpha_2$, we obtain a function $g_2: \mathscr{P}_1 \to \{-\alpha_2, 1 - \alpha_2\}$ such that for any $G = Y \in \mathscr{Y}_1$ and any $G = H \in \mathscr{H}_1$, we have

$$\left|\sum_{\mathbf{p}\in G} g_2(\mathbf{p})\right| \ll_\varepsilon \xi(A) b^{4+\varepsilon} + m_1^{\frac{1}{2}} b^{\frac{3}{2}},$$

where

$$m_1 = \max_{H\in\mathscr{H}_1} \#H = \max_{A_1\in\mathscr{A}_1} \#((A_1^{**}\backslash A_1)\cap\mathscr{P}_1).$$

Let $\mathscr{P}_2 = \{\mathbf{p}\in\mathscr{P}_1: g_2(\mathbf{p}) = 1 - \alpha_2\}$. We apply Lemma 8.7′ again with $X = \mathscr{P}_2$, $\mathscr{Y}_2 = \mathscr{Y}_{|\mathscr{P}_2}$, $\mathscr{Z}_2 = \mathscr{Z}_{|\mathscr{P}_2}$, $\mathscr{H}_2 = \mathscr{H}_{|\mathscr{P}_2}$ and $\alpha = \alpha_3$. By repeated application of Lemma 8.7′, we obtain a sequence of functions g_i, sets $\mathscr{P}_i$ and integers m_i $(1 \leqslant i \leqslant t)$ such that $g_i: \mathscr{P}_{i-1} \to \{-\alpha_i, 1 - \alpha_i\}$, $\mathscr{P}_i = \{\mathbf{p}\in\mathscr{P}_{i-1}: g_i(\mathbf{p}) = 1 - \alpha_i\}$ and

$$m_i = \max_{A_1\in\mathscr{A}_1} \#((A_1^{**}\backslash A_1)\cap\mathscr{P}_i),$$

and such that for any $E = B'(A_1)$ or $B''(A_1)$ or $(A_1^{**}\backslash A_1)$, where $A_1 \in \mathscr{A}_1$, we have

$$\left|\sum_{\mathbf{p}\in E\cap\mathscr{P}_{i-1}} g_i(\mathbf{p})\right| \ll_\varepsilon \xi(A) b^{4+\varepsilon} + m_{i-1}^{\frac{1}{2}} b^{\frac{3}{2}}. \tag{84}$$

Since $\alpha_1 = \cdots = \alpha_{t-1} = \frac{1}{2}$, $-g_i$ is also a function of the type $\mathscr{P}_{i-1} \to \{-\alpha_i, 1 - \alpha_i\}$ for $1 \leqslant i \leqslant t-1$. So replacing g_i by $-g_i$ if necessary, we can guarantee that

$$m_i \geqslant \tfrac{1}{2} m_{i-1} \quad (1 \leqslant i \leqslant t-1), \tag{85}$$

where $m = m_0$.

For any $E \subset \mathbb{R}^2$,

$$\sum_{\mathbf{p}\in E\cap\mathscr{P}_{i-1}} g_i(\mathbf{p}) = \sum_{\mathbf{p}\in E\cap\mathscr{P}_i} 1 - \alpha_i \sum_{\mathbf{p}\in E\cap\mathscr{P}_{i-1}} 1. \tag{86}$$

Using the relation $L^{-2} = \alpha_1\alpha_2\cdots\alpha_t$, we have, by (86), that

$$\begin{aligned}
&\left|\sum_{\mathbf{p}\in E\cap\mathscr{P}_t} 1 - L^{-2}\sum_{\mathbf{p}\in E\cap\mathscr{P}} 1\right| \\
&\leqslant \left|\sum_{\mathbf{p}\in E\cap\mathscr{P}_t} 1 - \alpha_t \sum_{\mathbf{p}\in E\cap\mathscr{P}_{t-1}} 1\right| + \alpha_t \left|\sum_{\mathbf{p}\in E\cap\mathscr{P}_{t-1}} 1 - \alpha_{t-1}\sum_{\mathbf{p}\in E\cap\mathscr{P}_{t-2}} 1\right| \\
&\quad + \alpha_t\alpha_{t-1}\left|\sum_{\mathbf{p}\in E\cap\mathscr{P}_{t-2}} 1 - \alpha_{t-2}\sum_{\mathbf{p}\in E\cap\mathscr{P}_{t-3}} 1\right| + \cdots \\
&\quad + \alpha_t\alpha_{t-1}\cdots\alpha_2\left|\sum_{\mathbf{p}\in E\cap\mathscr{P}_1} 1 - \alpha_1\sum_{\mathbf{p}\in E\cap\mathscr{P}} 1\right|.
\end{aligned} \tag{87}$$

It follows, on combining (84), (86) and (87), that for all $E = B'(A_1)$ or $B''(A_1)$ or $(A_1^{**}\backslash A_1)$, where $A_1 \in \mathscr{A}_1$, we have

$$\left|\sum_{p\in E\cap\mathscr{P}_t} 1 - L^{-2}\sum_{p\in E\cap\mathscr{P}} 1\right| \ll_\varepsilon \xi(A)b^{4+\varepsilon}\sum_{i=1}^{t}\prod_{j=i+1}^{t}\alpha_j + b^{\frac{3}{2}}\sum_{i=1}^{t} m_{i-1}^{\frac{1}{2}}\prod_{j=i+1}^{t}\alpha_j. \tag{88}$$

Since $\alpha_1 = \cdots = \alpha_{t-1} = \frac{1}{2}$ and $\frac{1}{2} \leqslant \alpha_t < 1$, elementary calculation gives

$$\sum_{i=1}^{t}\prod_{j=i+1}^{t}\alpha_j \ll 1. \tag{89}$$

By (84) and (86), we have the recursive relation

$$m_i \leqslant \alpha_i m_{i-1} + c_{16}(\xi(A)b^{4+\varepsilon} + m_{i-1}^{\frac{1}{2}}b^{\frac{3}{2}}), \tag{90}$$

where $c_{16} = c_{16}(\varepsilon)$ depends only on $\varepsilon > 0$. It follows that

$$\begin{aligned} m_i &\leqslant \left(1 + \frac{c_{16}\xi(A)b^{4+\varepsilon}}{\alpha_i m_{i-1}} + \frac{c_{16}b^{\frac{3}{2}}}{\alpha_i m_{i-1}^{\frac{1}{2}}}\right)\alpha_i m_{i-1} \\ &\leqslant \left(\prod_{v=1}^{i}\left(1 + \frac{c_{16}\xi(A)b^{4+\varepsilon}}{\alpha_v m_{v-1}} + \frac{c_{16}b^{\frac{3}{2}}}{\alpha_v m_{v-1}^{\frac{1}{2}}}\right)\right)\left(\prod_{v=1}^{i}\alpha_v\right)m. \end{aligned} \tag{91}$$

Since $\alpha_1 = \cdots = \alpha_{t-1} = \frac{1}{2}$ and $\frac{1}{2} \leqslant \alpha_t < 1$, it follows from (85) and elementary calculation that for all $i = 1, \ldots, t$,

$$\begin{aligned} &\prod_{\alpha=1}^{i}\left(1 + \frac{c_{16}\xi(A)b^{4+\varepsilon}}{\alpha_v m_{v-1}} + \frac{c_{16}b^{\frac{3}{2}}}{\alpha_v m_{v-1}^{\frac{1}{2}}}\right) \\ &\leqslant \prod_{\alpha=1}^{i}\left(1 + \frac{c_{16}\xi(A)b^{4+\varepsilon}}{\alpha_v 2^{1-v}m} + \frac{c_{16}b^{\frac{3}{2}}}{\alpha_v 2^{\frac{1}{2}(1-v)}m^{\frac{1}{2}}}\right) \\ &\ll_\varepsilon 1 + \frac{\xi(A)b^{4+\varepsilon}}{2^{-t}m} + \frac{b^{\frac{3}{2}}}{2^{-\frac{1}{2}t}m^{\frac{1}{2}}} \ll 1 + \frac{\xi(A)b^{4+\varepsilon}}{mL^{-2}} + \frac{b^{\frac{3}{2}}}{(mL^{-2})^{\frac{1}{2}}}. \end{aligned} \tag{92}$$

Combining (91), (92) and the relation $\alpha_1\alpha_2\cdots\alpha_t = L^{-2}$, we have

$$\begin{aligned} &\sum_{i=1}^{t} m_{i-1}^{\frac{1}{2}}\prod_{j=i+1}^{t}\alpha_j \leqslant 2\sum_{i=1}^{t} m_{i-1}^{\frac{1}{2}}\prod_{j=i}^{t}\alpha_j = 2\sum_{i=1}^{t}\left(m_{i-1}\prod_{j=i}^{t}\alpha_j\right)^{\frac{1}{2}}\left(\prod_{j=i}^{t}\alpha_j\right)^{\frac{1}{2}} \\ &\ll_\varepsilon \sum_{i=1}^{t}\left(\left(1 + \frac{\xi(A)b^{4+\varepsilon}}{mL^{-2}} + \frac{b^{\frac{3}{2}}}{(mL^{-2})^{\frac{1}{2}}}\right)\frac{m}{L^2}\right)^{\frac{1}{2}}\left(\prod_{j=i}^{t}\alpha_j\right)^{\frac{1}{2}} \\ &\ll \left(\frac{m}{L^2} + \xi(A)b^{4+\varepsilon} + \left(\frac{m}{L^2}\right)^{\frac{1}{2}}b^{\frac{3}{2}}\right)^{\frac{1}{2}}. \end{aligned} \tag{93}$$

For any convex set $E \subset [0, M)^2$, we have

$$\left|\sum_{\mathbf{p}\in E\cap\mathscr{P}} 1 - L^2\mu(E)\right| \ll L\sigma(\partial E) \ll LM,$$

where $\sigma(\partial E)$ denotes the length of the perimeter of E, and so

$$\left|L^{-2}\sum_{\mathbf{p}\in E\cap\mathscr{P}} 1 - \mu(E)\right| \ll ML^{-1} = k^{-1}. \tag{94}$$

Since the set $A_1^{**}\backslash A_1$ $(A_1\in\mathscr{A}_1)$ has the representation

$$A_1^{**}\backslash A_1 = (B''(A_1)\backslash B'(A_1))\backslash A_1,$$

where $B'(A_1), B''(A_1)\in\mathscr{B}(k)$, it follows from (94) that for any $E = B'(A_1)$ or $B''(A_1)$ or $(A_1^{**}\backslash A_1)$, where $A_1\in\mathscr{A}_1$, we have

$$\left|L^{-2}\sum_{\mathbf{p}\in E\cap\mathscr{P}} 1 - \mu(E\cap[0,M)^2)\right| \ll 1. \tag{95}$$

On combining (88), (89), (93) and (95), we see that for all $E = B'(A_1)$ or $B''(A_1)$ or $(A_1^{**}\backslash A_1)$, where $A_1\in\mathscr{A}_1$, we have

$$\begin{aligned}
&\left|\sum_{\mathbf{p}\in E\cap\mathscr{P}_t} 1 - \mu(E\cap[0,M)^2)\right| \\
&\leqslant \left|\sum_{\mathbf{p}\in E\cap\mathscr{P}_t} 1 - L^{-2}\sum_{\mathbf{p}\in E\cap\mathscr{P}} 1\right| + \left|L^{-2}\sum_{\mathbf{p}\in E\cap\mathscr{P}} 1 - \mu(E\cap[0,M)^2)\right| \\
&\ll_\varepsilon \xi(A)b^{4+\varepsilon} + b^{\frac{3}{2}}\left(\frac{m}{L^2} + \xi(A)b^{4+\varepsilon} + \left(\frac{m}{L^2}\right)^{\frac{1}{2}} b^{\frac{3}{2}}\right)^{\frac{1}{2}} + 1,
\end{aligned} \tag{96}$$

where $b = \max\{\log M, \log(d(A)+2)\}$.

We need an upper bound for m/L^2. By (79) and (80),

$$\begin{aligned}
\mu((A_1^{**}\backslash A_1)\cap[0,M)^2) &\leqslant \mu(A_1^{**}\backslash A_1) \leqslant \mu(A_1^{**}\backslash A_1^{*}) + \mu(A_1^{*}\backslash A_1) \\
&\ll (\xi(A))^2 + \xi(A)\log(d(A)+2) \ll (\xi(A))^2 b.
\end{aligned} \tag{97}$$

Combining this with (95), we have

$$mL^{-2} = L^{-2}\max_{A_1\in\mathscr{A}_1}\sum_{\mathbf{p}\in(A_1^{**}\backslash A_1)\cap\mathscr{P}} 1 \ll (\xi(A))^2 b + 1 \ll (\xi(A))^2 b. \tag{98}$$

It now follows from (96) and (98) that for any $E = B'(A_1)$ or $B''(A_1)$ or $(A_1^{**}\backslash A_1)$, where $A_1\in\mathscr{A}_1$, we have

$$\left|\sum_{\mathbf{p}\in E\cap\mathscr{P}_t} 1 - \mu(E\cap[0,M)^2)\right| \ll_\varepsilon \xi(A)b^{4+\varepsilon}. \tag{99}$$

Since

$$A_1 = A_1^{**} \backslash (A_1^{**} \backslash A_1) = (B''(A_1) \backslash B'(A_1)) \backslash (A_1^{**} \backslash A_1),$$

it follows from (99) that

$$\left| \sum_{\mathbf{p} \in A_1 \cap \mathscr{P}_t} 1 - \mu(A_1 \cap [0, M)^2) \right| \ll_\varepsilon \xi(A) b^{4+\varepsilon} \tag{100}$$

for all $A_1 \in \mathscr{A}_1$.

The underlying set $[0, M)^2$ belongs to the family $\mathscr{A}_1$. It follows from (100) that

$$|\#\mathscr{P}_t - M^2| \ll_\varepsilon \xi(A) b^{4+\varepsilon}. \tag{101}$$

Let $S^{(1)}$ be an arbitrary $(\min\{M^2, \#\mathscr{P}_t\})$-element subset of $\mathscr{P}_t$, and let $S^{(2)}$ be an arbitrary $(M^2 - \min\{M^2, \#\mathscr{P}_t\})$-element subset of $[0, M)^2 \backslash \mathscr{P}_t$. Now $S = S(A, M^2) = S^{(1)} \cup S^{(2)}$ is the desired M^2-element subset of $[0, M)^2$. Indeed, by (78), (100) and (101),

$$\left| \sum_{\mathbf{p} \in A(\lambda, \mathbf{v}) \cap S} 1 - \mu(A(\lambda, \mathbf{v}) \cap [0, M)^2) \right| \ll_\varepsilon \xi(A) b^{4+\varepsilon}$$

for all $-1 \leqslant \lambda \leqslant 1$ and $\mathbf{v} \in \mathbb{R}^2$. Lemma 8.21 follows.

Part C: More problems!

9
Miscellaneous questions

9.1 Arithmetic progressions

There are various limitations to the extent to which a sequence of natural numbers can be well distributed simultaneously among and within all congruence classes, unless the sequence is, in some sense, 'nearly' the sequence of all natural numbers or the empty sequence. We start with a theorem of Roth which gives some information on the discrepancy of an arbitrary sequence with respect to congruence classes.

Theorem 25A (Roth (1964)). *Let N be a natural number and let $\mathscr{A}$ be a set of distinct natural numbers not exceeding N. Let $\eta = N^{-1}\#\mathscr{A}$ denote the density of $\mathscr{A}$. Then there are natural numbers q_0, m_0 and h_0 such that*

$$\left| \sum_{\substack{n\in\mathscr{A}, n\leqslant m_0 \\ n\equiv h_0 (\mathrm{mod}\, q_0)}} 1 - \frac{\eta m_0}{q_0} \right| > c_1(\eta(1-\eta))^{\frac{1}{2}} N^{\frac{1}{4}}.$$

This theorem exhibits a limitation to the accuracy of approximation of the type

$$\sum_{\substack{n\in\mathscr{A}, n\leqslant m \\ n\equiv h (\mathrm{mod}\, q)}} 1 = \frac{\eta m}{q} + \Delta(q, m)$$

if such approximation is to be valid for all congruence classes of modulus $q \leqslant N$ and for all $m \leqslant N$. Theorem 25A can also be considered as a converse of the famous 'large sieve' of Roth (1965) and Bombieri (1965).

One can easily derive from Theorem 25A the following 2-colouring result.

Corollary 25B (Roth). *Let*

$$R(N) = \min_{f} \max_{q,h} \left| \sum_{j} f(h + jq) \right|,$$

where the maximum is taken over all arithmetic progressions in $\{1, 2, \ldots, N\}$ and the minimum is taken over all 2-colourings $f: \{1, 2, \ldots, N\} \to \{-1, +1\}$. Then $R(N) > c_2 N^{\frac{1}{4}}$.

In the opposite direction, Roth observed that a random sequence yields the upper bound $R(N) \ll N^{\frac{1}{2}}(\log N)^{\frac{1}{2}}$, and suspected that for every $\varepsilon > 0$, $R(N) \gg N^{\frac{1}{2}-\varepsilon}$. Sárközy disproved this conjecture by showing that $R(N) \ll N^{\frac{1}{3}+\varepsilon}$ (see Erdös and Spencer (1974)). Our next result shows that Corollary 25B is in fact essentially best possible.

Theorem 25C (Beck (1981*a*)). *Let $R(N)$ be defined as in Corollary* 25B. *Then for any $N \geqslant 2$,*

$$R(N) < c_3 N^{\frac{1}{4}}(\log N)^3.$$

We shall prove Theorems 25A and 25C at the end of this section.

The family of arithmetic progressions possesses the following remarkable property: Given any 2-colouring of $\mathbb{Z}$, one can find monochromatic arithmetic progressions of arbitrary finite length. It remains true for multicolourings as well. This is the famous van der Waerden's theorem, which can be considered as a 'super discrepancy' phenomenon. We state below the finite version of the theorem.

Van der Waerden's theorem. *Let $q \geqslant 2$ and $l \geqslant 3$. There exists a natural number $W = W(q, l)$ such that if the integers $\{1, 2, \ldots, W(q, l)\}$ are partitioned into q sets, one of these contains an arithmetic progression of length l.*

This result can be extended to partitions of the lattice $\mathbb{Z}^K$, as shown by Gallai (see Rado (1945)): Let $\mathbb{Z}^K = B_1 \cup \cdots \cup B_q$ be an arbitrary partition of the K-dimensional lattice. If $A \subset \mathbb{Z}^K$ is an arbitrary finite configuration, then one of the sets B_j contains a translation of a dilatation of A, i.e. $B_j \supset mA + \mathbf{x}$ for some $m \in \mathbb{N}$ and $\mathbf{x} \in \mathbb{Z}^K$.

It was conjectured by Erdös and Turán that there is a measure of the size of a set in $\mathbb{Z}$ that will guarantee the van der Waerden property. More precisely, their conjecture asserts that any set of positive upper density in $\mathbb{Z}$ possesses arithmetic progressions of arbitrary finite length. This conjecture, announced in the 1930s, was established in stages. Roth (1952), using analytic methods, showed that a set of positive upper density contains arithmetic progressions of length 3. Szemerédi (1969) showed that such sets contain arithmetic progressions of length 4. Finally, Szemerédi (1975) proved the full conjecture of Erdös and Turán, using complicated combinatorial arguments.

Szemerédi's theorem. *Let $B \subset \mathbb{Z}$ be such that for some sequence of intervals $[a_n, b_n]$ with $(b_n - a_n) \to \infty$,*

$$\frac{\#(B \cap [a_n, b_n])}{b_n - a_n} \to \delta > 0.$$

Then B contains arbitrarily long arithmetic progressions.

The multidimensional extension of Szemerédi's theorem was proved by Furstenberg and Katznelson (1978) (see also the excellent book of Furstenberg (1981)). This result serves as an illustration of a deep combinatorial theorem first established by ergodic-theoretic arguments (note that so far there is no alternative proof of the Furstenberg–Katznelson theorem). To formulate the theorem we need the notion of positive upper density of higher-dimensional sets. By an aligned box in $\mathbb{Z}^K$, we mean the cartesian product of K intervals. By the width of an aligned box, we mean the length of its shortest edge. A set $S \subset \mathbb{Z}^K$ is said to have positive upper density if for some sequence of aligned boxes $\mathscr{B}_n$ whose widths $\to \infty$,

$$\frac{\#(S \cap \mathscr{B}_n)}{\mu(\mathscr{B}_n)} > \delta$$

for some $\delta > 0$.

Furstenberg–Katznelson theorem. *If $S \subset \mathbb{Z}^K$ has positive upper density and A is a finite subset of $\mathbb{Z}^K$, then S contains a translation of a dilatation of A; i.e. $S \supset mA + \mathbf{x}$ for some $m \in \mathbb{N}$ and $\mathbf{x} \in \mathbb{Z}^K$.*

We now return to the van der Waerden function $W(q, l)$. Let $W(l) = W(2, l)$, i.e. the number of colours is 2. The best-known lower bound to $W(l)$ is due to Berlekamp (1968), who showed that for primes p, we have $W(p) \geqslant p2^p$. Our knowledge in the opposite direction is extremely poor. We cannot even prove that $W(l)$ is a primitive recursive function. So the basic problem is

Problem 1. *Find a 'reasonable' upper bound to $W(l)$.*

The next problem is more than 50 years old.

Problem 2 (Erdös). *Let $f: \mathbb{N} \to \{-1, +1\}$ be an arbitrary 2-colouring of the natural numbers. Is it true that*

$$\sup_{n \geqslant 1} \sup_{d \geqslant 1} \left| \sum_{i=1}^{n} f(id) \right| = +\infty?$$

We conclude this section by proving Theorems 25A and 25C.

We first of all prove Theorem 25A. For all integers n, we denote by $g(n) = g_{\mathscr{A}}(n)$ the characteristic function of $\mathscr{A}$ and by $h(n) = h_\eta(n)$ the corresponding expectation, i.e.

$$g(n) = \begin{cases} 1 & (n \in \mathscr{A}), \\ 0 & (\text{otherwise}), \end{cases}$$

and

$$h(n) = \begin{cases} \eta & (1 \leqslant n \leqslant N), \\ 0 & (\text{otherwise}). \end{cases}$$

Let

$$\chi_q(n) = \begin{cases} 1 & (n = lq, 0 \leqslant l < L, l \in \mathbb{Z}), \\ 0 & (\text{otherwise}), \end{cases}$$

where the parameter $L \geqslant 1$ will be specified later. Let

$$F_q = \chi_q * (g - h), \tag{1}$$

where $*$ denotes the convolution operation. More explicitly,

$$F_q(n) = \sum_{k \in \mathbb{Z}} \chi_q(n-k)(g(k) - h(k)) = \sum_{\substack{0 \leqslant l \leqslant L-1 \\ n - lq \in \mathscr{A}}} 1 - \eta \sum_{\substack{0 \leqslant l \leqslant L-1 \\ 1 \leqslant n - lq \leqslant N}} 1. \tag{2}$$

Next, let

$$Q = [N^{\frac{1}{2}}] \quad \text{and} \quad L = [\tfrac{1}{2}Q], \tag{3}$$

and consider the expression

$$E = \sum_{q=1}^{Q} \sum_{n \in \mathbb{Z}} |F_q(n)|^2.$$

Lemma 9.1. *We have*

$$E \gg \eta(1-\eta)N^2.$$

Proof. We recall some most elementary facts from harmonic analysis. If $F \in L^2(\mathbb{Z})$, let

$$\hat{F}(t) = \sum_{n \in \mathbb{Z}} F(n) e^{2\pi i n t}$$

denote its Fourier transform. Here $i = \sqrt{-1}$. Clearly

$$\int_0^1 |\hat{F}(t)|^2 \, dt = \sum_{n \in \mathbb{Z}} \sum_{m \in \mathbb{Z}} F(n)\overline{F(m)} \int_0^1 e^{2\pi i (n-m)t} \, dt.$$

Since

$$\int_0^1 e^{2\pi i (n-m)t} \, dt = \begin{cases} 1 & (n = m), \\ 0 & (n \neq m), \end{cases}$$

we have (Parseval identity)

$$\sum_{n \in \mathbb{Z}} |F(n)|^2 = \int_0^1 |\hat{F}(t)|^2 \, dt. \tag{4}$$

Moreover, if $f_1, f_2 \in L^2(\mathbb{Z})$ and $F = f_1 * f_2$, then

$$\hat{F}(t) = \hat{f}_1(t)\hat{f}_2(t). \tag{5}$$

By (1), (4) and (5), we have

$$E=\sum_{q=1}^{Q}\sum_{n\in Z}|F_q(n)|^2=\sum_{q=1}^{Q}\int_0^1|\hat{F}_q(t)|^2\,\mathrm{d}t$$
$$=\int_0^1\left(\sum_{q=1}^{Q}|\hat{\chi}_q(t)|^2\right)|(\hat{g}-\hat{h})(t)|^2\,\mathrm{d}t. \tag{6}$$

Note that

$$\int_0^1|(\hat{g}-\hat{h})(t)|^2\,\mathrm{d}t=\sum_{n\in Z}|g(n)-h(n)|^2=\eta(1-\eta)N. \tag{7}$$

On the other hand, for every real number t, there exist integers q_0, satisfying $1\leqslant q_0\leqslant Q$, and h_0 such that

$$|q_0t-h_0|\leqslant Q^{-1},$$

and so we have, with $\beta=q_0t-h_0$, that $|L\beta|\leqslant\frac{1}{2}$ and

$$\sum_{q=1}^{Q}|\hat{\chi}_q(t)|^2\geqslant|\hat{\chi}_{q_0}(t)|^2=\left|\sum_{l=0}^{L-1}\mathrm{e}^{2\pi il\beta}\right|^2$$
$$=L^2\left(\frac{\sin(\pi L\beta)}{\pi L\beta}\right)^2\left(\frac{\pi\beta}{\sin(\pi\beta)}\right)^2\geqslant\left(\frac{2L}{\pi}\right)^2. \tag{8}$$

Lemma 9.1 follows on combining (3) and (6)–(8). ■

We now deduce Theorem 25A. Note from (2) and (3) that for every q satisfying $1\leqslant q\leqslant Q$, we have that $F_q(n)=0$ whenever $n>2N$ or $n\leqslant 0$. It follows from Lemma 9.1 that there are integers q_1 and n_1, satisfying $1\leqslant q_1\leqslant Q$ and $1\leqslant n_1\leqslant 2N$, such that

$$|F_{q_1}(n_1)|^2\geqslant\frac{E}{2NQ}\gg\eta(1-\eta)N^{\frac{1}{4}}. \tag{9}$$

Since

$$\{n_1-lq_1\in[1,N]:l=0,1,\ldots,L-1\}$$
$$=\{k\in[1,\min\{n_1,N\}]:k\equiv n_1(\mathrm{mod}\,q_1)\}$$
$$\backslash\{k\in[1,\min\{n_1-Lq_1,N\}]:k\equiv n_1\,(\mathrm{mod}\,q_1)\}$$

and

$$\left|\sum_{\substack{k\in[1,m]\\k\equiv h(\mathrm{mod}\,q)}}1-\frac{m}{q}\right|<1$$

for all $m\geqslant 1$, $q\geqslant 1$ and $h\geqslant 1$, Theorem 25A follows from (2) and (9).

Remark. Roth's proof of Theorem 25A given here is in fact the motivation for the Fourier transform approach described in Chapters 6 and 7.

To prove Theorem 25C, we shall use the following general lemma.

Lemma 9.2. *Let X be a finite set and let $\mathscr{G}$ be a family of subsets of X. Let m and d be natural numbers such that*

$$\deg(\{A\in\mathscr{G}:\#A>m\})\leqslant d.$$

Then

$$\delta(\mathscr{G})\ll(d\log(d+2)\log(\#\mathscr{G}+2)+m\log(\#\mathscr{G}+2))^{\frac{1}{2}}\log(\#X+2).$$

Here the notation is as in §8.2.

Proof. The lemma follows immediately from Lemma 8.7 by choosing $\mathscr{H}=\{A\in\mathscr{G}:\#A\leqslant m\}$ and $\mathscr{Y}=\mathscr{X}=\{A\in\mathscr{G}:\#A>m\}$. ■

We now derive Theorem 25C from Lemma 9.2.

For integers i and j satisfying $i\leqslant j$, let

$$AP(a,q,i,j)=\{a+kq:i\leqslant k\leqslant j\},$$

i.e. $AP(a,q,i,j)$ denotes the arithmetic progression with difference q, starting from $a+iq$ and terminating at $a+jq$. We shall say that an arithmetic progression is elementary if it is of the type

$$AP(b,q,i2^s,(i+1)2^s-1),$$

where $q\geqslant 1$, $1\leqslant b\leqslant q$, $i\geqslant 0$ and $s\geqslant 0$. In other words, its length is a power of 2, say 2^s, and it has difference $q\geqslant 1$ and starts from $b+i2^sq$.

Let $\mathscr{G}_N$ denote the family of elementary arithmetic progressions contained in $\{1,2,\ldots,N\}$, i.e.

$$\mathscr{G}_N=\{AP(b,q,i2^s,(i+1)2^s-1)\subset[1,N]:q\geqslant 1,1\leqslant b\leqslant q,i\geqslant 0,s\geqslant 0\}.$$

Let $l\geqslant 2$ be an integer. By definition,

$$\begin{aligned}\deg(\{A\in\mathscr{G}_N:\#A\geqslant l\})&=\max_n\#\{A\in\mathscr{G}_N:\#A\geqslant l\text{ and }n\in A\}\\&\leqslant\max_n\sum_{1\leqslant q\leqslant(N-1)/(l-1)}\ \sum_{\substack{1\leqslant b\leqslant q\\ b\equiv n(\mathrm{mod}\,q)}}\ \sum_{\substack{s\\ 2^s\geqslant l\\ b+(2^s-1)q\leqslant N}}1.\end{aligned}$$

Simple calculation shows that the innermost sum is $\ll\log(N/ql)$. It follows that

$$\begin{aligned}\deg(\{A\in\mathscr{G}_N:\#A\geqslant l\})&\ll\max_n\sum_{1\leqslant q\leqslant(N-1)/(l-1)}\ \sum_{\substack{1\leqslant b\leqslant q\\ b\equiv n(\mathrm{mod}\,q)}}\log\left(\frac{N}{ql}\right)\\&=\sum_{1\leqslant q\leqslant(N-1)/(l-1)}\log\left(\frac{N}{ql}\right)\ll\frac{N}{l}.\end{aligned}$$

More explicitly, there is an absolute constant c_4 such that for any $l \geqslant 2$,

$$\deg(\{A \in \mathscr{G}_N : \#A \geqslant l\}) \leqslant c_4 N/l. \tag{10}$$

We now apply Lemma 9.2 to $\mathscr{G}_N$ with $m \gg\ll N^{\frac{1}{2}}$ and $d \gg\ll N^{\frac{1}{2}}$. Then

$$\delta(\mathscr{G}_N) \ll N^{\frac{1}{4}}(\log N)^2, \tag{11}$$

since clearly $\#\mathscr{G}_N \leqslant N^2$ and $\#X = \#(\{1, 2, \ldots, N\}) = N$. We claim that

$$R(N) \leqslant (2 \log_2 N)\delta(\mathscr{G}_N). \tag{12}$$

To see this, first observe that any arithmetic progression $a, a+q, \ldots, a+lq$ in $[1, N]$ is representable in the form

$$AP(b, q, 0, p_1) \backslash AP(b, q, 0, p_2),$$

where $a = b + (p_2 + 1)q$, $1 \leqslant b \leqslant q$ and $p_1 = l + p_2 + 1$. Moreover, both $AP(b, q, 0, p_i)$ $(i = 1, 2)$ are disjoint unions of not more than $(\log_2 N)$ elementary arithmetic progressions, i.e. elements of $\mathscr{G}_N$. Thus (12) follows.

Combining (11) and (12), we conclude that

$$R(N) \ll N^{\frac{1}{4}}(\log N)^3.$$

This completes the proof of Theorem 25C.

9.2 Unsolved problems on point distributions

The most famous question is to determine the L^∞-norm of the discrepancy of aligned boxes. We use the notation of §1.1.

Problem 3 (Great open problem). *Is it true that*

$$D(K, \infty, N) \gg_K (\log N)^{K-1}?$$

The case $K = 2$ was completely solved by Schmidt (see Theorem 3B). The L^1-norm is also a mystery.

Problem 4. *Is it true that*

$$D(K, 1, N) \gg_K (\log N)^{\frac{1}{2}(K-1)}?$$

In the case $K = 2$, the answer is affirmative. See Halász's theorem (Theorem 1C).

Note also that there is no non-trivial lower bound for $D(K, W, N)$ where $0 < W < 1$.

There are many 'random constructions' in this field. It would be desirable to find explicit constructions. For example, consider Roth's upper bound of the L^2-norm (see Theorem 2C and §§3.1–3.3).

Problem 5. *Find an explicit construction $\mathscr{P} \subset U_0^K$ such that $\#\mathscr{P} = N$ and that*

$$\|D(\mathscr{P})\|_2 \ll_K (\log N)^{\frac{1}{2}(K-1)}.$$

Again, the case $K = 2$ is solved. See Davenport's theorem (Theorem 2A).

In Chapter 5, we noted the existence of balls B with large $|D^{\text{tor}}[\mathscr{P}; B]|$. It is an open question to prove 'one-sided' estimates.

Problem 6. *Let $\mathscr{P} \subset U_0^K$ be an N-element distribution. Does there exist a ball B of diameter $\leqslant 1$ such that*

$$D^{\text{tor}}[\mathscr{P}; B] > 0 \quad \text{and} \quad |D^{\text{tor}}[\mathscr{P}; B]| \text{ is 'large'?}$$

Does there exist a ball B of diameter $\leqslant 1$ such that

$$D^{\text{tor}}[\mathscr{P}; B] < 0 \quad \text{and} \quad |D^{\text{tor}}[\mathscr{P}; B]| \text{ is 'large'?}$$

Finally, we mention three old problems which are, in a broader sense, related to our topic.

One day in the 1940s, Heilbronn looked out of his window and saw a group of soldiers. They were very dispirited and seemed not to be marching in formation. Heilbronn therefore set out to investigate how badly they could do.

Problem 7 (Heilbronn's triangle problem). *Let $\mathscr{P} = \{\mathbf{p}_1, \ldots, \mathbf{p}_n\}$ be a distribution of n ($\geqslant 3$) points in the unit square, and let $\Delta(\mathscr{P})$ denote the minimum of the areas of the triangles with vertices $\mathbf{p}_i, \mathbf{p}_j, \mathbf{p}_k$ where $1 \leqslant i < j < k \leqslant n$. Put*

$$\Delta(n) = \sup_{\mathscr{P}} \Delta(\mathscr{P}),$$

where the supremum is taken over all distributions $\mathscr{P}$ of n points in the unit square. Determine the order of magnitude of $\Delta(n)$.

It was shown by Erdös (see Roth (1951)) that

$$\Delta(n) \gg n^{-2}. \tag{13}$$

On the other hand, the inequality $\Delta(n) \ll n^{-1}$ is trivial. The first non-trivial upper bound for $\Delta(n)$ was obtained by Roth (1951) who showed that

$$\Delta(n) \ll n^{-1} (\log \log n)^{-\frac{1}{2}}.$$

This was improved to

$$\Delta(n) \ll n^{-1} (\log n)^{-\frac{1}{2}}$$

by Schmidt (1971), and then by Roth (1972*a*, *b*) to

$$\Delta(n) \ll n^{-\mu + \varepsilon} \tag{14}$$

where $\mu = 2 - (\frac{4}{5})^{\frac{1}{2}} = 1.105\ldots$ and $\mu = (17 - (65)^{\frac{1}{2}})/8 = 1.117\ldots.$

In fact, Heilbronn conjectured that $\Delta(n) \ll n^{-2}$, so that in view of (13), if the conjecture were true, then it would be best possible. However, the conjecture was shown to be false by Komlós, Pintz and Szemerédi (1982) who proved that

$$\Delta(n) \gg n^{-2} \log n. \tag{15}$$

Komlós, Pintz and Szemerédi (1981) also improved the upper estimate (14) to $\mu = \frac{8}{7} = 1.142\ldots$. However, there remains a gap between this and (15).

For further references of the problem, see also Roth (1973, 1976*a*).

We next state a problem on ellipses due to Danzer.

Problem 8 (Danzer's problem). *Let $\mathscr{P}$ be a set of points on the plane $\mathbb{R}^2$ with the property that every ellipse of unit area contains a point of $\mathscr{P}$. Is it true that if $N(r)$ denotes the number of points of $\mathscr{P}$ in the disc of radius r, then*

$$\lim_{r \to \infty} \frac{N(r)}{\pi r^2} = \infty?$$

The final problem can be considered the dual of Danzer's problem.

Problem 9. *Let $M \geqslant 2$ be an integer and let $\mathscr{Q}$ be a discrete set on the plane. Assume that any ellipse of unit area contains at most M points of $\mathscr{Q}$. Is it true that $\mathscr{Q}$ has density zero?*

References

T. van Aardenne-Ehrenfest (1945) 'Proof of the impossibility of a just distribution of an infinite sequence of points over an interval', *Proc. Kon. Ned. Akad. v. Wetensch.*, **48** (1945), 266–71.

(1949) 'On the impossibility of a just distribution', *Proc. Kon. Ned. Akad. v. Wetensch.*, **52** (1949), 734–9.

R.C. Baker (1978) 'On irregularities of distribution', *Bull. London Math. Soc.*, **10** (1978), 289–96.

J. Beck (1981*a*) 'Roth's estimate of discrepancy of integer sequences is nearly sharp', *Combinatorica*, **1** (1981), 319–25.

(1981*b*) 'Balanced two-colourings of finite sets in the square. I', *Combinatorica*, **1** (1981), 327–35.

(1983) 'On a problem of K.F. Roth concerning irregularities of point distribution', *Invent. Math.*, **74** (1983), 477–87.

(1984*a*) 'Some upper bounds in the theory of irregularities of distribution', *Acta Arith.*, **43** (1984), 115–30.

(1984*b*) 'Sums of distances between points on a sphere – an application of the theory of irregularities of distribution to discrete geometry', *Mathematika*, **31** (1984), 33–41.

(1984*c*) 'New results in the theory of irregularities of point distributions', *Number Theory Noordwijkerhout 1983*, pp. 1–16 (Lecture Notes in Mathematics 1068, Springer, 1984).

(TA*a*) 'Irregularities of distribution. I', *Acta Math.* (to appear).

(TA*b*) 'On the discrepancy of convex plane sets', *Monats. für Math.* (submitted).

(TA*c*) 'Irregularities of distribution. II', *Proc. London Math. Soc.* (to appear).

J. Beck and T. Fiala (1981) 'Integer-making theorems', *Discrete Applied Math.*, **3** (1981), 1–8.

J. Beck and J.H. Spencer (1984) 'Well-distributed 2-colourings of integers relative to long arithmetic progressions', *Acta Arith.*, **43** (1984), 287–94.

E.R. Berlekamp (1968) 'A construction for partitions which avoid long arthmetic progressions', *Canadian Math. Bull.*, **11** (1968), 409–14.

E. Bombieri (1965) 'On the large sieve', *Mathematika*, **12** (1965), 201–25.

W.W.L. Chen (1980) 'On irregularities of distribution', *Mathematika*, **27** (1980), 153–70.

(1983) 'On irregularities of distribution. II', *Quart. J. Math. Oxford (2)*, **34** (1983), 257–79.

(1985) 'On irregularities of distribution and approximate evaluation of certain functions', *Quart. J. Math. Oxford (2)*, **36** (1985), 173–82.

(TA) 'On irregularities of distribution and approximate evaluation of certain functions. II', *Proceedings of Conference on Analytic Number Theory at Stillwater 1984* (to appear).

J.G. van der Corput (1935*a*) 'Verteilungsfunktionen. I', *Proc. Kon. Ned. Akad. v. Wetensch.*, **38** (1935), 813–21.

(1935*b*) 'Verteilungsfunktionen. II', *Proc. Kon. Ned. Akad. v. Wetensch.*, **38** (1935), 1058–66.

H. Davenport (1956) 'Note on irregularities of distribution', *Mathematika*, **3** (1956), 131–5.
P. Erdös (1964) 'Problems and results on diophantine approximation', *Compositio Math.*, **16** (1964), 52–66 (Nijenrode lecture 1962).
P. Erdös and J.H. Spencer (1974) *Probabilistic Methods in Combinatorics* (Akadémiai Kiadó, Budapest, 1974).
H. Faure (1982) 'Discrépance de suites associées à un système de numération (en dimension *s*)', *Acta Arith.*, **41** (1982), 337–51.
L. Fejes-Tóth (1953) *Lagerungen in der Ebene, auf der Kugel und in Raum* (Springer, Berlin, 1953).
(1956) 'On the sum of distances determined by a point set', *Acta Math. Acad. Sci. Hung.*, **7** (1956), 397–401.
H. Furstenberg (1981) *Recurrence in Ergodic Theory and Combinatorial Number Theory* (Princeton University Press, Princeton, 1981).
H. Furstenberg and Y. Katznelson (1978) 'An ergodic Szemerédi theorem for commuting transformations', *J. d'Analyse Math.*, **34** (1978), 275–91.
P.M. Gruber and P.S. Kenderov (1982) 'Approximation of convex bodies by polytopes', *Rend. Circ. Mat. Palermo (2)*, **31** (1982), 195–225.
H. Hadwiger (1955) 'Volumschätzung für die einen Eikörper überdeckenden und unterdeckenden Parallelotope', *Elem. Math.*, **10** (1955), 122–4.
(1957) *Vorlesungen über Inhalt, Oberfläche und Isoperimetrie* (Springer, Berlin, 1957).
G. Halász (1981) 'On Roth's method in the theory of irregularities of point distributions', *Recent Progress in Analytic Number Theory*, vol. 2, pp. 79–94 (Academic Press, London, 1981).
(1985) Unpublished manuscript.
J.H. Halton (1960) 'On the efficiency of certain quasirandom sequences of points in evaluating multidimensional integrals', *Num. Math.*, **2** (1960), 84–90.
J.H. Halton and S.K. Zaremba (1969) 'The extreme and L^2 discrepancies of some plane sets', *Monats. für Math.*, **73** (1969), 316–28.
J.M. Hammersley (1960) 'Monte Carlo methods for solving multivariable problems', *Ann. New York Acad. Sci.*, **86** (1960), 844–74.
G. Harman (1982) 'Sums of distances between points of a sphere', *Internat. J. Math. & Math. Sci.*, **5** (1982), 707–14.
J. Komlós, J. Pintz and E. Szemerédi (1981) 'On Heilbronn's triangle problem', *J. London Math. Soc. (2)*, **24** (1981), 385–96.
(1982) 'A lower bound for Heilbronn's problem', *J. London Math. Soc. (2)*, **25** (1982), 13–24.
L. Kuipers and H. Niederreiter (1974) *Uniform Distribution of Sequences* (Wiley-Interscience, New York, 1974).
H.L. Montgomery (1985) 'Irregularities of distribution by means of power sums' (manuscript).
A. Ostrowski (1927) 'Mathematische Miscellen. IX. Notiz zur Theorie der Diophantischen Approximationen', *Jber. Deutsch. Math. Ver.*, **36** (1927), 178–80.
R. Rado (1945) 'Note on combinatorial analysis', *Proc. London Math. Soc.*, **48** (1945), 122–60.
K.F. Roth (1951) 'On a problem of Heilbronn', *J. London Math. Soc.*, **26** (1951), 198–204.
(1952) 'Sur quelques ensembles d'entier', *C.R. Acad. Sci. Paris*, **234** (1952), 388–90.
(1954) 'On irregularities of distribution', *Mathematika*, **1** (1954), 73–9.
(1964) 'Remark concerning integer sequences', *Acta Arith.*, **9** (1964), 257–60.
(1965) 'On the large sieves of Linnik and Renyi', *Mathematika*, **12** (1965), 1–9.
(1972*a*) 'On a problem of Heilbronn. II', *Proc. London Math. Soc. (3)*, **25** (1972), 193–212.
(1972*b*) 'On a problem of Heilbronn. III', *Proc. London math. Soc. (3)*, **25** (1972), 543–9.
(1973) 'Estimation of the area of the smallest triangle obtained by selecting three out of

n points in a disc of unit area', *Proceedings of Symposia in Pure Mathematics*, vol. 24, pp. 251–62 (American Mathematical Society, Providence, 1973).
(1976*a*) 'Developments in Heilbronn's triangle problem', *Adv. in Math.*, **22** (1976), 364–85.
(1976*b*) 'On irregularities of distribution. II', *Communications on Pure and Applied Math.*, **29** (1976), 749–54.
(1979) 'On irregularities of distribution. III', *Acta Arith.*, **35** (1979), 373–84.
(1980) 'On irregularities of distribution. IV', *Acta Arith.*, **37** (1980), 67–75.
(1985) 'On a theorem of Beck', *Glasgow Math. J.*, **27** (1985), 195–201.
E. Sas (1939) 'Über eine Extremumeigenschaft der Ellipsen', *Compositio Math.*, **6** (1939), 468–70.
W.M. Schmidt (1968) 'Irregularities of distribution', *Quart. J. Math. Oxford (2)*, **19** (1968), 181–91.
(1969*a*) 'Irregularities of distribution. II', *Trans. Amer. Math. Soc.*, **136** (1969), 347–60.
(1969*b*) 'Irregularities of distribution. IV', *Invent. Math.*, **7** (1969), 55–82.
(1971) 'On a problem of Heilbronn', *J. London Math. Soc. (2)*, **4** (1971), 545–50.
(1972*a*) 'Irregularities of distribution VI', *Compositio Math.*, **24** (1972), 63–74.
(1972*b*) 'Irregularities of distribution. VII', *Acta Arith.*, **21** (1972), 45–50.
(1974) 'Irregularities of distribution. VIII', *Trans. Amer. Math. Soc.*, **198** (1974), 1–22.
(1975) 'Irregularities of distribution. IX', *Acta Arith.*, **27** (1975), 385–96.
(1977*a*) 'Irregularities of distribution. X', *Number Theory and Algebra*, pp. 311–29 (Academic Press, New York, 1977).
(1977*b*) *Lectures on Irregularities of Distribution* (Tata Institute of Fundamental Research, Bombay, 1977).
R. Schneider and J.A. Wieacker (1981) 'Approximation of convex bodies by polytopes', *Bull. London Math. Soc.*, **13** (1981), 149–56.
I.M. Sobol (1967) 'On the distribution of points in a cube and the approximate evaluation of integrals' (Russian), *Ž. Vyčisl. Mat. i Mat. Fiz.*, **7** (1967), 784–802; English translation in *U.S.S.R. Comp. Math. and Math. Phys.*, **7** (1967), 86–112.
V.T. Sós (1983) 'On strong irregularities of the distribution of $\{n\alpha\}$ sequences', *Studies in Pure Mathematics in Memory of Paul Turán*, pp. 685–700 (Akadémiai Kiadó, Budapest, 1983).
K.B. Stolarsky (1973) 'Sums of distances between points on a sphere. II', *Proc. Amer. Math. Soc.*, **41** (1973), 575–82.
W. Stute (1977) 'Convergence rates for the isotrope discrepancy', *Ann. Prob.*, **5** (1977), 707–23.
E. Szemerédi (1969) 'On sets of integers containing no four elements in arithmetic progression', *Acta Math. Acad. Sci. Hung.*, **20** (1969), 89–104.
(1975) 'On sets of integers containing no *k* elements in arithmetic progression', *Acta Arith.*, **27** (1975), 199–245.
R. Tijdeman and G. Wagner (1980) 'A sequence has almost nowhere small discrepancy', *Monats. für Math.*, **90** (1980), 315–29.
I.V. Vilenkin (1967) 'Plane nets of integration' (Russian), *Ž. Vyčisl. Mat. i Mat. Fiz.*, **7** (1967), 189–96; English translation in *U.S.S.R. Comp. Math. and Math. Phys.*, **7** (1967), 258–67.
G. Wagner (1980) 'On a problem of Erdös in diophantine approximation', *Bull. London Math. Soc.*, **12** (1980), 81–8.
G.N. Watson (1958) *A Treatise on the Theory of Bessel Functions* (Cambridge University Press, Cambridge, 1958).
S.K. Zaremba (1970) 'La discrépance isotrope et l'intégration numerique', *Ann. di Mat. Pura et Appl. (4)*, **87** (1970), 125–36.

Index of theorems and corollaries

If a result in stated more than once in this book, the page number here corresponds to its first appearance.

Index of names

Index